Proceedings of the First European Astronomical Meeting Athens, September 4-9, 1972

Volume 3

Galaxies and Relativistic Astrophysics

Edited by

B. Barbanis

J. D. Hadjidemetriou

Springer-Verlag
Berlin Heidelberg New York 1974

With 61 Figures

ISBN-13: 978-3-642-65697-2 e-ISBN-13: 978-3-642-65695-8
DOI: 10.1007/978-3-642-65695-8

Editors' Note

The present third volume of the Proceedings of the First European Astronomical Meeting held in Athens, Greece, under the auspices of the IAU, from 4th to 9th September 1972, contains the papers presented at this meeting which are related to recent observational and theoretical work in the field of galaxies and high energy astrophysics.

The editors would like to thank all those who have contributed to this volume. Also, the Executive Commitee of the IAU that sponsored the Meeting, the Greek Ministry of Culture and Science for its financial support and the Scientific Organizing Committee and the Local Organizing Committee for their efforts in organizing this Meeting.

B. BARBANIS
J. D. HADJIDEMETRIOU

Introductory Remarks

A need for closer contacts between the astronomers in Europe has been increasingly felt in the course of the past years. This need is well understood by anyone who is aware of the necessity of internationalization of research on our continent. Progress in astronomical research requires pooling of manpower and financial means in order that efforts in Europe may lead to achievements comparable to those elsewhere, particularly to those in the United States. Typical for this development are such international projects as ESRO, ESO, and possibly in the near future, JOSO and CESRA. At the same time, we observe a trend for internationalization of large national projects in both radio astronomy and optical astronomy.

Whereas in these large projects specially oriented groups have already arrived at an encouraging degree of integration across the national borders in Europe, it is felt by many an astronomer that the relations with colleagues in other countries outside these large projects still leave much to be desired. In most of the astronomical institutes in the European countries there is relatively little acquaintance with the research programmes on a small and intermediate scale in progress in the institutes of other countries. The language barrier undoubtedly is an important cause of this regrettable situation, but it would not seem to be the only one. Equally important seems to be the circumstance that there is little opportunity, especially for the younger, not yet internationally recognized, astronomers, to have contacts with their collagues of other countries outside the occasional symposia and colloquia. With the tendency for restricted participation, which is desirable in order to render such meetings fruitful for a follow-up by participants, they certainly do not suffice for the establishment of good contacts between the coming generation of astronomers.

Whereas in the United States the American Astronomical Society plays an essential role in promoting such contacts, such a common organization is lacking on the European continent. Therefore, a most important step towards remedying this situation was the initiative of the International Astronomical Union to create the institute of Regional Meetings. This initiative was supplemented admirably by the Greek National Committee for Astronomy, when early in 1972 it proposed to the IAU to hold a first European regional meeting in Athens. The present three volumes containing the proceedings of these meetings represent the fruit of this excellent effort.

For those who were present at the meetings in Athens, the memory will remain of an inspiring series of lectures and discussions, many of these latter very lively, and with that participation from astronomers from all over Europe (and also from across the ocean) which the International Astronomical Union's Executive

Committee must have had in mind when it initiated these meetings. Hopefully, it may induce colleagues from other states in Europe to organize similar meetings in the future. For the Scientific Organizing Committee, it was a privilege to be able to share in this first "regional" effort. It is a pleasure, on behalf of this committee, to acknowledge the excellent help it has had from the part of the Greek National Committee.

A. Blaauw
Chairman,
Scientific Organizing Committee

Contents

Observational and Theoretical Aspects of Galaxies

Relativistic and High Energy Astrophysics

List of Contributors to Volume 3

ALLADIN, S. M., Centre of Advanced Study in Astronomy, Osmania University, Hyderabad 500007, India

BALLABH, G. M., Centre of Advanced Study in Astronomy, Osmania University, Hyderabad 500007, India

BARBANIS, B., University of Patras, Department of Astronomy, Patras, Greece

BONOMETTO, S., Instituto di Fisica "Galileo Galilei", Via Marzolo 8, I-35100 Padova, Italy

BURBIDGE, G. R., Department of Physics, University of California, San Diego, La Jolla, CA 92037, USA

CAZZOLA, P., Instituto di Fisica "Galileo Galilei", Via Marzolo 8, I-35100 Padova, Italy

CHANDRASEKHAR, S., Laboratory for Astrophysics and Space Research, 933 East 56th Air, Chicago, Illinois 60637, USA

CONTOPOULOS, G., University of Thessaloniki, Department of Astronomy, Thessaloniki, Greece

DALLAPORTA, N., Osservatorio Astronomico, Vicolo dell' Osservatorio 5, I-35100 Padova, Italy

DAVIES, R. D., University of Manchester, Nuffield Radio Astronomy Laboratories Jodrell Bank, Cheshire, Great Britain

DUIN, R. M., Sterrewacht te Leiden, Kaiserstraat 57, Leiden 2401, The Netherlands

EMERSON, D. T., Mullard Radio Astronomy Observatory, Cambridge, Great Britain

HADJIDEMETRIOU, J. D., Department of Theoretical Mechanics, University of Thessaloniki, Thessaloniki, Greece

KALINKOV, M. P., Bulgarian Academy of Sciences, Department of Astronomy, 7th November Street 1, Sofia, Bulgaria

LUCARONI, L., Instituto di Fisica "Galileo Galilei", Via Marzolo 8, I-35100 Padova, Italy

LUCCHIN, F., Instituto di Fisica "Galileo Galilei", Via Marzolo 8, I-35100 Padova, Italy

LUKAČEVIĆ, I., Faculty of Sciences and Institute for Geomagnetism, Belgrade, Yugoslavia

LYNDEN-BELL, D., University of Cambridge, Institute of Astronomy, Madingley Road, CB3 OHA, Cambridge, Great Britain

MACKAY, C. D., Cavendish Laboratory, Free School Lane, Cambridge, Great Britain

MAKSUMOV, M. N., Astrophysical Institute of the Academy of Sciences of the Tadjik SSR, Str. Swiridenko 22, Dushanbe, USSR

MCVITTIE, G. C., 74 Old Dover Road, Canterbury Kent, Great Britain

MONNET, G., Observatoire de Marseille, 2, Place Le Verrier, F-13004 Marseille, France

NORTHOVER, K. J. E., Mullard Radio Astronomy Observatory, University of Cambridge, Department of Physics, Cambridge, Great Britain

ÖGELMAN, H., Physics Department, METU, Orta Dogu Teknik Üniversite, Ankara, Turkey

ÖZEL, M. E., Physics Department, METU, Orta Dogu Teknik Üniversite, Ankara, Turkey

OORT, J. H., President Kennedylaan 169, Oestgeest, The Netherlands

OZERNOY, L. M., I. E. Tamm Department of Theoretical Physics, Lebedev Physical Institute, Leninsky Prospect 53, Moscow, USSR

PACINI, F., Laboratorio di Astrofisica Spaciale, Casella Postale 67, I-00044 Frascati (Roma), Italy

PERSIDES, S., University of Thessaloniki, Astronomy Department, Thessaloniki, Greece

PIŞMIŞ, P., Instituto de Astronomia, Universidad Nacional de Mexico, Apartado Postal 70-264, Mexico 20, D. F., Mexico

REES, M. J., Astronomy Center, University of Sussex, Falmer, Brighton, Great Britain

SASTRY, K. S., Centre of Advanced Study in Astronomy, Osmania University, Hyderabad 500007, India

SEIELSTAD, G. A., Owens Valley Radio Observatory, P.O.Box 387, Big Pine, California 93513, USA

STEPHENSON, R. J., University of Manchester, Nuffield Radio Astronomy Laboratories, Jodrell Bank, Cheshire, Great Britain

STROM, R. G., Sterrewacht te Leiden, Kaiserstraat 57, Leiden 2401, The Netherlands

ULRICH, M.-H., The University of Texas at Austin, Department of Astronomy, Austin, Texas 78712, USA

UNSÖLD, A., Neue Universität, Bau 13/I, D-2300 Kiel, Ohlshausenstr. 40–60, Germany

WARNER, P. J., Mullard Radio Astronomy Observatory, Cambridge, Great Britain

WEILER, K. W., Radiosterrewacht Westerbork, Post Hooghalen, The Netherlands

WESTERLUND, B. E., European Southern Observatory, Casilla 16317 Correo 9, Santiago de Chile, Chile

WILSON, A. S., Sterrewacht te Leiden, Kaiserstraat 57, Leiden 2401, The Netherlands

Recent Radio Work in Nearby Galaxies

(Invited Lecture)

By J. H. Oort
Sterrewacht, Leiden, The Netherlands

With 7 Figures

I. Introduction

What are the main problems into which high-resolution *radio* observations of galaxies provide new insight?

Continuum Observations. At wavelengths longer than 10 cm almost all emission is synchrotron radiation, the thermal radiation being generally too weak to be observed except in supergiant H II regions. The synchrotron emission is likely to be proportional to a fairly high power of the density, and may thus give information on the pressure differences in the gas caused by the gravitational waves in spiral galaxies. This is clearly a new datum of considerable importance for the spiral problem. In the second place the intensity of the general non-thermal emission may provide a rough measure of the comparative strength of the magnetic field in different galaxies, and, with future more sensitive observations, also of the general pattern of the field.

A problem of no less interest is the activity of the *nuclei* of galaxies. In the past decennium there has come more and more evidence indicating that eruptive activity is a common occurrence, even in nuclei of galaxies which are otherwise quite normal. Because eruptive activity appears to be generally accompanied by increased synchrotron emission, radio observations can provide unique information on the frequency and nature of this activity and of its relation to other properties of the galaxies.

Observations of the 21-cm Line. These should furnish the first clear data on the gas flow in the spiral waves and on the density distribution of the gas in a spiral arm. The only data we had previously on the wave motion came from the Galactic System, where already the unraveling of the spiral structure itself is difficult because of the edge-on view, and it is of course even more difficult to study the dynamics of the spiral arms. Observations of external galaxies are therefore certain to give unique new information.

From the line observations we hope also to find out whether large-scale *radial* motions, like those shown by the 3-kpc arm and the $+135$ km s^{-1} expanding arm, exist in other galaxies.

A third important field of research is the *overall* density distribution of neutral hydrogen in galaxies, and the extension of the rotation curves to rather larger distances from the centre than can be reached by optical observation.

II. Requirements and Instruments

The beam diameter should not exceed the average breadth of a spiral arm, or about 0.5 kpc. As we have to go out to at least 5 Mpc in space to reach a reasonable number of galaxies this implies that in angular measure the beam should not be much larger than 20″. At the same time the sensitivity should be sufficient to measure quite low brightness temperatures. For the 21-cm line as well as for the continuum around 20 cm one must go down to 0.5 K, and it is in many cases desirable to reach even considerably lower brightness temperatures. At present the only places where such a combination of resolution and sensitivity has been attained and applied to these problems are the synthesis instrument at the Mullard Radio Astronomy Observatory in Cambridge and the Westerbork Synthesis Radio Telescope in the Netherlands. Observations with a 4′ beam have been made on a few galaxies at the Owens Valley Radio Observatory. In Cambridge data have been obtained for the two nearest spirals M 31 and M 33, and recently in M 101. At Westerbork some 50 spiral systems have been observed in the continuum; for about a third of these *line* measurements have been made as well. The programme included among others all spiral galaxies above +15° declination which are brighter than $10^{m}.5$ on the Shapley-Ames scale, and the complete list of Seyfert galaxies above +10°.

The Westerbork telescope consists of ten fixed and two movable 25-m dishes arranged along an East-West line; the ten fixed telescopes are spaced at 144 m, the other two can move on tracks having a total length of 300 m. The output of all fixed telescopes is correlated with that of each of the movable telescopes, so that twenty interferometers are used simultaneously. Until the summer of 1972 the operating frequency for the continuum observations has been 1415 MHz. The effective beamwidth was 23″ in right-ascension and 23″/sin δ in declination. Most of the galaxies have been observed during 12 hours. In some ten cases, a.o. M51, M31, NGC4258, integration was extended to four times 12 hours. The r.m.s. noise after 12 hours observation is about $\pm 1^{\circ}\!.0$ for full resolution.

Line observations with a sufficiently narrow beam have been made in Cambridge for M 33 and M 31 and in Westerbork for some 15 spiral galaxies. The latter observations have been but partly reduced, and I can report only on a few provisional results.

III. Nuclei of Spiral Systems

The nuclei show a large range of intrinsic radio brightness. Table 1 shows a few examples. The radio brightness at 1400 MHz is given as log P, where P is the power in $W\,Hz^{-1}$ sterad^{-1}; it is also given in the intrinsic brightness of Sgr A as a unit, the power of which is 6×10^{17} for the region within 500 pc (3°) from the centre (for the inner components, within $0^{\circ}\!.5$ from the centre, it is about 1.3×10^{17}). The galaxies are arranged roughly in the order of advancing type, or of increasing open-ness of the arms.

For comparison the average intrinsic brightness of the nuclei of Seyfert galaxies, as well as their average type, are shown in the last line. The other columns of the table will be discussed later.

Table 1. Radio data for some bright galaxies

Galaxy	Type	r Mpc	Nucleus		Arms		Base disk		Opt diam. kpc
			$\log P$ W Hz^{-1} ster^{-1}	$\frac{P}{P_{\mathrm{Sgr}\,A}}$	T_b	width kpc	T_b	S (5 Mpc) m.f.u.	
NGC 4736	Sb	4.0	18.4	4			0.8	130	8.6
NGC 4258	Sb	6.6	~19.0	~16	4	~1.5			36
M 81	Sb	3.3	19.0	14	~0.3	~1.5	0.1	109	23
M 31	Sb	0.69	17.7	1[a]	0.3	~2.0	0.1	80	32
The Galaxy	Sbc		17.8	1	2	0.5			30
M 51	Sbc	4.0[b]	19.1	20	8	0.3	2.5	~300	12.6
M 101	Sc	4.6	<17	<0.4	~1				33
NGC 6946	Sc	10.5[c]	20.0[d]	180[d]	3.5	~3	~2		32
M 33	Scd	0.72	<16	<0.02	<0.04			56	13.0
Seyfert galaxies	Sb		~20.5	~500					

[a] For the region within 35 pc from the centre M 31 is intrinsically 17 times fainter than the corresponding part of Sgr *A*.
[b] A recent investigation by SANDAGE and TAMMANN has indicated that the distance is much larger than what has been assumed in the present article. The actual distance appears to be about 9.7 Mpc (private communication from Dr. TAMMANN).
[c] Distance from private communication from Dr. SANDAGE, based on angular sizes of HII regions.
[d] These quantities refer to an unresolved central region. It should be noted that there is *no* truly small nucleus. WADE (1968) failed to find a component of less than about 2″ (100 pc).

The radio power is probably connected with eruptive activity in the nucleus. This is clearly the case for the Seyfert galaxies, where the intrinsic brightness is on the average at least two orders of magnitude brighter than in "normal" Sb galaxies. In M 51 as well as in NGC 4258, with radiopowers about 20 times that in M 31, the nuclear emission lines are intense and wide, though not nearly as wide as in Seyfert nuclei, while in the radio-weak M 31 the optical emission is also quite weak. In M 81, however, where there is a pronounced radio nucleus the emission lines are inconspicuous. For NGC 6946 I have found no mention of nuclear emission. The central region emits strong radiation at 20 cm; it should be noted, however, that WADE found *no small* component (of ~2″) in its nucleus (WADE, 1968).

Very late-type spirals, which have no distinct optical nucleus, appear to have no radio nucleus either.

One might perhaps surmise that all nuclei of galaxies are latent volcanos which only from time to time develop such an activity as we see for instance in the Seyfert nuclei.

Among the bright galaxies there are three in which we clearly see the effects of an expulsion of very large masses, viz. M 82, where masses of the order of at least 10^6 M$_\odot$ were expelled about 10^6 years ago, NGC 1275, where at an equally recent epoch a mass of more than 10^8 M$_\odot$ has been expelled at a velocity of several thousand km s^{-1}, and NGC 4258, where nearly 10^8 M$_\odot$ appears to have been thrown out, in this case considerably longer ago (about 18 million years).

A fourth case of a large ejection may be seen in our own Galaxy, where there is rather direct evidence that about 10^6 $M_\odot$ was ejected 6 million years ago.

There are at least three more among the brightest galaxies which show clear evidence of eruptive activity, viz. NGC 253, NGC 4736 and NGC 5128. NGC 253 and 5128 are fairly strong radio sources. The central region of NGC 253 emits 2.4 f.u. at 1425 MHz corresponding with log $P=20.6$, NGC 5128 has a moderately strong radio nucleus, but a very extensive radio source connected with it, with several separate components which must have been expelled at various epochs in the nearby past. NGC 4736 is a comparatively weak radio emitter, but shows two radio components at about 1 kpc on either side of the nucleus, from which they have presumably been ejected. We see therefore that activity of some sort is a very common property of nuclei of large spirals.

In the two cases of very recent ejection, M 82 and NGC 1275, the intrinsic radio brightness of the nucleus is exceptionally high: about 10^3 and 10^6 times the luminosity of Sgr *A*, respectively. In the much older NGC 4258 eruption it is only about 15 times brighter than Sgr *A*. In the Galactic System the nucleus has apparently become rather quiescent, judging from its moderate radio brightness.

In Seyfert galaxies, with their present extreme activity, no signs have as yet been found of large quantities of expelled matter outside the small nuclei. This indicates that their activity might be rather young, and might not be a permanent attribute of these galaxies.

Of the 8 Seyfert galaxies for which data are available 6 have a radio luminosity between 250 and 1500 times that of Sgr *A*. Only NGC 4051 and NGC 1068 fall outside this range, with 20 and 10000 times Sgr *A*, respectively.

Among the roughly 50 bright spirals observed there are two with *multiple* nuclei, viz. NGC 4736 and NGC 3726. The former has three components, of which the central one coincides with the optical nucleus of the multi-arm spiral, and has a luminosity 4 times that of Sgr *A*. The other two, each with 3 times the brightness of Sgr *A*, lie at distances of about 1 kpc from the centre. The Sc galaxy NGC 3726 has two components of about 30 and 15 times the luminosity of Sgr *A* at 4 kpc from the centre, while the nucleus itself is below the sensitivity limit of the Westerbork observation. In both galaxies the multiplicity of the nucleus appears to indicate past activity, possibly analogous to that which NGC 5128 (Centaurus *A*) shows on a very much larger scale.

It should be noted that practically all radiation from external galaxies at 1400 MHz is non-thermal, probably of the synchrotron type, and that most of the nuclear sources seem to be unresolved.

IV. Synchrotron Emission from Spiral Arms and Disk

Except in the latest-type spirals (like M 33), where thermal radiation may give an important contribution, the radiation observed at 1400 MHz is largely synchrotron emission. The emissivity must vary with a rather high power of the magnetic field strength, probably between 2.5 and 3, and therefore with at least the same power of the gas compression. The synchrotron intensity can thus provide a sensitive means for measuring the density variation of the interstellar gas at its passage through the gravitational spiral wave. It may ultimately show the entire pattern of the density wave in the gas. At present, however, the available resolving power and sensitivity are not yet sufficient to study these details. But one *can* determine the general location of the compression regions, and the overall intensity of the spiral arm radiation.

The latter appears to vary greatly from one galaxy to another. The latest-type spirals, with very open and irregular structure, like M 33 and NGC 2403, have arms whose brightness temperatures are less than 0.1 K. Some Sc spirals, such as M 101 and NGC 628 have arms with T_b between 0.2 and 0.3, while in other Sc's, like NGC 6946, T_b reaches 3.5 K. Among the Sb's there is a similarly large range, from T_b about 0.3 K for M 31 and M 81 to about 8 K for M 51, NGC 4321 and NGC 5055. In the two Sa galaxies investigated the disk radiation is unobservably small.

The width of the arms in the spiral plane shows likewise a considerable range (cf. Table 1), varying from about 0.5 kpc in the Galaxy to about 1.5 kpc in galaxies like M 31.

It is not yet clear what causes these differences. It has been suggested that the arm brightness may be correlated with the shape of the rotation curve, and that it would be small in M 33, for instance, because the main body of the galaxy has roughly a solid-body rotation, in which case the amplitude of the density wave would be small. But this is certainly not the main cause of the differences, since it is not only the brightness temperature of the arms, but equally that of the disk outside the arms which shows this large variation. It seems more likely that the determining factor is the average magnetic field strength in the various spirals.

There is no clear correlation between disk brightness and radio luminosity of the nuclear region.

Detailed studies have been made of the spiral arm radiation in M 31 (POOLEY, 1969; VAN DER KRUIT, 1972), M 51, NGC 6946, IC 342, and NGC 4258. The relation between the synchrotron arms in M 51 and its optical arms, as defined principally by the H II regions and *OB* stars, is shown in Figs. 1 and 2 (MATHEWSON *et al.*, 1972). The white curves in Fig. 1 indicate the maxima of the non-thermal radiation in various position angles. These ridge lines coincide remarkably closely with the places where the maximum amount of dust is visible. They lie clearly inside the bright optical arms. From the amount over which they would have to be rotated in order to centre them on the latter an estimate can be made of the time it takes to form stars from the compressed gas. This is found to be 6 million years if the inclination of the spiral on the sky is 35°. If an inclination of 20°, as advocated by TULLY (1972), is adopted this period would be reduced to about $3\frac{1}{2}$ million years.* The inner radio arms in this very condensed spiral

* As noted in footnote b of Table 1 the distance of M 51 should be revised. With the new distance the formation time is increased, and may roughly be estimated at 10 million years. The optical diameter of the galaxy becomes 30 kpc, so that the system is to be ranked among the large spirals. The corrected thickness of the arms varies from less than 600 to about 800 pc. The brightness of the nucleus becomes about 100 times that of Sgr *A*.

Fig. 1. Ridges of 1415 MHz emission superposed on a blue plate of M51 taken by HUMASON with the 200-inch telescope. The small dotted circle around the nucleus shows the diameter of the central radio source, 16″ of arc. The short full lines indicate regions where polarization could be measured. They show the direction of the E-vectors, the length indicating the brightness temperature of the polarized radiation, the scale of which is indicated at the top of the figure

galaxy (diameter about 9 kpc) are quite thin (less than 250 pc); in the outer regions the half-width becomes about 350 pc. There is a pronounced radio

Fig. 2. Isophotes of 1415 MHz emission from M 51 and NGC 5195. The contour unit is 0.8 K brightness temperature. The contours drawn are Nos. 1 to 15, then 18, 20, 25, 30, 40, 50 and 60. The contours around isolated minima are "spider-legged". The positions of the nuclei of the two galaxies are denoted by crosses and the peaks of the associated radio sources by dots

nucleus; the surface brightness in the arms decreases generally with increasing distance from the centre. There is a strong asymmetry in the radio picture, the Northern half, directed towards NGC 5195, is about twice as strong as the Southern half. The companion is also connected with M 51 by a radio bridge. NGC 5195 is itself a fairly strong source, with a diameter less than 7″. Linear polarization has been found in several regions in M 51.

NGC 6946, in which the general appearance of the arms is rather different from that in M 51, shows a similar relation between radio emission and optical arms (Fig. 3).

In NGC 4258 we see a very different picture. The strong optical arms, which in their compactness resemble the arms in M 51, emit synchrotron radiation comparable in strength with that observed in M 51. But a considerably stronger and much more extensive radiation is seen in regions which appear to have no connection with the optical spiral pattern (Fig. 4). This emission seems to start from the nucleus in two opposite ridges of remarkably smooth structure, which run across the ordinary spiral arms and extend over a very large area. The inner parts of these ridges co-incide precisely with the remarkable set of

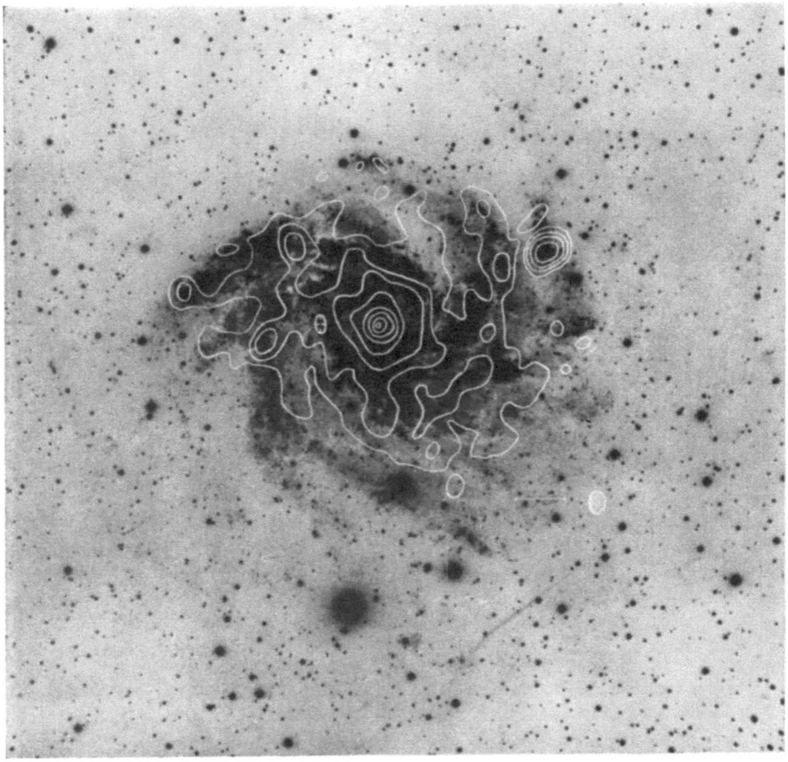

Fig. 3. Synchrotron radiation from NGC 6946. The outer contour corresponds to $T_b = 2.3$ K; the other contours are for 4.6, 6.9, 11.6, 23.1, 46.2, 69.3 and 92.4 K. The beamwidth is 25″ × 28″. The horizontal line is 1′. The optical picture is from ARP's Atlas

Fig. 4. Synchrotron radiation from NGC4258. Contour units are 1 milliflux unit per synthesized beam, corresponding to 0°.9 K in brightness temperature. The optical nucleus is indicated by a white cross. The optical picture has been reproduced from a 200-inch plate taken by SANDAGE

filamentary H_α-arms discovered in 1961 by COURTÈS and CRUVELLIER (1961), who noted particularly that there were no blue supergiants in these arms and presumably therefore the ionization was due to a mechanism other than that of ionization by stellar radiation. They noted also that the arms seemed to partake in the rotation of the galaxy, were remarkably smooth, and showed no large velocities beside the rotation. It seems plausible to assume as a working hypothesis that these anomalous arms, both optical and radio, lie in the equatorial plane of the galaxy. The most striking features of the anomalous *radio* arms, beside their co-incidence with the filamentary H_α-arms, are their very steep preceding edges and the remarkable smoothness of these edges over the entire disk of the galaxy. The sharpness of the edges suggest that these are due to interaction between two streams of gas. We have therefore tried to design a model based on the hypothesis that the ridges are the sites where gas clouds expelled from the nucleus interact with the disk gas of the spiral (VAN DER KRUIT *et al.*, 1972). In the specific model considered – which contains various quite uncertain

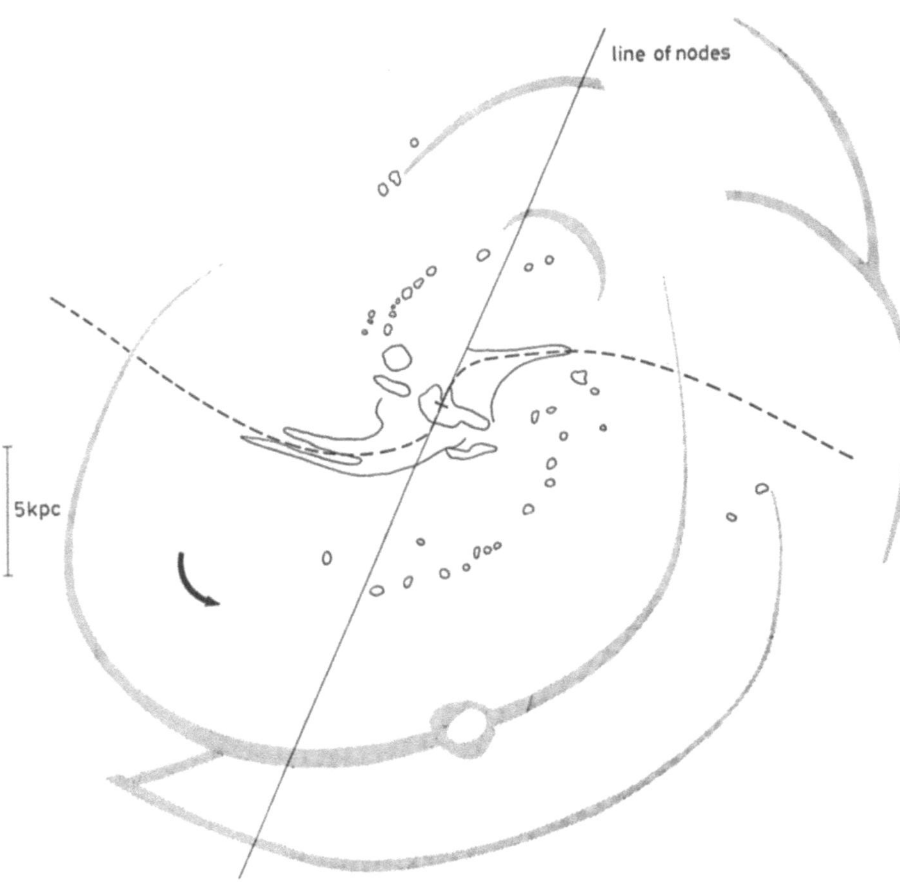

Fig. 5. Radio arms, anomalous H_α-arms and ordinary spiral arms in NGC 4258 drawn in its equatorial plane on the supposition that all arms lie in this plane

parameters—the gas clouds were supposed to have been expelled 18 million years ago in two opposite directions in the disk at velocities between about 800 and 1600 km s^{-1}. The ejected clouds were supposed to have swept up the disk gas, and to have concentrated it near the ridge lines.

Behind the ridge lines there are broad plateaus of somewhat weaker, but still exceptionally strong synchrotron emission. In our model these are ascribed to a spread in the position angle of the ejection.

If the radio ridges lie in the plane of the spiral their shape in this plane is as shown in Fig. 5, in which the run of the ordinary arms has also been sketched. The shape of the anomalous arms is entirely different from that of the normal optical arms; they are also very much more regular. In the quadrants where the expelled clouds would have passed the normal arms these seem to have been swept away. The total mass of ejected gas must have been at least $5 \times 10^7\ M_\odot$. The present ridge will develop into a large-scale spiral in less than one revolution.

It may be that we witness in this galaxy the formation process of new spiral arms, or possibly just the sweeping-up of the irregular old *gas* arms into a new regular pattern.

Extensive observations have been made of M 31, mostly in Cambridge, and partly in Westerbork. The Cambridge observations, at 408 and 1407 MHz, showed that the emission from the spiral arms is non-thermal and quite faint (about 0.3 K) at 1407 MHz (Pooley, 1969). The extensive Cambridge measures

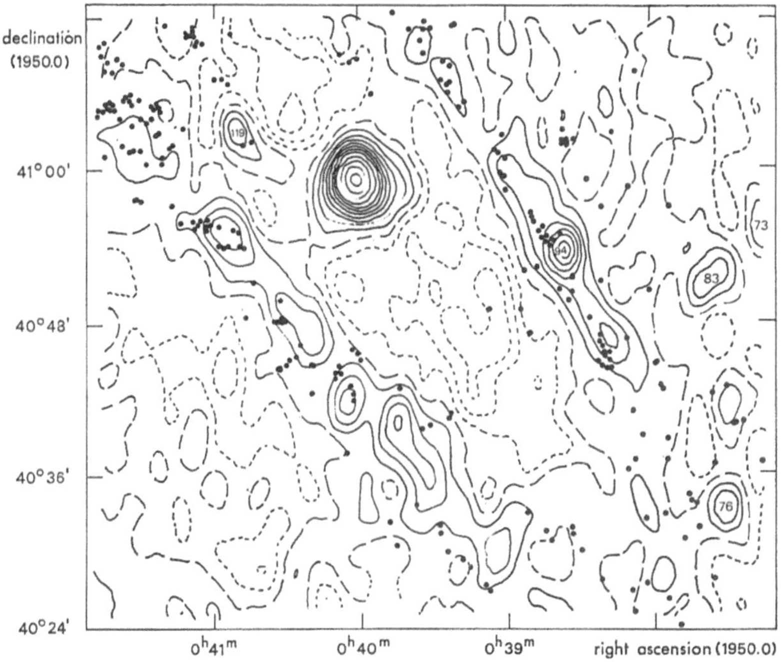

Fig. 6. Contour map of the brightness distribution in the Andromeda Nebula with a beam of 400 × 500 pc. Contour units are 0.1 K. Dots indicate positions of emission nebulae from the survey by Baade and Arp

of the hydrogen line will be reported on by Mr. EMERSON. In Westerbork the continuous radiation was studied with about twice the resolution used in the Cambridge analysis (VAN DER KRUIT, 1972). These observations show considerable structure in the arms (Fig. 6).

In a nuclear region of about 300 pc radius the radiation is considerably stronger. In contrast to the nuclear region of our Galaxy that of M 31 appears to be roughly spherical in shape. Observations with an 80×120 pc beam show the presence of a number of small condensations, one of which coincides with the optical nucleus.

V. Neutral Hydrogen

Neutral hydrogen has been observed in M 101 with a beam of 4′ at the Owens Valley Radio Observatory (ROGSTAD, 1971), in M 31 and M 33 by BALDWIN a.o. with the Cambridge half-mile telescope, while during the past 3/4 year observations with high-resolution were made at Westerbork in some 15 spirals. The latter observations were made with a provisional line-receiver, having only 8 frequency channels, of 25 km s^{-1} width each. Ten interferometers were used simultaneously. Extensive observations, comprising between 8 and 20 half days of observing per galaxy, were made of NGC 224 (M 31), 891, 1300, 3031 (M 81), 3034 (M 82), 4258, 4594, 4736, 5194 (M 51), 5383 and 5457 (M 101). In addition, IC 342, NGC 7320, II Zwicky 40 and Maffei II were observed for 1 to 4 half-day periods, while exploratory observations were made on a few other systems. Reductions and discussions, which are made in Groningen and Leiden, have not yet been completed, so that I can only show a few very provisional results.

Complete analyses have been made at Cambridge for M 31 and M 33, among others, but these have not yet been published. NGC 891 will be discussed in a separate communication during this meeting.

I will report especially on the distribution of the H I column density in M 81, M 51 and NGC 4258. All three show a remarkable concentration in the outermost parts of the arms. This is shown most clearly in M 81 (Fig. 7). Quite dense hydrogen is seen in regions where the optical arms have become almost invisible. The hydrogen is in those places still very clearly concentrated in the arms. The large surface intensity extends to a part of the arms at a distance R 11 kpc from the centre. The surface density reduced to a column perpendicular to the spiral plane averages about 6×10^{20} atoms per cm^2, very nearly the same as the column density in the Galactic System in the general surroundings of the Sun. The high density begins in the same arms around $R = 5$ kpc. Inside this radius the column density is practically everywhere less than about $2 \times 10^{20}\,cm^{-2}$. The curious phenomenon of the concentration of neutral hydrogen in outer regions of galaxies has been known for some time for our own Galaxy and the Andromeda nebula. ROBERTS was the first to draw attention to it as a general property of spiral galaxies. The present observations show how the phenomenon is related to the spiral structure.

Much the same picture is shown by NGC 4258 and M 51. In NGC 4258 all the hydrogen which is clearly concentrated in arms lies in the faint outer arms

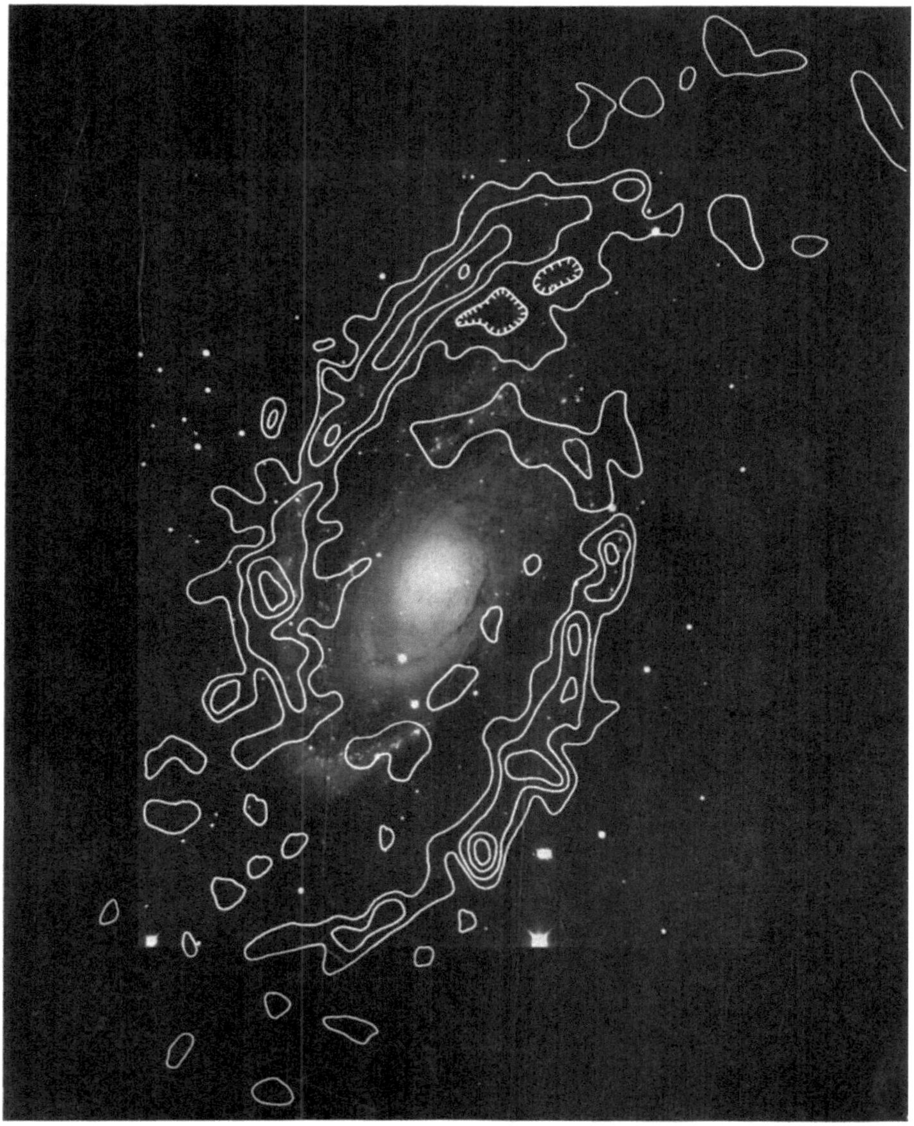

Fig. 7. Column density of neutral atomic hydrogen in M81. The contour interval is 0.4×10^{21} atoms per cm^2. The observations have been smoothed to a beamwidth of 50″. The underlying optical picture is from a 200-inch plate

where it extends with full density to about 20 kpc from the centre on both ends of the major axis. The column density reduced to face-on view is again about $5 \times 10^{20}\ cm^{-2}$ in these regions.

Also in M51 the higher HI density starts only at some distance from the centre, though sooner than in the other two galaxies. It begins roughly at a point where the arms have made 3/4 revolution and are still bright optically. But high density continues to outer regions where the arms have become invisible. The column density is about 7×10^{20} cm^{-2}.

It is clear from the above that the HI observations will give the possibility to extend the measurement of the rotation velocity to very much greater distances from the centre than can be reached optically. They will therefore give extremely valuable data on the density distribution in the far outer regions and will provide the means for getting better estimates of the total masses.

A very provisional determination has been made of the rotation curve of M 51, and a first attempt has been made to relate waves in this curve with the places where a shock occurs in the gas. A preliminary study has also been made of the velocity field in the Southern half of M 81. In both cases the observations seem to indicate a rough agreement with the wave theory, but it would be premature to draw any firm conclusions from the very fragmentary reductions that we have yet been able to make.

Many astronomers in Leiden and Groningen have taken part in the Westerbork work on which I have reported. The continuum results have been mainly discussed by VAN DER KRUIT, partly with MATHEWSON, both in Leiden. The results on the nuclei are entirely due to VAN DER KRUIT. NGC 6946 and M 101 are from a programme by ROTS and ALLEN in Groningen. Work on the line observations has been done by ALLEN, BAJAJA, BURKE, DEKKER, ROTS, SANCISI, SHANE, and VAN WOERDEN (Groningen and Leiden). The telescope group under the direction of RAIMOND, WEILER and BAARS, as well as the computer group led initially by BROUW and presently by VAN SOMEREN GRÉVE, have given an enormous amount of time and ingenuity to all these programmes.

References

COURTÈS, G., CRUVELLIER, P.: Compt. Rend. Acad. Sci. Paris **253**, 218 (1961).
KRUIT, P. C. VAN DER: Astrophys. Letters **11**, 173 (1972).
KRUIT, P. C. VAN DER, OORT, J. H., MATHEWSON, D. S.: Astron. Astrophys. **21**, 169 (1972).
MATHEWSON, D. C., VAN DER KRUIT, P. C., BROUW, W. N.: Astron. Astrophys. **17**, 468 (1972).
POOLEY, G. G.: Monthly Notices Roy. Astron. Soc. **144**, 101 (1969).
ROGSTAD, D. H.: Astron. Astrophys. **13**, 108 (1971).
TULLY, R. B.: The Kinematics and Dynamics of M 51, University of Maryland, (Ph. D. Thesis).
WADE, C. M.: Astron. J. **73**, 876 (1968).

Neutral Hydrogen Observations of External Galaxies with the Mark IA Radio Telescope

By R. D. DAVIES and R. J. STEPHENSON
University of Manchester, Nuffield Radio Astronomy Laboratories, Jodrell Bank, Cheshire, U.K.

With 1 Figure

Abstract

The resurfaced Mark IA radio telescope has been used to make surveys of a number of extragalactic systems with an angular resolution of $12' \times 12'$ arc. These include the M 81/M 82/NGC 3077 group, the NGC 5194/5 system and the companion galaxies of M 101.

Widely distributed intergalactic neutral hydrogen is found throughout the M 81/M 82/NGC 3077 group. Several speculations on its origins are given.

I. The Observing System

These observations were amongst the first made at 21-cm wavelength with the resurfaced Mark IA radio telescope. At the time of the observations the surface had not been adjusted finally. Significant improvements have been made subsequently to the surface structure which are expected to improve the aperture efficiency beyond the value of 0.40 found in the present experiments.

The beamwidth at this wavelength is $12' \times 12'$; sidelobes are less than 1–2 per-cent of the main beam response. Observations of continuum sources gave an aerial temperature of 1 K for 1.6 flux units (1 f.u. $= 10^{-26}$ W m^{-2} Hz^{-1}).

Observations of external galaxies are made using a parametric amplifier with a system temperature of ~ 100 K. Spectra are taken with the Jodrell Bank 256 channel autocorrelation spectrometer set to receive an overall bandwidth of 5 MHz (~ 1000 km s^{-1}). The velocity resolution of the spectra is 7 km s^{-1}. Reference spectra on a position adjacent to the galaxy being observed are used to remove instrumental baseline curvature from the spectra taken at the position of the galaxy.

II. Neutral Hydrogen in the M 81/M 82/NGC 3077 Group

A complete survey has been made of the M 81/M 82/NGC 3077 group on a $6' \times 6'$ arc grid. This provides a fully sampled map of the system. A first interesting result of this work is the discovery of neutral hydrogen emission distributed throughout the group. It was found over an area of $2° \times 1.5°$ centred on M 81. The brightness temperature at positions away from the main galaxies was typically 0.1 to 0.2 K, or about 10 percent of the averaged value seen in the present beamwidth when centred on the galaxies themselves. This work confirms and extends the observations reported by ROBERTS (1972).

Such widespread intergalactic neutral hydrogen within a cluster is unique and it is interesting to speculate upon its origin. One possibility is that the peculiar galaxy M 82 may be responsible for this gas since it is undergoing violent disruption at the present epoch. If this neutral hydrogen had been ejected from M 82 one might expect it to be more or less uniformly distributed around M 82. This explanation runs into a major difficulty – the intergalactic neutral hydrogen is centred on M 81 rather than M 82.

A second possibility is that the neutral hydrogen may have been swept out of M 81 by tidal (gravitational) forces during a close passage of M 82. This phenomenon can be seen optically in many systems shown in ARP'S (1966) Atlas of Peculiar Galaxies. It is shown to be feasible theoretically (WRIGHT, 1972; TOOMRE and TOOMRE, 1972). This seems an attractive explanation of the gas in the M 81/M 82/NGC 3077 system since it is centred on M 81. Further, the velocity of the gas external to M 81 has velocities which are consistent with bound rotation about M 81. ARP (1965) has published a long exposure photograph of M 81 which shows an extended outer envelope. This envelope indicates the presence of what appears to be stellar material at great distances from the centre of M 81. Such a situation could also be the result of a close passage of M 82.

A third possible explanation is that the intergalactic gas never condensed from the original protocluster. The close proximity of the three major galaxies in the cluster may inhibit condensation of gas lying in the more tenuous outer parts of the cluster. Careful calculations are required to show if this is a feasible explanation.

The Jodrell Bank data on the M 81/M 82/NGC 3077 group is being used in conjunction with observations by the Westerbork synthesis instrument to provide a high angular resolution map of the group. This is a collaborative programme with the Leiden and Groningen groups; it also includes a study of the NGC 5194/5195 system.

III. The Neutral Hydrogen Companions of M 101

NGC 5474, 5477 and A1353 (Ho IV) are three close companions of M 101. Their neutral hydrogen content has been studied with the Mark IA system. Fig. 1 shows the grid of 5 spectra centred on NGC 5474. It has a narrow velocity profile which is 40 km s^{-1} wide at half intensity. It also produces a measurable broadening of the telescope beamwidth. The optical and neutral hydrogen parameters of M 101 and its companion galaxies are given in Table 1. Also included are the luminosity and the distance independent ratio M_H/L.

The value of M_H for the galaxies NGC 5474 (Type Scd) and NGC 5477 (Type Sd) are similar to those for the Magellanic Clouds. A1353 appears to be a different class of object – it has the lowest optical surface luminosity of any galaxy in which neutral hydrogen has been seen so far. The values of M_H/L are similar to those found in other Scd, Sd, and Im galaxies.

Table 1. Optical and neutral hydrogen parameters of M 101 and its companion galaxies

Galaxy	m_{pg}	Type	M_H[a]	L/L_0[a]	M_H/L
NGC 5457 (M 101)	8.2	Scd	80×10^8	160×10^8	0.51
NGC 5474	11.2	Scd	3.1	10	0.31
NGC 5477	13.9	Sd	3.0	0.8	3.7
NGC 1353 (Ho IV)	13.0	Im	0.63	2.0	0.31

[a] Calculated for an assumed distance of 4 Mpc.

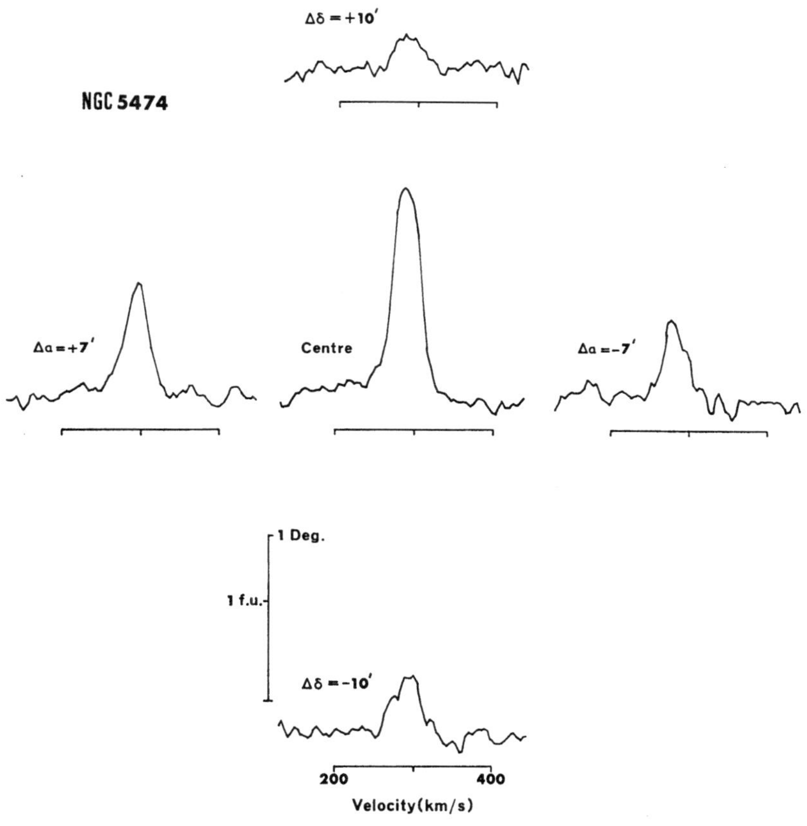

Fig. 1. A grid of 5 neutral hydrogen spectra centred on NGC 5474. The vertical scale is given in aerial temperature and flux density (1 f.u. $= 10^{-26}$ W m^{-2} Hz^{-1}). The velocity is relative to the l.s.r.

References

Arp, H. C.: Science **148**, 363 (1965).
Arp, H. C.: Atlas of Peculiar Galaxies, California Inst. Tech., Pasadena (1966).
Roberts, M. S.: in D. S. Evans, (ed.), External Galaxies and Quasi-stellar Objects, D. Reidel Publ. Co., Dordrecht, p. 12 (1972).
Toomre, A., Toomre, J.: Astrophys. J. **178**, 623 (1972).
Wright, A. E.: Monthly Notices Roy. Astron. Soc. **157**, 309 (1972).

Discussion

Bertola, F.:
Did you get evidence of rotation in NGC 3077 at 21-cm?

Davies, R. D.:
We find a clear indication of rotation in NGC 3077 of 80 to 90 $km\,s^{-1}$ along the major axis. The precise value of the rotation is being derived from the observational data. There is some confusion by Milky Way hydrogen and M 81 hydrogen which has to be corrected for.

The H I Distribution in Nearby Galaxies

By D. T. EMERSON
Mullard Radio Astronomy Observatory, Cambridge, U.K.

Abstract

The Cambridge half-mile radio telescope has been used at 21-cm to map the integrated H I distribution in a number of nearby galaxies; the following results are presented for M 31, with a resolution of 90″ of arc, which corresponds to 300 pc along the major axis of the galaxy.

1) Peaks of H I emission are found to be coincident with dust features.

2) H II regions (BAADE and ARP, 1964) tend to be clustered around peaks of H I emission, rather than being exactly coincident. This may be due to H II regions within the dust being heavily obscured.

3) Unpublished 408 MHz observations of M 31 (POOLEY, 1969) with similar resolution (80″ arc) show peaks of continuum emission coincident with peaks of H I, and with dust. H I, dust, continuum, and H II regions in M 31 are thus all very closely associated.

4) Within 4 kpc of the nucleus, the H I density is not more than ~5% of that found associated with some dust lanes further out. This excludes a small region of anomalous velocity hydrogen very close to the nucleus.

5) The so-called "S.W. Companion" is found to be linked to the main galaxy on both the *N* and *S* of the major axis, and is almost certainly part of an outer arm; this arm can be traced in H I along the *SE* edge of the galaxy.

The H I distribution in M 101 has been observed by J. M. PIGGINS, and shows features strongly associated with the spiral arms, but consistent with the correlation of H I, H II, and dust found for M 31.

References

BAADE, W., ARP, H.: Astrophys. J. **139**, 1027 (1964).
POOLEY, G. G.: Monthly Notices Roy. Astron. Soc. **144**, 101 (1969).

Discussion

MEZGER, P. G.:
Can H I brightness temperatures be interpreted in terms of H I column densities? If yes, is it possible to correlate these column densities with the extinction of the dust lanes? Do you think that opacity of some dense and cool clouds in the spiral arms of M 31 may affect the interpretation of your observations?

EMERSON, D. T.:
Ignoring optical depth effects the H I intensities (in K km s^{-1}) are directly proportional to H I column densities. I think it unlikely that corrections due to opacity within the arms would be as great as 50%. As regards the extinctions of the dust lanes this is difficult to investigate quantitatively. But the brightest peaks of H I certainly seem to correlate with the darkest regions of the dust lanes.

OORT, J. H.:
The Westerbork measures give column densities (perpendicular to the equatorial plane) between 5 and 10×10^{20} atoms per cm^2 over the entire strong parts of the arms in M 81, M 51 and NGC 4258. This is very much the same as the column densities in our own Galaxy in the vicinity of the Sun.

EMERSON, D. T.:
In M 31, column densities (perpendicular to the equatorial plane) range from just over 10×10^{20} atoms per cm^2 along the brightest H I arm to about 2 or 3×10^{20} along the weaker outer arm.

CONTOPOULOS, G.:
How does your work compare with that of Westerbork?

EMERSON, D. T.:
The Westerbork telescope has not yet looked at the H I in M 31 and the large extent of the H I emission (about 5°) means that it would be difficult for Westerbork to survey the whole of the Andromeda galaxy. Their observations of M 51 and M 81, however, appear to be entirely consistent with the H I emission being coincident with the dust lanes as in M 31.

COURTÈS, G.:
The contour map is given with an interval of 600 K km s^{-1} but what is in fact your accuracy on the radial velocity of a specific point?

EMERSON, D. T.:
The integrated H I map has a noise level less than about 100 K km s^{-1}. The velocity filter used in these observations has a bandwidth of 39 km s^{-1}, enabling velocities at a specific point to be determined within a few km s^{-1}.

Compact Structures Associated with the Radio Galaxies 3 C 66, 264 and 315

By K. J. E. NORTHOVER
Mullard Radio Astronomy Observatory, Cambridge, U.K.

Abstract

The radio galaxies 3 C 66, 264 and 315 have been mapped at frequencies of 5.0 and 2.7 GHz by the Cambridge One-Mile radio telescope. The resolution at 5.0 GHz is 6″.5 in right ascension and 6″.5 cosec δ in declination. All three sources consist of an unresolved component, coincident with the nucleus of the galaxy associated with the source, and extended regions of low brightness radio emission. A feature of the map of 3 C 66 is a narrow jet of radio emission, unresolved across its width, extending for approximately 40″ from the nucleus of the galaxy. The jet eventually ends in an extended low-brightness region of emission. There is some evidence for a similar structure in the map of 3 C 264.

The maps suggest that there is continuous activity in the nuclei of the galaxies and that energy is transported continuously from the nuclei to the outer regions of the radio sources. The jet in 3 C 66 provides evidence that the activity in the galactic nucleus is highly directed.

Discussion

ÖGELMAN, H.:
Are the diffusion lengths you have calculated, based on just the radiative loss of the electrons or have you included the diffusion across magnetic lines?

NORTHOVER, K. J.:
I have only considered radiative loss effects.

SWARUP, G.:
Do you think that there is a similarity between the origin of radio emission from bridges seen in many extra – galactic double radio sources and the cases described by you?

NORTHOVER, K. J.:
The bridges seen in double radio sources are weak and well resolved, the components are usually bright and compact. In 3 C 66 the jet is bright at radio wavelengths and the outer components are weak and well resolved. I don't think that there is any similarity between the two types of objects.

Recent Optical Studies on Nearby Galaxies

(Invited Lecture)

By G. MONNET
Observatore de Marseille, Marseille, France

With 7 Figures

Most of our knowledge of the detailed structure of galaxies comes from the local group, to which belong about 15 dwarf elliptical galaxies, four dwarf and two larger irregular galaxies, a moderate luminosity Scd spiral (M 33), two giant spirals (M 31 and our Galaxy) – but unfortunately no giant elliptical.

Much observational data have been obtained in recent years, mainly by the use of new types of instruments – for instance image tube echelle spectrographs, Perot-Fabry interferometers, spectrum scanners, etc.

I. Models of the Stellar Population

The best known galaxy is by far the Andromeda nebula, where extensive results on the stellar population content have been recently obtained.

a) The Nucleus (Radius: 5 pc). A synthetic stellar model has been obtained from spectroscopic observations by SPINRAD and TAYLOR (1971). The minimum M/L ratio in the V band is 45, with most of the mass coming from M 5–M 8 super metal-rich dwarf stars. The upper limit from $V-K$, $K-L$ color observations by SANDAGE *et al.* (1969) is 65. From the virial theorem EINASTO (1972) has obtained a dynamical (M/L) ratio of about 40, in good agreement with the photometric result. Oddly enough, the position angle of the major axis of the nucleus is 14° greater than that of the main body of M 31 – as discovered by JOHNSON (1961) – and confirmed in 1971 by Princeton observers from a Stratoscope flight. This result probably implies that the nucleus is dynamically isolated.

b) The Bulge (Radius: 0.4 kpc). The bulge of M 31 has been observed by SANDAGE *et al.* (1969) in $U-B-V-R-I-H-K-L$ colours. There is no colour gradient (less than 0.02 mag) except in $U-B$ which varies from 0.8 at the center to 0.6 at 200 pc. This single color change seems rather difficult to explain by a variation in the stellar luminosity function or in the metal abundances and is perhaps related to the UV excess found by CODE from the OAOII satellite. COHEN (1970) has made narrow band photometry in 11 band-passes from 5300 Å to 3448 Å, and found no color changes from the center to 1.7 kpc. Narrow band analysis by SPINRAD *et al.* (1971) gives a rather similar population model to that in the nucleus but with metal abundances close to the solar level. The mass to luminosity ratio is somewhat higher than 45, a rather high value, probably difficult to reconcile with the smaller values obtained from analysis of the RUBIN and FORD (1970) rotation curve.

c) The Halo Component (Mean Radius: 4.5 kpc). Since BAADE (1944), the globular clusters were universally considered as the prototypes of old metal-poor population II. However these concepts have been recently shaken by the fundamental photometric and spectroscopic survey of the 44-th brightest globular clusters in M 31 made by VAN DEN BERGH (1969). He found a rather large average metallicity, higher than in our Galaxy, with no dependence on position relative to the disk. Radial velocities of the clusters give a high dispersion of $\sim 110\,\mathrm{km\,s^{-1}}$, so the globular clusters are not members of the disk population, as would otherwise seem to be implied by the high metallicity. These striking abundance results have been confirmed from narrow band photometry by SPINRAD and SCHWEIZER (1972), who found that some of the clusters are even super metal-rich. The conclusion obtained by VAN DEN BERGH is that much of the heavy element formation took place in M 31 before the collapse of the proto-galaxy contrary to our Galaxy.

Similar observations on the globular clusters of the Fornax dwarf galaxy by VAN DEN BERGH (1969), show that they are all metal-poor.

d) The Disk Component (Mean Radius: 10 kpc). It is known since BAADE's work that the spiral arms are clearly outlined against the bright central bulge from its dust component, to about 1.5′ of arc from the nucleus. JOHNSON (1972) has photographed a dust arm coming to within 6″ of arc which seems to imply that there is no inner Lindblad resonance. There is a well-known ionized gas emission in the inner 400 pc (2′ of arc) discovered by MÜNCH (1960) from the [O II] doublet. An extensive study has been made by RUBIN and FORD (1971) in the [N II] 6584 Å line. They found the inner part of the disk in fast rotation, but with deviations as large as $100\,\mathrm{km\,s^{-1}}$, which may indicate a bar phenomenon. The excess positive velocities are correlated with the position of JOHNSON's absorbing clouds. The emissivity of the gas is extremely low ($\sim 9\,\mathrm{cm^{-6}\,pc}$), but the [S II] ratios indicate a very high density ($\sim 10^4\,\mathrm{cm^{-3}}$), so the clouds are extremely clumpy. At about 1.6 kpc there is a dip in the radial velocity of the gas, which is also exhibited by the E Fraunhöfer line (λ 5269) in the stellar component (RUBIN and FORD, 1972).

In spite of this enormous body of observational data, there is still no fully satisfactory model of M 31, particularly the agreement between spectral photometric and kinematical results is rather poor.

For the other nearby spiral galaxies, the available information is much poorer. However much work has been devoted to the study of the mass to luminosity variation in the outer parts of a number of spirals, computed from the rotation curve and the luminosity distribution. In general, large increases with radius have been found except in very early type galaxies, i.e. Sd, Sm. But $(B-V)$ colour observations in a large number of spirals show practically no variation with the radius. So, the M/L variation, if real, must come from extremely subluminous bodies; i.e. very low mass stars. This result seems fairly reliable for NGC 300 (LEWIS, 1972) an M 33 (BOULESTEIX and MONNET, 1970) with a steep rise from about 1 near the nucleus to 30 or more in the outer parts. The most likely explanation would be that the mass distribution is dominated by a large radius, old-star subsystem, in other words a spheroidal population; but much caution is still needed as these results rely heavily on rotational motions in the outer parts, where the data are rather scarce.

II. Chemical Composition

M 86 = NGC 4406, in the Virgo Cluster, is the nearest giant elliptical galaxy. Multicolor observations have been made by Mc Clure and van den Bergh (1968), and they have derived a population model of the central part, independently confirmed by Spinrad (quoted by van den Bergh, 1972).

Most of the light in the core comes from a metal rich population with a minimum Mass/Luminosity ratio, of about 13, and most of the mass from *M* dwarfs. The same holds for M 31, as found in its pioneering study by Baum (1959) – from star counts – and later confirmed from spectrophotometric measures by Greene (1968).

M 32 = NGC 221 is by far the best observed elliptical galaxy of intermediate luminosity. Multicolor photometry has been made by Wood (1966) and Mc Clure and van den Bergh (1968). They have derived a population model consisting of a dwarf-enriched cluster population, and definitely not by adding main sequence stars to a metal-poor globular cluster base. The minimum Mass/Luminosity ratio is about 5.

On the other hand, observations of the colour magnitude diagrams for dwarf spheroidal systems: *Draco*, Baade and Swope (1961) – *Sculptor*, Hodge (1965) – *Ursa Minor*, van Agt (1967), certainly show that these dwarf galaxies are exceedingly metal-poor.

In conclusion, the metal abundance results for dwarf-intermediate and giant ellipticals show a progressive metal enrichment which suggest that only one generation of stars are observed in the low mass systems and that the opposite is true for high mass systems (van den Bergh, 1972).

In early type galaxies, i.e. Sb, Sc, and Irregulars, the metal abundances can be studied from the line ratios in H II regions over the whole disk of the galaxy, not only in the central part. A radial composition gradient might be expected if, as believed, the heavy elements come from Supernovae explosions, as the ratio of stellar mass versus gaseous mass decreases with increasing distance from the nucleus. Peimbert (1968) found an excess of nitrogen in the very strong nuclear regions of M 51 and M 81, as did Rubin and Ford (1971) in the much fainter nuclear region of M 31 (in nitrogen and sulphur). Searle (1971) has found a decrease of the N/O ratio from the inner part of some spiral galaxies (M 51, M 101, M 33) to the outer parts, implying an overabundance of nitrogen in the center. The Scd galaxy M 33 is close enough to show blue and red supergiants. Near the nucleus there are 15 blue for 1 red, while in the outer region the ratio is 5 blue for 1 red (Walker, 1964). van den Bergh (1968) has suggested that this variation comes from a gradient of metal abundance in the gas.

On the other hand, a dwarf irregular galaxy of the local group, NGC 6822 has been found highly deficient in *N* and deficient in *O* (Peimbert and Spinrad, 1970), but not in helium.

The study of the chemical composition of the gas in spirals and irregulars thus gives the same result as that of the stars in the ellipticals, an increase of of the heavy element abundances with increasing mass.

III. The Spiral Arms

a) Gas Distribution. Of particular interest is the Triangulum nebula M 33, as it is the nearest early type spiral and the first external galaxy where a detailed H I distribution has been obtained (WRIGHT *et al.*, 1972). The distribution of neutral gas, ionized gas, and young stars, shows a large asymmetry, with much of these zero age components in the southern half of M 33 – a phenomenon which is especially clear from the 21-cm distribution (Fig. 1).

On the other hand there is no such asymmetry in the distribution of the underlying continuum of old stars (DE VAUCOULEURS, 1961) which is thus strictly restricted to population I components.

Such an asymmetry is in fact a very usual phenomenon; it is observed optically in many spirals (M 101, NGC 6946, NGC 2903, M 83, etc.) and it will be particularly

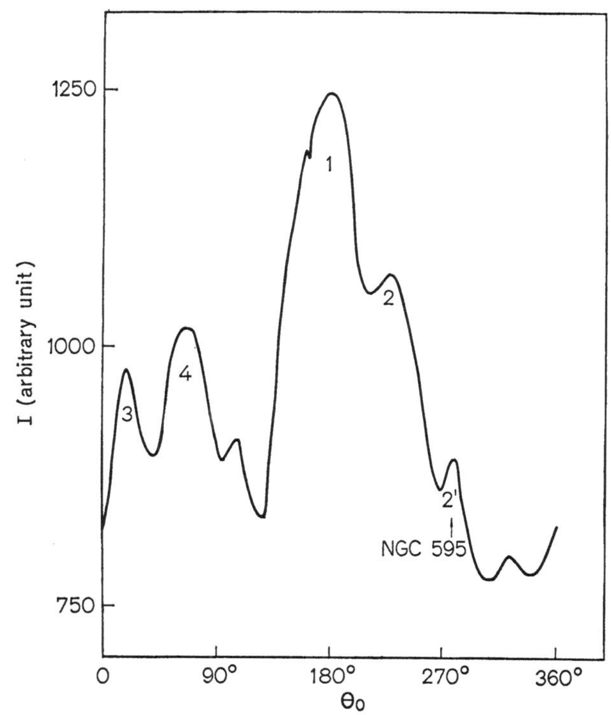

Fig. 1. The geometrical shape of the optical arms has been taken from the $H\alpha$ photograph (Fig. 2b) and swept in the plane of M 33. The integrated amount of neutral hydrogen has been computed from the H I density map (Fig. 2a) and is given here in ordinate versus the position angle Θ_0 of the rotating pattern. One sees the major Southern arm (number 1), it's symmetrical (number 3), two symmetrical other arms (number 2 and 4) all of them with clear optical counterparts on Fig. 2b. The bump in 2′ is entirely due to a large H II region, unconnected to the spiral arm pattern, NGC 595

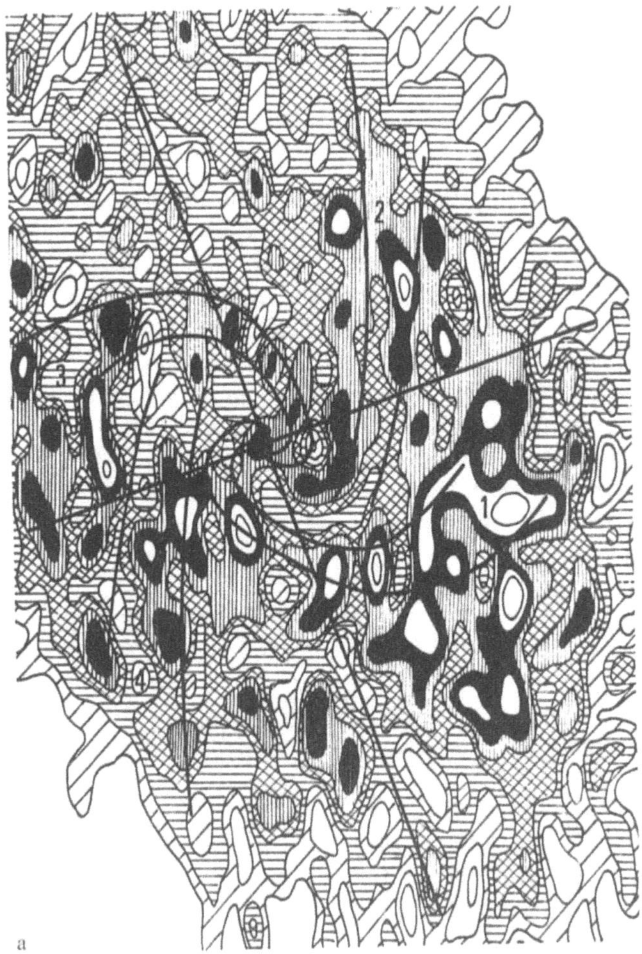

Fig. 2a. This is the H I density map in M33 according to WRIGHT *et al.* (1972). On a 1° diameter

interesting in the next few years to correlate the optical results with the detailed distributions in neutral gas which are being made at Cambridge and Westerbork.

In M33, comparison of the HI and HII distributions shows an optical counterpart of each major HI cloud (Fig. 2). But the HII regions are much fainter and more scarce in the outer parts of the galaxy, compared to the flat distribution of neutral gas. This is a well-known phenomenon in our Galaxy and in other external galaxies, which when integrated with a larger beam give the rings in neutral hydrogen found by ROBERTS (1966).

This shows that star formation is more active in the inner part of the disk of spirals, perhaps because the volume density of HI is higher owing to the strong variation with z of the stellar gravitational potential.

From WRIGHT'S *et al.* results, the ratio of neutral gas density in and between the arms is ~ 3–5. But there is also a large amount of ionized gas in the interarm

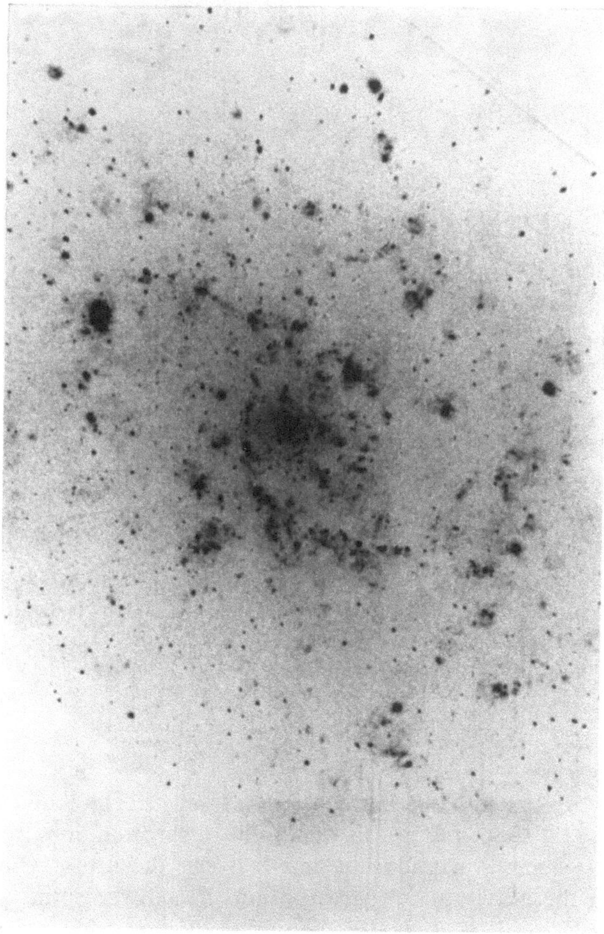

Fig. 2b. This give the H II distribution on the same scale and same orientation that in Fig. 2a. – This Hα photograph has been taken through a 15 cm diameter, 15 Å wide interference filter, at the 77 inch telescope of Haute Provence Observatory

region, in the disk, which was detected by MAYALL and ALLER (1939) in the [O II] lines, then studied in the $H\alpha$ and [N II] 6584 Å lines over the whole galaxy by CARRANZA *et al.* (1968) (Fig. 3). This disk emission has been frequently seen on smaller galaxies by different observers, the BURBIDGES, DE VAUCOULEURS, etc., and systematically studied in the nearest Sc galaxies by the COURTÈS' group since 1967. It appears as a very common phenomenon among Sc galaxies, with emission measures of $\sim 30\ cm^{-6}$ pc, and much fainter in Sbc or Sb galaxies (MONNET, 1971). It is only $8\ cm^{-6}$ pc in M 51, the only Sbc galaxy where an emission has been detected. A probable guess is that this emission comes from partially ionized gas (typically 30%) excited by low energy protons.

b) Kinematics. The radial velocity field is now fairly well known in most of the large nearby spirals – M 33, M 31, M 51, M 101, etc. – mainly through the use

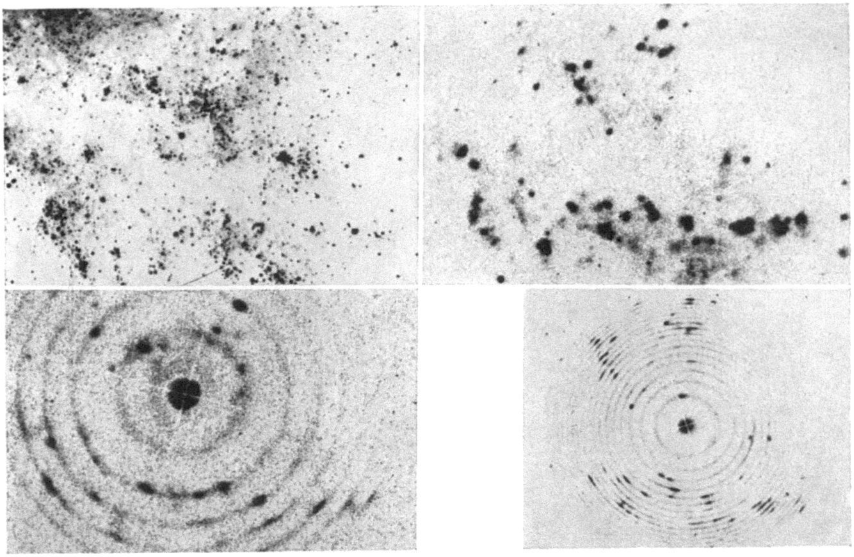

Fig. 3. M33 (NGC598) field: 6′ of arc. All four photographs refer to the same part of the Southern arm of M33, with a length of about 1.5 kpc. Top, left: taken in blue continuum light; Top, right: taken in $H\alpha$ light; Bottom, left: $H\alpha$ Perot-Fabry rings projected on the galaxy (77″ Telescope – Haute Provence Observatory); Bottom, right: $H\alpha$ Perot-Fabry rings projected on the galaxy (200″ Telescope – Hale Observatory). A smooth $H\alpha$ emission can be seen, in and between the arms

of Perot Fabry or Image Tube Echelle spectrographs. The study of the ionized disk emission by the COURTÈS' group, and of the absorption lines by SIMKIN (1970) in M51, show velocity variations in and between the arms, which are clearly associated with the gravitational perturbations due to the spiral arms.

As was shown by LIN *et al.* (1969) from angular momentum consideration, one expects *smaller* velocities at the inside of the spiral arms, and *higher* velocities at the outside.

This is clearly shown for M33 (Fig. 4) and NGC6946 (Fig. 5). Somewhat more limited evidence comes from observations across the spiral arms in M51 (TULLY, 1972); and in a piece of spiral arm in M31 (Fig. 6). The first radio results from synthesis techniques at 21-cm give the same result in M101 (ROGSTAD, 1971). The qualitative agreement is thus fairly impressive.

Quantitative computations are much less easy and there is still no reliable dynamical model of any spiral galaxy, hence no fully satisfactory test of LIN's Quasi-Stationnary-State Hypothesis – i.e., that the shape of the spiral arms are maintained against the strong differential rotation by collective motions of the stars and gas. In M33, from WRIGHT'S HI distribution, it is easy to compute the gravitational perturbation due to the gas itself. The iso-potential curves are given in Fig. 7. They are highly asymmetrical as is the distribution of the gas density, of about 4% of the total potential of gas plus stars. In this early type galaxy, we are thus rather far from the popular linear LIN's theory and fully non linear computations are needed.

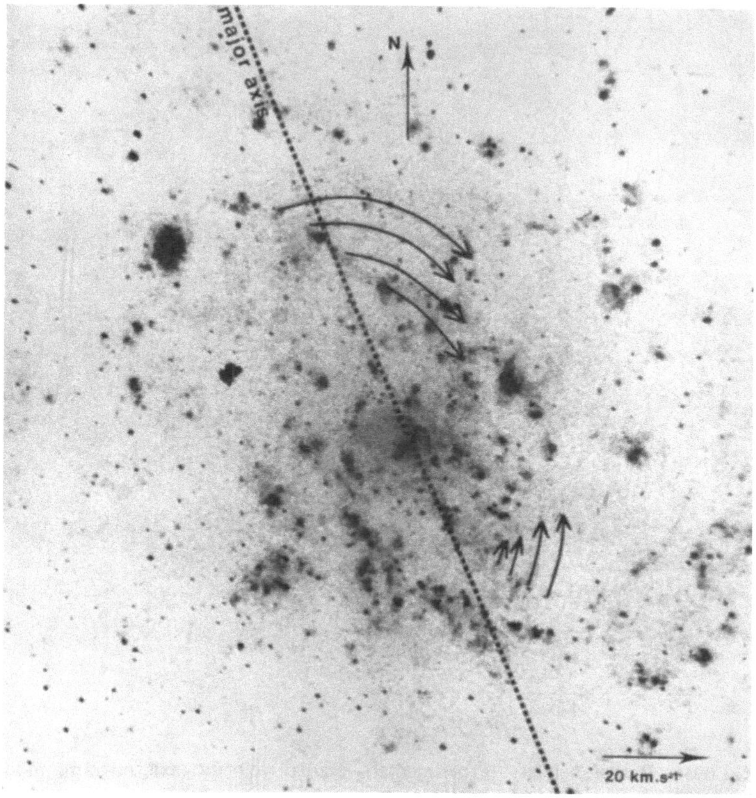

Fig. 4. The same with Fig. 2b. – The arrows indicate residual velocities of ionized gas in the plane of M33 – i.e. the difference between the local radial velocity and the theoretical velocity given by the mean rotation curve. One sees rather small negative residual velocities (~ 5 to $15\ \mathrm{km\,s^{-1}}$), in the concavity of the major Southern arm (no 1 on Fig. 2a) and large positive residual velocities ($\sim 30\ \mathrm{km\,s^{-1}}$) in the Northern spiral arm no 3

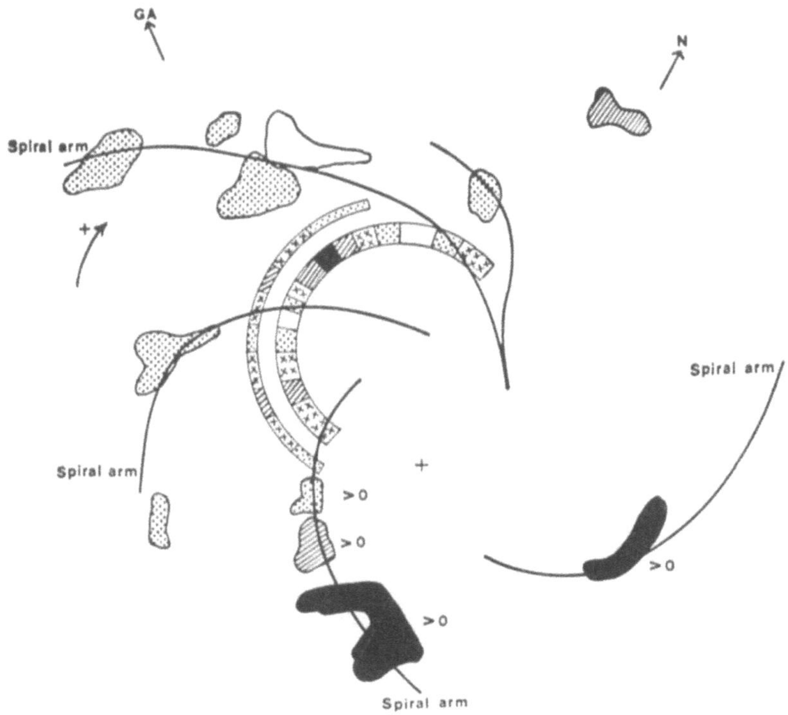

Fig. 5. NGC 6946. Radial velocity (experimental) – Radial velocity (axisymmetric model). [dotted] $\sim 0\,\mathrm{km\,s^{-1}}$, [crosses] -7 to $-21\,\mathrm{km\,s^{-1}}$, [hatched] -21 to $-35\,\mathrm{km\,s^{-1}}$, [black] $\leqq -35\,\mathrm{km\,s^{-1}}$. The same symbols are used for positive residuals, but with a >0 subscript. Residual velocities in the plane of NGC 6946 are shown by symbols. They have been computed as the difference between the observed radial velocity and the theoretical radial velocity given by the best axisymmetric model indicated by the velocity pattern. The following model has been adopted: Systemic velocity: $V_s = +10\,\mathrm{km\,s^{-1}}$. Major axis: P. A = 60°. Inclination: $i = 60°$. Rotation Law: $\Theta = 72\,\tilde{\omega}^{1/2}$. Θ rot. velocity in $\mathrm{km\,s^{-1}}$. $\tilde{\omega}$ vector radius in minute of arc. One sees large positive residual velocities in the Southern part, and in the Northern part residual velocities increase in the inside of the spiral arms and decrease in the outside

Fig. 6. M 31. N_4 spiral arm (on a 15′ of arc length). Systemic velocity: $V_s = -313\ \mathrm{km\,s^{-1}}$. Inclination: $i = 13°$. (Observed radial velocity) − (Theoretical radial velocity). [black] $\leqq 15\ \mathrm{km\,s^{-1}}$, [hatched] -15 to $-5\ \mathrm{km\,s^{-1}}$, [crosses] -5 to $+5\ \mathrm{km\,s^{-1}}$, [dotted] $+5$ to $+15\ \mathrm{km\,s^{-1}}$, [open] $\geqq +15\ \mathrm{km\,s^{-1}}$

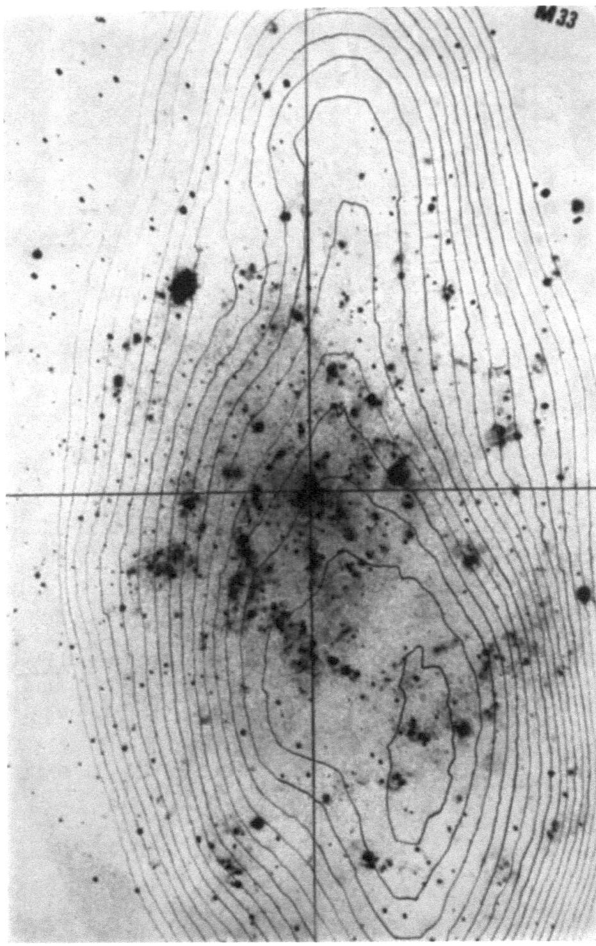

Fig. 7. This gives the iso-gravitational potential Φ_G curves of the neutral hydrogen mass distribution. The central maximum potential is $2200\,\mathrm{km^2\,s^{-2}}$ – and each iso curve step is $-60\,\mathrm{km^2\,s^{-2}}$. The mean asymmetry is given by:

$$\frac{\Phi_G\,(\text{South}) - \Phi_G\,(\text{North})}{\text{Stellar Potential}} = 4\%.$$

The curves have been superposed on the $H\alpha$ photograph given on Fig. 2b

References

AGT, S. VAN: Bull. Astron. Inst. Neth. **19**, 275 (1967).

BAADE, W.: Astrophys. J. **100**, 137 (1944).

BAADE, W., SWOPE, H.: Astron. J. **66**, 300 (1961).

BAUM, W.: Publ. Astron. Soc. Pacific **71**, 106 (1959).

BERGH, S. VAN DEN: J. Roy. Astron. Soc. Can. **62**, 145 and 205 (1968).

BERGH, S. VAN DEN: Astrophys. J. Suppl. no **171**, 145 (1969).

BERGH, S. VAN DEN: In D. EVANS (ed.), External Galaxies and Quasi-Stellar Objects, IAU Symp. **44**, 6 (1972).

BOULESTEIX, J., MONNET, G.: Astron. Astrophys. **9**, 350 (1970).
CARRANZA, G., COURTÈS, G., GEORGELIN, Y., MONNET, G., POURCELOT, A.: Ann. Astrophys. **31**, 63 (1968).
COHEN, G.: Publ. Astron. Soc. Pacific **82**, 760 (1970).
DE VAUCOULEURS, G.: Amer. Phil. Soc. Yearbook 1961, 268 (1961).
EINASTO, J.: In D. EVANS (ed.), External Galaxies and Quasi-Stellar Objects, IAU Symp. **44**, 37 (1972).
GREENE, T.: Astron. J. **73**, S. 15 (1968).
HODGE, P.: Astrophys. J. **142**, 1390 (1965).
JOHNSON, H.M.: Astrophys. J. **133**, 309 (1961).
JOHNSON, H.M.: Astrophys. J. Letters **174**, L 71 (1972).
LEWIS, B.: Astron. Astrophys. **16**, 165 (1972).
LIN, C., YUAN, C., SHU, F.: Astrophys. J. **155**, 721 (1969).
MAYALL, N., ALLER, L.: Publ. Astron. Soc. Pacific **51**, 112 (1939).
MCCLURE, R., BERGH, S. VAN DEN: Astron. J. **73**, 313 (1968).
MONNET, G.: Astron. Astrophys. **12**, 379 (1971).
MÜNCH, G.: Astrophys. J. **131**, 250 (1960).
PEIMBERT, M.: Astrophys. J. **154**, 33 (1968).
PEIMBERT, M., SPINRAD, H.: Astron. Astrophys. **7**, 311 (1970).
ROBERTS, M.: Astrophys. J. **144**, 639 (1966).
ROGSTAD, D.: Astron. Astrophys. **13**, 108 (1971).
RUBIN, V., FORD, W.: Astrophys. J. **159**, 379 (1970).
RUBIN, V., FORD, W.: Astrophys. J. **170**, 25 (1971).
RUBIN, V., FORD, W.: In D. EVANS (ed.), External Galaxies and Quasi-Stellar Objects, IAU Symp. **44**, 49 (1972).
SANDAGE, A., BECKLIN, E., NEUGEBAUER, G.: Astrophys. J. **157**, 55 (1969).
SEARLE, L.: Astrophys. J. **168**, 327 (1971).
SIMKIN, S.: In W. BECKER and G. CONTOPOULOS (eds.), The Spiral Structure of Our Galaxy, IAU Symp. **38**, 79 (1970).
SPINRAD, H., GUNN, J., TAYLOR, B., MCCLURE, R., YOUNG, J.: Astrophys. J. **164**, 11 (1971).
SPINRAD, H., SCHWEIZER, F.: Astrophys. J. **171**, 403 (1972).
SPINRAD, H., TAYLOR, B.: Astrophys. J. Suppl. **22**, 445 (1971).
TULLY, R.: University of Maryland, (Ph. D. Thesis) (1972).
WALKER, M.: Astron. J. **69**, 744 (1964).
WOOD, D.: Astrophys. J. **145**, 36 (1966).
WRIGHT, M., WARNER, P., BALDWIN, J.: Monthly Notices Roy. Astron. Soc. **155**, 337 (1972).

Discussion

MEABURN, J.:
To avoid airglow why not use $H\beta$ instead of $H\alpha$?

MONNET, G.:
The gain is not obvious specially in a dust rich galaxy where $H\beta$ may be much fainter than $H\alpha$, but we are trying it.

Gas Motions in the Nuclei of Seyfert Galaxies

By M.-H. ULRICH
The University of Texas at Austin, Department of Astronomy
Austin, Texas, U.S.A.

Abstract

Recent observations of profiles of the permitted lines in the nuclei of some Seyfert galaxies have shown that the broad wings of the permitted lines constitute evidence for mass motions with velocities in the range 1000–4000 km s^{-1}. Consequences of these high-velocity motions are discussed.

In the optical spectra of most Seyfert galaxies the forbidden lines and the permitted lines have different widths. The width of the permitted lines can be as large as 100 Å, whereas the width of the forbidden lines does not exceed 20 Å. This shows that in Seyfert nuclei, the ionized gas is present under two different phases: the broad component of the permitted lines originates in a very dense phase $N_e > 10^7$ cm^{-3}; the forbidden lines and the center of the permitted lines originate in a less dense phase $N_e < 10^7$ cm^{-3}.

In the nearby Seyfert galaxies NGC 1068, NGC 4151 and NGC 7469, the region emitting the forbidden lines is spatially resolved into a few clouds extending over several hundred parsecs and with velocities in the range 200–600 km s^{-1} with respect to the center of the galaxy (WALKER, 1968a, 1968b; ULRICH, 1972a, 1973). On the contrary, the broad wings of the hydrogen lines in NGC 4151 and NGC 7469 originate in a region which is seeing-limited and which, in projection on the sky, coincides with the seeing-limited region emitting the non-thermal continuum. This shows that the permitted line region is relatively close to the center of the galaxy, and is significantly smaller than the forbidden line region.

Detailed observations of the profiles of hydrogen lines in a few Seyfert nuclei have shown that the broad wings cannot be explained by electron scattering alone and therefore the broad wings constitute evidence for the presence of mass motions with velocities in the range 1000 to 4000 km s^{-1} (ANDERSON, 1971; ULRICH, 1972b). Some implications of these high-velocity mass motions are discussed below.

The intensity of a hydrogen line, for example $H\beta$, emitted by a gas cloud of volume V and mass M is given by

$$L(H\beta) \propto N_e^2 V \tag{1}$$

or

$$L(H\beta) \propto N_e M. \tag{2}$$

Consequently, for a given luminosity $L(H\beta)$ observed at a certain time, the mass of gas necessary to emit $L(H\beta)$ is proportional to N_e^{-1}. Since in the region emitting the broad component of the permitted line the density is very high, it is easy to see that only a small mass and therefore a small volume is sufficient to produce the observed intensity of the broad wings of the hydrogen lines (cf. formula 1). Since high velocity mass motions are present in such small volumes, it seems reasonable to expect that rapid expansion of the dense clouds can occur. As the gas expands, the density decreases and the luminosity in the hydrogen line of a cloud of constant mass decreases following formula (2).

Table 1. Expansion of a dense gas cloud[a]

N_e (cm^{-3})	10^7	10^8
l (cm)[b]	1.6×10^{17}	3.4×10^{16}
M ($M_\odot$)[b]	130	13
t (year)	57	12
Rate ($M_\odot$/year)	2.3	1

[a] It is assumed that the expansion velocity is 1000 km s^{-1} and that the $H\beta$ luminosity is 10^{41} erg s^{-1} when the expansion starts.

[b] The numerical values of l and M are for a homogeneous sphere of gas at $T\simeq15000$ K. The gas is assumed to be optically thin in the Balmer lines.

In Table 1, we consider a dense gas cloud of constant mass expanding at a constant velocity $v=1000$ km s^{-1}, and assume that at the time the expansion starts, the cloud emits 10^{41} erg s^{-1} in $H\beta$; this latter figure is the typical $H\beta$ luminosity of the Seyfert nuclei which have been studied in some detail.

The mass M of the gas cloud depends on the initial value of the electron density but remains constant during the expansion. The linear dimension l is the initial value of the cloud radius assuming the gas is contained in a homogeneous sphere. The time t is the time necessary for a decrease by a factor 10 of the $H\beta$ luminosity of the cloud of mass M.

Obviously a decrease of $L(H\beta)$ by a factor 10 would be easily noticeable and in some instances would make the $H\beta$ line undetectable. The rate given at the bottom line of Table 1, is M/t i.e. the mass of gas whose $H\beta$ luminosity in average decreases by a factor 10 in one year; it is also the mass of gas which each year has to be injected in the dense region if its overall $H\beta$ luminosity remains constant.

Table 1 has two important consequences:

(i) If, in a given Seyfert nucleus, the luminosity and the profile of the broad wings of the hydrogen lines remain constant over a few years, it means that either there is some mechanism preventing the high-velocity gas to expand, or that the dense gas expands but is constantly replaced by new dense gas.

(ii) If the gas does expand with the assumed velocity of 1000 km s^{-1} then smaller is the initial density, slower is the decrease of $L(H\beta)$ but higher is the rate at which the gas becomes undetectable and has to be replaced.

Evidently regular observations over several years of the profiles and of the total intensities of the permitted lines of some Seyfert nuclei will be of great interest.

References

ANDERSON, K. S.: Astrophys. J. **169**, 449 (1971).
ULRICH, M.-H.: Astrophys. J. **171**, L37 (1972a).
ULRICH, M.-H.: Astrophys. J. **174**, 483 (1972b).
ULRICH, M.-H.: Astrophys. J. **181**, 51 (1973).
WALKER, M. F.: Astrophys. J. **151**, 71 (1968a).
WALKER, M. F.: Astron. J. **73**, 854 (1968b).

Electronographic Observations of the Optical Structure of Radio Galaxies

By C. D. MACKAY
Institute of Astronomy, Cambridge, U.K.

Abstract

A "Spectracon" electronographic image tube (MCGEE *et al.*, 1969) has been used to make observations of a number of the brighter 3C radio galaxies as part of a programme to search for structural features in the radio galaxies and others in the field which may be associated with the presence of radio emission. Observations were made in the *B* and *V* bands, and the linearity of the electronographic process allowed accurate measurement of the distribution of luminosity and colour over the objects in the field.

Although a number of radio galaxies have been observed already only the results of the observations of 3C465 and some field galaxies were described in detail. It was found that the outer envelope of the central galaxy was extended very asymmetrically along a direction which approximately bisects the angle between the two bright ridges of radio emission (MACDONALD *et al.*, 1968). This envelope was detected to a diameter in excess of 80 kpc (with $H = 100\,\mathrm{km\,s^{-1}\,Mpc^{-1}}$) and found to be very red in colour. The three galaxies between 50 arc sec and 150 arc sec to the east and south of the central galaxy were also measured. The galaxy approximately 90 arc sec from the central galaxy was found to have a much steeper brightness gradient on the side nearest to the radio source. The colour of the galaxy was redder nearer the radio source (to the west of the galaxy) and became progressively bluer along the minor axis.

A more substantial account of these and other results is in the course of preparation. This work is part of a cooperative programme between the Mullard Radio Astronomy Observatory, Cambridge, and the Astronomy Department of Imperial College, London. The observations were made at the Royal Greenwich Observatory, Herstmonceux.

References

MACDONALD, G. H., KENDERDINE, S., NEVILLE, A. C.: Monthly Notices Roy. Astron. Soc. **138**, 259 (1968).

MCGEE, J. D., MCMULLAN, D., BACIK, H., OLIVER, M.: Adv. Electronics Electron Phys., **28 A**, 61, Academic Press, London (1969).

Discussion

OORT, J. H.:
Variation of axial ratio with decreasing surface brightness of the type you found in the galaxy *C* near 3C456 has been observed in some brighter elliptical galaxies like NGC3115 which contain a disk population as well as an elliptical population.

MACKAY, C. D.:
This is true. Indeed, many of the galaxies we have studied show substantial deviations from the simple shape of constant ellipticity as was found by Professor OORT many years ago.

DANKS, A. C.:
What is the spectral sensitivity of the photocathode used in "Spectracon" and which emulsion was used?

MACKAY, C. D.:
The photocathod had an S11 response and was used with *B* and *V* band filters. The emulsion was Ilford L4.

HAWKINS, M.:
What steps did you take to correct for the non-uniformity in the sensitivity of the photocathode?

MACKAY, C. D.:
None, but the photocathode was uniform to within 2%, over the area covered by the galaxies described.

HAWKINS, M.:
But even this would distort the outer contours at least.

MACKAY, C. D.:
That is true, but the contour map should not be believed to more than one contour level. The measurements of ellipticity and colour were done using one-dimensional microphotometer scans which were checked for background non-uniformity.

The Magellanic Clouds

(Invited Lecture)

By B. E. Westerlund
European Southern Observatory, Santiago de Chile

Abstract

Recent observational data are summarized, and the distances, structure and motions of the two Magellanic Clouds are discussed.

I. Introduction

Since the Symposium on the Magellanic Clouds in Santiago de Chile in March, 1969, a large number of investigations of great variety has been carried out in this field. Our knowledge has increased appreciably in certain areas. However, many fundamental problems are still as far from their solution as in 1969; this is in particular true for the structure of and motions in the two galaxies; and the study of the older stars is still in its infancy.

In the following, some of the recent observational results will be described. The structure of the two galaxies is discussed with the emphasis on types of objects for which new information has become available, and the motions in the Clouds are considered.

Among research areas omitted here are the interrelations between the Galaxy and the Clouds and the structure of and motions in the giant H II region 30 Doradus; both topics deserve extensive discussion, separately.

II. New Observational Data

a) The Brightest Stars

During the last few years a large number of bright stars in the region of the Magellanic Clouds have been catalogued. Sanduleak (1968) listed 169 proven or possible members of the Small Magellanic Cloud (SMC) and followed this (1969 b) with a list of 47 members of the Wing of the SMC and (1970) with a catalogue of 1272 possible members of the Large Magellanic Cloud (LMC), all from objective-prism spectral classification. Fehrenbach and Duflot (1970) listed 495 members of the LMC together with 1830 galactic stars found by radial-velocity measurements according to Fehrenbach's objective-prism technique. The large velocity shifts (about $+275\ \mathrm{km\,s^{-1}}$ in LMC and $+165\ \mathrm{km\,s^{-1}}$

in SMC) make this method very efficient, though some galactic high-velocity stars may be included. CAROZZI *et al.* (1971) have applied FEHRENBACH's technique to search for Cloud members in the Wing of the SMC. They report to have found 29 new probable members. ARDEBERG *et al.* (1972) published UBV data, spectral types and radial velocities for 409 LMC stars, 42 possible LMC stars and 132 galactic foreground stars. The majority of the objects were chosen from the FEHRENBACH-DUFLOT catalogue. An additional list of 47 LMC stars and 40 foreground stars has just been completed by BRUNET *et al.* (1972). BIGAY *et al.* (1972) have observed 100 supergiants in the LMC in SANDULEAK's (1970) catalogue photoelectrically in the U, B, V system.

It is of interest to note that the FEHRENBACH survey has led to an appreciable increase in the known number of LMC supergiants of types A–G. For the still later classes the infrared objective-prism survey by WESTERLUND (MULLER, 1971, p. 31) resulted in the identification of about 600 M supergiants and 400 carbon stars for which, however, no additional data are available yet.

b) X-Ray Radiation from the Clouds

X-ray radiation from the Clouds had just been detected at the time of the Symposium on the Magellanic Clouds in Santiago de Chile in 1969 (MULLER, 1971, p. 95). Now 4 sources are identified in the LMC and one in the SMC (GIACCONI *et al.*, 1972); their positions, from the Uhuru catalogue, are given in Table 1. Both Clouds appear also to have a diffuse emission region which, however, may consist of several unresolved discrete sources (PRICE *et al.*, 1971; GIACCONI *et al.*, 1972). None of the sources have so far been definitely identified with an optical object; the proposed identification of 30 Dor with a strong X-ray source (PRICE *et al.*, 1971) is not supported by the Uhuru data for LMC X-1.

Table 1. Position of X-ray sources in the Magellanic Clouds

Source name	Position (1950)						Previous name
2U 0115-73	1^h	15^m	2^s	$-73°$	41′	24″	SMC X-1
2U 0521-72	5	21	36	−72	1	12	LMC X-2
2U 0532-66	5	32	19	−66	37	12	LMC X-4
2U 0539-64	5	39	22	−64	4	48	LMC X-3
2U 0540-69	5	40	58	−69	48	0	LMC X-1

The source in the SMC is located in the Wing. It shows no evidence of complex structure (LEONG *et al.*, 1971). Its intensity is highly variable on a time scale of hours; this rapid variation indicates that the source must be no bigger than the solar system. Its greatest observed intensity corresponds to a power in the 2–7 keV band of 1.2×10^{38} erg s^{-1}. Of the four sources in the LMC three are point sources ($<0^\circ\!.25$) whereas the fourth, LMC X-2, is extended with a width of $1^\circ\!.1$. At the distance of the LMC (50 kpc) this corresponds to 0.9 kpc and the source may well consist of a cluster of sources.

LMC X-3 and LMC X-1 may be two of the brightest X-ray sources observed with luminosities in the energy range of 2–7 keV about 7×10^{37} erg s^{-1}; this is 10 times greater than that of the Crab nebula. The total power in X-rays emitted by the LMC of about 3×10^{38} erg s^{-1} is about 0.1 of that estimated for our Galaxy. This is comparable to the ratio of masses. Most of the emission is probably concentrated in small clusters of sources.

c) The Variable Stars

The variable stars in the Clouds continue to attract interest and new observational material has been presented in a large number of papers. It is impossible here to give more than a brief summary of recent results for a few selected types of variables.

Variable Supergiant Stars. The stability of the brighter supergiants in the Magellanic Clouds has been questioned for a long time. Until very recently, no short term variations had been detected. Thus, BOK *et al.* (1966) noted that the five LMC stars studied by them appeared to be stable well within 0.05 mag over longer periods with no evidence for regular or semi-regular pulsations in brightness.

ROSENDHAL and SNOWDEN (1971) concluded from a study of five of the most luminous supergiants in the LMC that the maximum amplitude of variability on short time scales (minutes to hours) is less than ± 0.015 mag in both blue and visual wavelength ranges for all five stars, but that all five stars appeared to be variable when observed over a time interval of a month or more with the variability ranging from 0.02 to 0.07 mag in the visual and 0.04 to 0.08 mag in the blue. They propose that pulsation is a possible cause of the variability. Some of the stars studied have emission-lines in their spectra and are apparently losing mass; as a consequence of this the nuclear-energized pulsations would not be stabilized due to the increasing central condensation during evolution away from the main sequence.

The stars studied by ROSENDHAL and SNOWDEN are rather evolved supergiants with types ranging from B8 to F8. Recently, APPENZELLER (1972) has observed 18 bolometrically very bright O and B stars in the LMC. Almost all the observed stars were found to be variable on a time scale of several days to several weeks. Only for the two hottest stars (O5 and O6) and for one star with a very peculiar spectrum no variations could be detected. The amplitudes of the variations (0.005 to 0.226 in V) seemed to depend strongly on the spectral type with a maximum at about B2.5.

Nightly variations were detected in five of the 18 stars; four of these stars have P Cyg characteristics. In three of these stars the observed variations looked like slow irregular drifts with no indication of any periodicity. In the two other the variations are not inconsistent with a period of 8 hours. This is the fundamental pulsation period of a 95 solar masses zero-age main sequence (ZAMS) star; the two stars may thus be vibrationally unstable.

Almost all the stars in APPENZELLER'S programme show easily detectable long term variations. The amplitudes are highly different, but all the stars show maximum variations in V and in $U-B$ and very little in $B-V$. Most striking is HDE 269128 with V varying by 0.40 and $U-B$ by 0.15 during the observing run, whereas $B-V$ stayed constant within ± 0.003 mag. For these long-term variations the correct physical explanation has yet to be found.

The Novae. Until the new survey for Magellanic Cloud novae, initiated by GRAHAM, started at Cerro Tololo Inter-American Observatory in September 1970 four novae were known in the SMC (TUC, 1897, 1927, 1951, 1952) and seven in the LMC (DOR, 1926, 1936, 1937, 1948; MENSAE, 1951, 1968, 1970a). GRAHAM'S survey (GRAHAM and ARAYA, 1971) has added three novae in the LMC (MENSAE, 1970b; DOR, 1971a, 1971b). Nova Mensae 1970b was a very fast nova and Nova Dor 1971a a moderately fast nova. The latter is the best observed of Magellanic Cloud novae. It was followed photoelectrically for over 400 days (ARDEBERG and DE GROOT, 1972) and it is believed that a few observations were obtained before its principal maximum which according to ARDEBERG and DE GROOT occurred on Julian Date 2440985.20 with $V=11.84$, $B-V=+0.02$ and $U-B=-0.86$. Spectra (disp. 73 Å/mm) were obtained of Nova Dor 1971a at the European Southern Observatory on La Silla as well as of Nova Mensae 1970b and have been described for the latter (HAVLEN *et al.*, 1972).

The Cepheids. Several papers dealing with this class of variables in the Magellanic Clouds have appeared, and the differences between the cepheids in the SMC, the LMC and the Galaxy have been discussed extensively.

The most disturbing effects existing between the cepheids in the SMC and in our Galaxy are (SANDAGE and TAMMANN, 1971): (1) the SMC contains many cepheids with periods between 1.5 and 4 days, whereas our Galaxy has few; (2) the $B-V$ colours of the SMC cepheids are bluer than those in the Galaxy by about 0.1 mag for periods shorter than 10 days; (3) the amplitude of these short-period, blue SMC cepheids are very large, reaching 1.6 mag. These three points may indicate that cepheids differ from galaxy to galaxy and that their use as precision distance indicators are suspect. However, SANDAGE and TAMMANN point out that the three points may be intimately related and "may be but different statements of a single special phenomenon, which is *unrelated to the form and zero point of the* $P-L-C$ (period-luminosity-colour) *relation*".

The Cepheid instability strip in the (M_{bol}, log Te) diagram is likely the same for all galaxies. However, the filling of the strip may differ from galaxy to galaxy due to differences in the evolutionary tracks that thread the strip. Thus, the bluer Cepheids in the SMC result from their penetrating deeper into the strip (on the second passage) than the galactic cepheids. This explains point (1) and (2) above. Point (3) would then follow if for this type of cepheids the amplitude of light variation were largest at the blue end of the strip. SANDAGE and TAMMANN conclude that this is true for cepheids with $0.40<\log P<0.86$ and derive for this group a period-luminosity-amplitude relation:

$$M^0_{\langle B\rangle} = -2.386 \log P - 1.406 f_B - 0.205$$

where the amplitude defect is

$$f_B = 10^{0.4(\Delta B - \Delta B_{max})};$$

ΔB = observed amplitude in magnitude, and
ΔB_{max} = maximum amplitude reached by any Cepheid at a given period.

In this connection it should be mentioned that the latest form of the $P-L-C$ relation is

$$M^0_{\langle B\rangle} = -3.534 \log P + 3.647\,(\langle B\rangle^0 - \langle V\rangle^0) - 2.469\,.$$

These relations are used to derive distance moduli for the Magellanic Clouds (see below).

It is of interest to note that SANDAGE and TAMMANN also find that by taking the amplitude differences into consideration "there is no evidence for intrinsic differences in the colour properties of cepheids in the four galaxies" (SMC, LMC, M 31, Galaxy). Thus, the difference in colour remaining between the SMC and the galactic cepheids is found to be 0.041 ± 0.020 mag which is well within the observational errors.

The RR Lyrae Stars. Four fields in the Magellanic Clouds are being studied by GRAHAM (1972), and he has obtained very interesting results for the field RR Lyrae variables in two of them: one near NGC 1783 in the LMC and one centered on NGC 121 in the SMC. A third field is centered on NGC 1835 in the LMC and the fourth between NGC 361 and NGC 362 near the SMC Bar.

There are now 3 cluster RR Lyrae stars with known periods in the SMC, all in NGC 121, and 22 in the LMC: 6 in NGC 1466, 10 in NGC 1835, and 6 in NGC 2257 (GRAHAM, 1972). The mean periods of the RR Lyrae *a b* type stars is 0.560 in the SMC cluster and 0.559 in the LMC clusters, the corresponding time averaged blue magnitudes, $\langle \bar{B} \rangle$, are 19.7 and 19.3, respectively. On the basis of the variable star data all four clusters fall into Oosterhoff type I.

In the field of NGC 1783 72 probable RR Lyrae stars were detected and periods have been determined for 50 of type *a b* and 1 of type *c*. Preliminary results from the field around NGC 121 are: 60 probable RR Lyrae stars of which periods are determined for 16 of type *a b* and 4 of type *c*. The periods and averaged magnitudes are compared in Table 2. We note that there is a shift of about $0^d.05$ towards a longer period in the case of the SMC variables; this may indicate that they are slightly metal poor compared with the LMC variables.

A comparison of the $\langle \bar{B} \rangle$ values of the cluster and field RR Lyraes show the former to be about $0^m.3$ brighter. This is most likely due to the background enhancement in the globular cluster areas and the field star data should be considered as more accurate.

GRAHAM notes that the dispersion in mean apparent blue magnitude $\langle B \rangle$ is surprisingly small for the field RR Lyrae stars in the LMC. This indicates (1) that the intrinsic dispersion of the absolute magnitudes of these stars is very small; (2) that the LMC RR Lyrae stars studied are concentrated in a disk system rather than spread through an extended halo system; and (3) that the

Table 2. Most frequent periods and time averaged blue magnitudes for field RR Lyraes in the SMC and the LMC

	SMC	LMC
Most frequent period between	$0^d.55$–$0^d.60$	$0^d.50$–$0^d.55$
$\langle \bar{B} \rangle$	20.0	19.56

interstellar absorption is very homogeneous or/and very small. From (1) follows that the range of metal abundance may be quite small among the old LMC population. This is supported by the absence of the short RR_{ab} variables with $P < 0\overset{d}{.}45$ and, at the other extreme, by the possible absence of Oosterhoff type II clusters.

d) Globular Clusters

WALKER (1970, 1971, 1972a, 1972b) has carried out electronographic photometry of globular clusters in the Magellanic Clouds to faint limiting magnitudes ($V > 22$).

So far, the results for 3 clusters, Kron 3 in the SMC; NGC 2209 and NGC 2257 in the LMC, have appeared and WALKER has also described NGC 419 in the SMC at a recent conference. He has concluded that Kron 3 with its curved giant branch, short horizontal branch (at $V = 19.1$) and subgiant branch extending down to the limit of the observations ($V = 22.46$) resembles the "metal-rich globular" clusters in our Galaxy although certain differences remain. One of these is the "short horizontal branch" referred to above which is far too luminous if $m - M = 19.2$ for the SMC. WALKER proposes that it may be instead an asymptotic giant branch and that a good fit (with $m - M = 19.0$) would then be obtained with M 13 for all branches; except that Kron 3 has none of the blue horizintal-branch of M 13.

Another aspect in which Kron 3 differs from the globulars in our Galaxy is in the occurrence in Kron 3 of extremely red giants ($B - V$ up to 2.4). The brightest red giants are much less concentrated to the center of the cluster than the fainter stars; this suggests that they may be older stars that have lost mass during an earlier passage through the red-giant stage and that they have now reevolved from the horizontal or asymptotic branch back into the red-giant region.

NGC 2209 has been studied previously by HODGE (1960) and GASCOIGNE (1966). Its colour-magnitude diagram is puzzling: it has a sequence extending from $V = 17.7$; $B - V = +1.4$ to about $V = 20.0$; $B - V = 0$ with a possible gap at $V = 19.5$; $B - V = 0.3$. WALKER's extention to fainter limit ($V = 21.9$) shows a vertical main sequence from about $V = 19.5$ to $V = 21.9$ at $B - V = +0.20$. This is very similar to the Hyades and WALKER concludes that the cluster is not an old metalpoor Population II globular cluster but rather an object similar to the Hyades. However, in the Hyades and in similar galactic clusters the yellow-giant branch begins at $B - V \geqq 0.9$ instead of at $B - V \sim +0.55$ as in NGC 2209. The unusual appearance of NGC 2209 is ascribed by WALKER to a difference in chemical composition between the stars in it and in those found in clusters of comparable age in our Galaxy.

NGC 2257 is, when studied to $V = 22.3$, basically similar to the classical "Population II" globular clusters in our Galaxy though also here differences appear. Most striking is a gap in the giant branch at the magnitude level of the horizontal branch; the star density is zero in this gap. From the colour distribution of the horizontal branch stars and the height of the red-giant branch at $(B - V)_0 = +1.4$: $\Delta V = 2.55$, WALKER proposes that NGC 2257 may be interpreted either as (1) a cluster with extremely low metal content like M 92 but with an abnormally low red-giant branch, or (2) a cluster of intermediate metal deficiency like M 3 or M 13, which shows the "M 13 anomaly" in the distribution of its horizontal branch stars and in which the giants above the horizontal branch lie mostly on the asymptotic-giant branch.

It is concluded from the study of these three clusters that the existing differences between the individual objects are not sufficiently well-known to make this class of objects suitable for high precision determinations of distances.

The influence of the Magellanic Cloud field stars on the colour magnitude diagram of globular clusters is not yet well established.

NGC 2209 is considered as one of the most outlying clusters in the LMC, at $6^h 09^m$, $-73° 50'$, or 6° SE of the centre. However, long exposures have revealed rather rich star fields at this distance from the centre and it appears important to investigate to what extent these stars may contribute to vertical main sequences of clusters.

e) The Dust Content

The *foreground absorption* in the direction of both Clouds is small. FEAST *et al.* (1960) determined $E_{B-V}=0.07$ for both Clouds with standard deviations of $0\overset{m}{.}01$ and $0\overset{m}{.}04$ for the SMC and the LMC respectively. Most other determinations have given lower values, and GRAHAM (1972) considers the reddening virtually negligible in front of both Clouds, basing this conclusion in particular on GASCOIGNE'S (1969) and PHILIP'S (1971) work. HODGE (1969) noted that $3\overset{\circ}{.}2$ from the center of the LMC reddening is undetected. Also a recent investigation by DACHS (1972) confirms the very low foreground absorption in the direction of the Clouds.

The Dust Content of the Clouds appears remarkably low in relation to their gas content. FEAST *et al.* (1960) found mean absorptions of $A_v=0.33\pm0.03$ for 48 stars in the LMC and $A_v=0.34\pm0.03$ for 15 stars in the SMC (foreground absorption of 0.21 included). The maximum value measured by them was $A_v=0\overset{m}{.}96$ in the LMC and $0\overset{m}{.}60$ in the SMC.

Recently PAYNE-GAPOSCHKIN (1971) has tabulated the total absorption, A_{pg}, over the LMC in the Harvard coordinate system (X, Y). The maximum value given is $A_{pg}=0.68$; the fluctuations over the LMC are rather smooth and the absorption at $3\overset{\circ}{.}5$ from the dynamical center was adopted as zero.

Indications of a higher amount of dust in the Small Cloud have been given by WESSELINK (1961) who from galaxy counts estimated an absorption of up to $A=2$ mag in the center of the SMC. However, it is possible that the crowding in the central area led to a systematic "loss" of galaxies with a following high absorption value.

Also PAYNE-GAPOSCHKIN and GAPOSCHKIN (1966) noted in their investigation of the variable stars in the SMC that the Cepheids in some regions are systematically faint. They consider the background effects but conclude that absorption within the SMC is probably a major factor.

HUTCHINGS (1966) studied the $\lambda 4430$ band absorption in both Clouds and concluded that dust existed to about the same and rather considerable amount in both of them. However, his interpretation has been reanalyzed for the SMC by DACHS (1970), who concludes that a major fraction of the diffuse interstellar absorption at 4430 Å observed in the spectra of SMC stars is originating within the Galaxy.

Differences in interstellar reddening between objects of different types or ages have been looked for by several authors. FEAST *et al.* (1960) found a higher reddening of stars within nebulosity than of those not in nebulosity. They found the differences in reddening between emission-line stars and supergiants without emission features not statistically significant, and DACHS' (1972) colour excesses $E_{B-V} = +0\overset{m}{.}24 \pm 0\overset{m}{.}15$ for 6 emission line stars, and $E_{B-V} = +0\overset{m}{.}07 \pm 0\overset{m}{.}07$ for 26 non-emission line stars do only indicate strong variations from star to star.

VAN DEN BERGH and HAGEN (1968) carried out integrated UBV observations of 39 clusters in the LMC, 13 in the SMC and 7 in our Galaxy. They derived a mean reddening for the LMC and SMC clusters of $E_{B-V} = 0.20 \pm 0.05$ for clusters with ages smaller than 3×10^7 years and 0.06 ± 0.01 for clusters with ages between 3×10^7 to 1×10^9 years.

The higher reddening of some of the youngest clusters might be due to a high content of red supergiants as for example in NGC 2100, NGC 2004 and L 72, whereas 30 Dor (NGC 2070) is undoubtedly intrinsically reddened. The high absorption in the surroundings of 30 Dor has been noted by LE MARNE (1968) and by GASCOIGNE (1969) and is also seen in the colour-magnitude diagram of the central cluster (WESTERLUND, 1966).

Dust-to-gas ratios in the Clouds have been given values between 4 and 100 times smaller than in our Galaxy. VAN GENDEREN (1969, 1970) derived N_H/E_{B-V} diagrams for the SMC and the LMC by calculating the number of H I atoms, N_H, in a line-of-sight column of 1 cm^2 cross-section in the fields of his objects from HINDMAN'S (1967) and MCGEE and MILTON'S (1966) data. He found the mean ratio in the central part of the SMC to be a factor of 5 and in the northern part a factor of 12 smaller than in the Galaxy. For the LMC he derived a corresponding factor of 4. However, by adopting the currently used (but rather uncertain) values for the tilt angles of the Clouds he concluded that the two Clouds have more or less the same dust-to-gas ratio.

VAN DEN BERGH and HAGEN (1968) have determined a mean ratio at least 10 times smaller in the SMC than in the Galaxy, and DACHS (1970) has concluded that the dust-to-gas ratio is at least hundred times smaller in the SMC than in the Galaxy.

The Extinction Law in the Magellanic Clouds resembles the Perseus rather than the Cygnus law. BRÜCK *et al.* (1970) concluded this from a comparison of the energy distribution in spectra of reddened Cloud stars with galactic stars of the same spectral types. Corrections were applied to eliminate the difference in luminosity between classes I (MC) and V (Galaxy). They also found that the change in slope in the neighbourhood of $1/\lambda = 2.3$ is similar to that found in the galactic extinction laws.

They concluded that the composition, size, and optical properties of the dust grains in the Clouds resemble those in the anticentre region of the Galaxy.

Magnetic-Field Structure. The photoelectric determinations of the distribution of dust in and in front of the Magellanic Clouds may be compared with the optical polarization data derived by MATHEWSON and FORD (1970) and by SCHMIDT (1970). In both investigations the existence of a large-scale regular magnetic field is shown; the lines of force are parallel to the line joining the two galaxies. The possibility that this field is of galactic origin is rejected by MATHEWSON and FORD because of the low galactic dust content in the direction of the SMC and because of a significant difference between the degrees and angles of polarization in the LMC and in the Galaxy.

SCHMIDT (1970) arrived by studying foreground stars to the same conclusions, but considers (private communication) that more observations are necessary to establish positively that the polarization really occurs in the SMC and not in the Galaxy.

The regions of strongest polarization occur in the Clouds in the "Population I" areas (rich in OB stars, H II regions, H I concentrations). The 30 Doradus complex is the largest of them and has also the strongest polarization. The magnetic field extends from 30 Dor to the east and south-east along the H I concentration in that area. However, there is some indication that the field extends into and along the Bar, though SCHMIDT's surface polarimetry does not show this. This descrepancy between star and surface polarization angles could possibly be understood by the introduction of two high albedo dust layers containing differently aligned magnetic fields; the large scale Magellanic field would then be located in the one nearer to us. It is of interest to note that HODGE (1972a) has found a tendency among the dark nebulae, identified by him in the LMC, to show alignment with the Bar. There is also a strong suggestion of an elongation of the dust clouds parallel to the magnetic field.

f) Chemical Composition of the Clouds

High-dispersion spectra of the three stars, HD 33579 (PRZYBYLSKI, 1968; WOLF, 1972a), and HD 32034 (PRZYBYLSKI, 1971) in the LMC, and HD 7583 (PRZYBYLSKI, 1972; WOLF, 1972b) in the SMC have been analyzed for their chemical composition. WOLF finds the metal abundances in HD 33579 as well as in HD 7583 to be within a factor of two of those in the Sun and in galactic supergiants, whereas PRZYBYLSKI finds HD 7583 to be metal deficient by a factor of 10 and HD 32034 by a factor of 6. Slight differences in adopted temperatures may cause such discrepancies, and it appears reasonable to conclude that until now no significant differences have been found between the extreme Population I in the Clouds and in the Galaxy. An indication of weaker OI 7774 lines in the A-type supergiants in the Clouds as compared with those in the Galaxy (OSMER, 1972) may likewise be explained as an effect of adopted temperature; OSMER used $T_e = 9\,500°$ whereas WOLF and PRZYBYLSKI have $T_e \sim 8\,100°$.

Differences in the chemical composition of the Clouds in relation to our Galaxy have been inferred from the sizes of the loops in the evolutionary tracks of the Cepheids during their second crossing of the instability strip. CHRISTY (MULLER, 1971; p. 136) proposes that small differences in He content or metal content are sufficient to shorten or lengthen these loops. The SMC cepheids are the bluest and the galactic cepheids the reddest; the latter make then the shortest loops.

Some of the colour-magnitude diagrams of clusters in the Magellanic Clouds have been interpreted as showing composition differences between the Clouds and the Galaxy. ARP (1967) has studied the positions of the red ends of the giant branches of NGC 458 (SMC) and NGC 1866 (LMC) and compared them with the red limit for giants in our Galaxy. It is clear from his Fig. 7 that there is a progression in the red limits of the giant stars in the order: SMC, LMC and the Galaxy (the latter being most to the red). Redward displacements indicate increasing metal content and it is concluded that the SMC has the lowest content.

Peculiar colour-magnitude diagrams may be found in the Magellanic Clouds for other reasons than abundance effects. NGC 330 in the SMC was found by ARP (1959) to have a distribution of the supergiants in the V, $B-V$ diagram completely different from that of galactic clusters of the same age. It was shown by FEAST (1964a) that possible blending effects could explain some of the peculiarities. Recently, FEAST (1972) has shown that practically all the main sequence stars in the cluster, with $-3\overset{m}{.}6$—M_v—$-4\overset{m}{.}2$, are *Be* stars. As a consequence the position of the main sequence of NGC 330 in the colour-magnitude diagram may require significant correction for the effects of high stellar rotational velocities before a comparison can be made with a main sequence of non-rotating stars.

It is naturally necessary to extend the comparisons to include other classes of objects before definite conclusions regarding the metal contents of the three galaxies can be drawn. So far, for instance, the red giant stars in the field of the LMC, in particular in the Bar, appear similar to those in the Galaxy (WESTERLUND, 1960; TIFFT and SNELL, 1971), whereas VAN DEN BERGH and HAGEN (1968) propose that the very red stars found in many globular clusters in the Clouds indicate that the oldest stellar population in the Clouds may differ from Population II in our Galaxy and instead resemble that in some dwarf spheroidal galaxies.

III. The Distances to the Clouds

Table 3 summarizes recent determinations of the distance moduli of the Magellanic Clouds.

Table 3. True distance moduli of the Magellanic Clouds

Objects	LMC	SMC	Reference
Cepheids	18.6	19.3	SANDAGE, TAMMANN (1971)
RR Lyrae ($\langle M_v \rangle = 0.5$)	18.7	19.1	GRAHAM (1972)
Globular clusters:			
NGC 1466, 2257	18.7		GASCOIGNE (1966)
NGC 121		19.1	TIFFT (1963)
Old giant population	18.5	19.1	TIFFT, SNELL (1971)
Supergiants (Pop I)	18.1		DIVAN (1972)
Nova Dor 1971 a	18.5		ARDEBERG, DE GROOT (1972)
Novae	18.8	19.4	VAN DEN BERGH (1968)
Eclipsing variables	18.4	19.2	GAPOSCHKIN (1970, 1972)
Adopted values	18.5	19.2	

The data used by SANDAGE and TAMMANN (1971) are from GASCOIGNE (1969); GASCOIGNE derived in fact $m-M=18.7$ for the LMC and 19.4 for the SMC from his data using a slightly different $P-L-C$ relation.

The RR Lyrae data are from GRAHAM'S (1972) investigation, discussed on p. 43.

We feel that the globular clusters at present do not qualify as distance indicators for high precision determinations. The results derived by WALKER (see above) show that the clusters differ in many ways that make a simple fitting of their colour-magnitude diagrams or horizontal branches rather uncertain.

The low value obtained by DIVAN (1972), using the BARBIER, CHALONGE, DIVAN (B, C, D)-method for the two-dimensional classification of some supergiants in the LMC, may be due to the still existing uncertainty in the calibration of the most luminous part of the B, C, D diagram. The distance determination using Nova Dor 1971a should probably be given higher weight than the mean value derived by VAN DEN BERGH (1968). GAPOSCHKIN'S determinations (1970, 1972) using the eclipsing variables are based on 53 stars in the LMC and 30 in the SMC.

The accepted mean moduli correspond to distances of 50 and 69 kpc for the LMC and the SMC, respectively.

IV. The Structure of the Clouds

a) The Magellanic System

The low-resolution surveys of the neutral hydrogen in the Clouds showed the existence of large H I envelopes around both Clouds but also (for references see KERR, 1971) that a bridge existed between the two Clouds. The results of the optical polarization studies, that a large-scale regular magnetic field exists with the lines of force parallel to the line joining the two galaxies, support the idea of a Magellanic System.

The sharp edges to the neutral atomic hydrogen envelope on the leading and trailing edges of the complex may indicate that the two Clouds are separating. More detailed studies of the inter-Clouds area show that the neutral atomic hydrogen exists in the form of rather isolated small clouds with velocities in some cases indicating ejection from either SMC or LMC (TURNER *et al.*, 1968). Optically the bridge between the SMC and LMC is so far incomplete. The Wing of the SMC appears to end rather abruptly at about $RA=2^h15^m$ (WESTERLUND and GLASPEY, 1971).

The existence of possible links between our Galaxy and the Clouds remains in doubt (cf. DE VAUCOULEURS and FREEMAN, 1969, p. 8). GAPOSCHKIN (1970) considers that the LMC, the SMC and the Galaxy form a physical triplet galaxy. The identified linking objects are: a mini-cluster (No. 1), 39 Link-Cepheids, 61 Link-Lyrae stars (of which two are in the mini-cluster No. 1), and 19 Link-Eclipsings. All these objects are found over the asymmetrical part of the Milky Way at Centaurus-Carina-Vela.

It appears highly desirable to obtain further information about these objects. Likewise, it is hoped that the FEHRENBACH objective-prism technique will lead to the identification of any objects having luminosities and radial velocities corresponding to a position between the Galaxy and the Clouds.

b) The Space Orientations

DE VAUCOULEURS and FREEMAN (1969) summarize in their Table 1 determinations of the inclination (i) and the position angle (p_0) of the line of nodes of the LMC. They adopt as mean values

$$i = 27° \pm 2°, \quad p_0 = 170° \pm 1°.$$

The latter value is probably not far from correct, whereas the determination of i cannot yet be considered final. It depends mainly on an analysis of red-light isophotes of the outermost loop and the assumption that it is circular in its own plane. A similar assumption for the system of faint open clusters in the outer parts of the LMC gave $i=45°$ (LYNGÅ and WESTERLUND, 1963). MCGEE and MILTON (1966) derived values of $i=27° \pm 5°$ and $i=25° \pm 9°$ from the distribution of neutral hydrogen and open clusters respectively. However, in the case of the neutral hydrogen the extension south of the Bar was disregarded, and we are not convinced that the cluster catalogue used by them is complete in the outer regions. If the outermost contours of the neutral hydrogen are considered (see MCGEE and MILTON, 1966, Fig. 8) a value of i near 45° is obtained.

It is attractive to assume that in systems such as the Magellanic System, having a common envelope of neutral hydrogen and probably a general magnetic field, there is some connection between the plane of symmetry of the more massive galaxy and the position in space of the two major components.

The presently accepted distances of the LMC and the SMC are 50 and 69 kpc respectively. It is easily seen that with a separation of 20° on the sky between the centres of the two galaxies a tilt of 45° of the LMC will make its plane of symmetry extend through the main body of the SMC. (1° at 50 kpc = 0.88 kpc; at 69 kpc = 1.21 kpc.)

For the SMC DE VAUCOULEURS and FREEMAN (1969) give $i=60° \pm 3°$ and $p_0 = 45° \pm 3°$. HINDMAN (1967) finds from the neutral hydrogen distribution $i > 70°$ (end-on Bar) and $p_0 = 55°$. GAPOSCHKIN (1970) has proposed that both Clouds are seen face-on; this results from the distribution of the variable stars.

We conclude that also for the SMC the definite value of i is still to be determined.

c) The Sub-Systems of the Clouds

We have found it convenient to describe each Cloud in terms of a small number of sub-systems (WESTERLUND, 1970). Tables 4 and 5 summarize their contents and dimensions with recent observational data taken into consideration. As compared with the previous version there are only minor changes, and we will here discuss only some classes of objects, mainly belonging to Population I in the LMC, which have recently been more thoroughly investigated.

However, firstly we wish to call attention to some of the fundamental differences between the two Clouds.

The Neutral Hydrogen in the Main Bodies of the two Clouds shows a striking difference in distribution. The LMC neutral hydrogen has a very patchy appearance, while the SMC gas is much smoother in its distribution. The LMC has

Table 4. The sub-systems of the Small Magellanic Cloud, $m-M=19.2$

Dimensions	Bar 2°5 × 1°	Wing 6° × 1°	Central System 7° (possibly flat)	Halo 7° (spherical)
Observed content	a) *Extreme Population I:* OB associations, young clusters, supergiants of class I a H II regions	*Extreme Population I:* OB associations, H II regions, blue and yellow supergiants of class I a and I–O long-period cepheids	a) *Population I:* Old open clusters planetary nebulae	*Population II:* Old globular clusters, RR Lyrae variables
	b) *Population I:* Supergiants of class I b Cepheids		b) *Population II:* Red giant stars, intermediate-age globular clusters	
Limiting magnitude	17 mag	20 mag	20 mag	22.5 mag

Note: One degree is 1.2 kpc. HINDMAN's (1967) results from H I observations indicate that the optical Bar should possibly be considered as an end-on Bar plus a superposed arc of extreme Population I objects. The center of the SMC central system is assumed to lie near NGC419; the length of the Wing has been measured from this point. The extent of the "central" system is based on GASCOIGNE's (1966) classification of the intermediate-age globular clusters. A division of the globulars of this group into two classes would lead to a much smaller central system. Interference by foreground stars influences the colour-magnitude diagrams of the central system severely.

Table 5. The sub-systems of the Large Magellanic Cloud, $m-M=18.5$

Dimensions	Bar 3° × 1°	Central System 6° (flat system)	Disc 14° (flat)	Halo 24° (spherical)
Observed content	a) *Extreme Population I:* OB associations and H II regions (superposed on Bar and belonging to Central System?)	*Extreme Population I:* OB associations, young clusters, supergiants of class I a and I–O. H II regions and most of the neutral hydrogen	*Population I:* Old open clusters, planetary nebulae, carbon stars	*Population II:* Old globular clusters. LMC-type globulars
	b) *Population II* and *Old Population I:* Red Giants		*Population II:* Field RR Lyrae stars	
Limiting magnitude:	$I=13.5$ mag $V=18$	$V=18$ mag	$V=17$ mag	$V=22.3$ mag

Note: One degree is 0.9 kpc. The center of the LMC has been taken to coincide with the centroid of the system of the planetary nebulae, R.A. $=5^h 24^m$, Dec $=-69°5$ (1950).

52 clearly defined concentrations, the SMC only three. However, the total amount of gas is roughly the same in the two galaxies, about 5×10^8 solar masses. A typical concentration in the LMC has a diameter of about 500 pc, a velocity dispersion of 11.5 km s^{-1}, a gas density of 1 atom/cm^{-3} and a total mass of 4×10^6 $M_\odot$. In both Clouds double peaked profiles are observed implying structure in depth.

In the LMC the H I complexes are closely correlated in position as well as in motion with the H II regions and with regard to motion with most of the *OB* stars in these areas.

In the SMC there is only a superficial correlation between the H I and H II as is to be expected from the smooth distribution of H I. However, FEAST (1970) finds a good agreement between the optical radial velocities of a number of H II regions in the SMC and the velocities of the most intense 21-cm peak in each case.

The Bars of the two Clouds are strikingly different. The Bar of the SMC is blue, dominated by Extreme Population I objects, whereas the Bar of the LMC contains a very strong concentration of faint red stars.

From an infrared spectral survey WESTERLUND (1960) concluded that the luminosity of the *M* giants detected in the LMC Bar, $M_I = -5.5$ $(m - M = 18.5)$, corresponded to Population II. It should be noted that these objects are detected because they have TiO bands strong enough to be seen in low dispersion; this indicates that they are not similar to the red giants in ordinary globular clusters.

WALKER *et al.* (1969) concluded from two-colour composite photographs also that the red Bar Population is of the right magnitude to be identified with BAADE'S Population II. They suggest, however, that it is more likely that the LMC Bar contains a mixture of evolved stars with a considerable spread of ages and chemical compositions. Also TIFFT and SNELL (1971) find a very strong old component and possibly a sprinkling of intermediate age stars.

The concentration of such a large number of very old stars to the Bar of the LMC shows that the earliest generations of stars were formed in the Bar, while at present star formation occurs in the Central System. From this point of view the core of the Central System of the SMC, i.e. a region centered on NGC 419 and having a radius of about 1°.5 (cf. WESTERLUND, 1970), corresponds to the Bar of the LMC, whereas the SMC Bar and Wing are the sites of present star formation.

Significant is in this connection also that the SMC Bar contains much neutral hydrogen (HINDMAN, 1967), whereas only very small quantities are found in the LMC Bar (MCGEE and MILTON, 1966). In fact, the point of maximum integrated brightness of H I falls in the SMC Bar. The high temperature and wide velocity spread there are consistent with a considerable depth of gas. HINDMAN suggests that the Bar is seen end-on, and there is evidence of gas streaming along it (see below). Similar conclusions cannot be drawn from the observations of the LMC. The extended disk of ionized hydrogen, centered at $5^h 30^m$, $-69° 30'$, with the size R.A. $2° \times$ Dec. $1°$ discussed by GEORGELIN and MONNET (1970), possibly explains the "hole" in the H I distribution around Constellation II (cf. MCGEE and MILTON, 1966), but does not explain the lack of H I in the Bar.

The System of Stellar Associations in the LMC. HODGE and LUCKE (1970) have studied 122 recognized stellar *O* associations in the LMC. Their mean diameter is found to be 65 pc ($m-M=18.5$) which is very similar to the mean diameter of 60 pc for stellar associations in the solar neighbourhood. Fifteen "star clouds" are also recognized; with a mean diameter of 225 pc they correspond to the stellar associations identified in M 31.

The density of very luminous stars is much higher in the LMC associations than in the galactic ones. We find 13 stars per 10^3 pc^2 in the LMC with $M_v < -4$ mag; the corresponding value in the Galaxy is 3. This confirms the generally accepted idea that the Central System of the LMC has recently experienced a burst of star formation. The luminosity functions of these very young associations agree well with the initial luminosity function, provided the very fast evolution of the massive stars is taken into consideration (WESTERLUND, 1961). This has been confirmed recently by LUCKE (1972) from a study of 16 associations in the LMC.

The centroid of the associations in the LMC is at R.A. $= 5^h 24^m$, Dec $= -68° 39'$ (1975). Their radial distribution shows

1) a very low density at the center,

2) a double hump with maxima at 40′ (590 pc) and 80′ (1180 pc) from the center, and

3) a gradual tailing off beyond 1250 pc to zero density at 3500 pc.

Points (1) and (2) stress merely the very irregular distribution of the Extreme Population I in the Central System and its grouping into SHAPLEY'S Constellations (or Super-Associations). The distribution of H II regions shows similar features.

It should be noted that the density of stellar associations decreases more rapidly with distance from the center than the density of the bright stars as well as of neutral hydrogen. This may be interpreted as the bright stars in the outer areas of the Central System being slightly older, or as indicative of associations forming preferentially in the areas of highest density. In low density areas only small groups of stars or individual stars may form. The latter explanation appears preferable in view of the fact that a sprinkling of young stars exists between the Super-Associations and that pronounced age differences have also been observed in some young groupings such as NAC 2100 (WESTERLUND, 1961) and Anon b4 (WALKER and MORRIS, 1968).

The System of Clusters in the LMC. About 1600 clusters are known in the LMC, and it is estimated that about 6000 clusters may exist with the brightest stars having $M_B \leqq 2^m.0$. HODGE (1972b) has estimated the ages of the 509 brightest clusters; the limiting magnitude is $V \leqq 15.5$.

The brightest blue star in each cluster was used for the age determination. HODGE discusses the errors that may arise if the star is not at the turn-off point, and concludes that they may be of the order of one million years.

The distribution of clusters as a function of age over the entire LMC is obtained and the evolutionary history of the system during the last 14 million years is illustrated. HODGE concludes that the clusters appear to form in aggregates of typically 24 clusters, with group formation spanning 1.5 kpc and 2×10^6 years. The average rate of formation of clusters is one per 3×10^4 years.

As a rule the turn-off point in the colour-magnitude diagram of a young LMC cluster appears to be about 2 mag below the brightest blue star if the latter is brighter than $V=11$ and one magnitude below if it has $V=12$ (WESTERLUND, 1961). It is, however, possible that the turn-off point should be still lower in most of the clusters. (One exception is the 30 Dor cluster.) The U, B, V photoelectric photometer by WALKER and MORRIS (1968) of the cluster Anon b4 and NGC 2081 showed the supergiants to begin at about $V=13$ in good agreement with WESTERLUND'S (1961) result for the latter. It appears thus doubtful that many clusters younger than 6×10^6 years exist in the LMC. Apart from this systematic effect on the age determination for the youngest clusters, the evolutionary history as given by HODGE should be essentially correct. It is to be expected that the extension of this study to fainter limiting magnitudes will eventually establish if star formation has occurred in well separated bursts (WESTERLUND and SMITH, 1964) or if it has been a more continuous procedure.

The Planetary Nebulae. The planetary nebulae belong to the "Central System" of the SMC and to the "Disk System" of the LMC (Tables 4 and 5).

In the SMC their surface distribution is symmetrical around a line through the optical center ($0^h 48^m -73.^\circ 3$) and with a position angle of 16° (WESTERLUND, 1968). If the SMC is virtually seen end-on (HINDMAN, 1967), this line may define the true equatorial plane.

In the LMC the surface distribution of the planetary nebulae is approximately circular with the centroid at $5^h 22^m -69.^\circ 5$ (WESTERLUND and SMITH, 1964). 50 per cent of the planetaries fall within 2 kpc of this center. SANDULEAK *et al.* (1971) propose that the radial distribution of the planetaries may be represented by the relationship

$$\rho(r)=4.8\times e^{-r^2/7.0},$$

where $\rho(r)$ is the number of planetaries per square degree and r the distance in degrees from the center. They propose further that the exponential term may represent the distribution of the disk population of stars and hence the bulk of the total mass in the LMC. Furthermore, they propose that the near circular distribution of the planetaries suggests that the LMC is seen nearly face on. However, the system of planetary nebulae in the LMC is not yet sufficiently understood for such interpretations. There is a puzzling difference in distribution between the low-luminosity planetaries, which generally have high excitation, and the high-luminosity ones. The former group appears to have a distribution markedly elongated along the line of nodes, and no member of it appears to the east of R.A. $=5^h 40^m$. The high luminosity objects are more evenly distributed.

The possibility must also be considered that some of the planetary nebulae are halo objects. WEBSTER (1969) suggests that three planetaries move in highly elliptical orbits, whereas the remaining 15 objects in her study have low eccentricity orbits.

The Variable Stars. 1830 variables in the LMC are catalogued by PAYNE-GAPOSCHKIN (1971) and 1541 in the SMC by PAYNE-GAPOSCHKIN and GAPOSCHKIN

(1966). Dominating are the Cepheids; 1111 are known in the LMC and 1155 in the SMC.

GAPOSCHKIN (1972) has studied the structure of the LMC as displayed by the variables. He divides the LMC into 30 bright regions (Ridges) and 20 dark areas (Lanes). The Bar is divided into four parts following the grouping of several hundreds of the faint Cepheids ($P<7$ days).

As a rule the Ridges are the luminous parts of known Super-Associations, and GAPOSCHKIN groups in fact most of them into six larger units. The lanes frequently form the borders of these units. With this in mind it becomes easier to understand the distribution of some of the classes of variables: the bright Cepheids ($P>12$ days) fall in between the bright regions and the lanes; the red variables occur in the bright regions, and the eclipsing variables tend to be in the lanes.

The faint Cepheids ($P<7$ days) form large groupings, in particular in the Bar, and the moderate Cepheids ($7^d<P<12^d$) show no marked pattern at all. It should be recalled that a cepheid with $P=10$ days has an age of about 10^8 years.

The geometrical center of the LMC, determined from the distribution of the variable stars, falls at 5^h22^m, $-68\overset{\circ}{.}5$. This is close to the centroid of the associations and to the so-called "radio center of rotation".

The Cepheids with the largest periods conform in the SMC closely to the distribution of the most luminous stars and the H II regions. It is of particular interest to note that all the Cepheids in the Wing have periods greater than 7 days and most of them have periods greater than 15 days (age less than 5.5×10^7 years).

The area covered by the short period variables approaches the elliptical distribution shown in near-infrared photographs of the Central System. In this way they confirm the non-concentrical displacement of the region of star formation discussed previously (p. 52).

Spiral Structure? The LMC and the SMC are classified SB(s)m and SB(s)mp respectively, by DE VAUCOULEURS in his system (see DE VAUCOULEURS and FREEMAN, 1969), and it is inferred that they are typical of a definite stage of the barred spiral sequence. It is possible that there is some support for this classification of the LMC in some remnants of a rather ill-defined outer arm. In the SMC there is no indication of an organized spiral structure, but this might be due to the SMC being seen end-on.

No spiral structure has been proposed in the SMC on the basis of the H I distribution.

HINDMAN (1967) interprets the form of the contours in the H I velocity–Right Ascension plane as showing expanding shell-structures. He identifies three shells with hydrogen masses 1.9; 1.0; and 0.3×10^7 solar masses and expansion velocities of the order $20\,\mathrm{km\,s^{-1}}$. The age of these formations would be about 10 years. Most of the neutral atomic hydrogen remains in the smooth distribution (the total H I mass is estimated to 4.8×10^8 solar masses).

Recent studies of the SMC supergiants (SANDULEAK, 1969a) show a good correlation between their distribution and the H I pattern with no special structure apparent. The clusters in the SMC show an ellipsoidal distribution (LINDSAY,

1958). Too little is known about their ages and motions for further interpretation.

JOHNSON (1961) attempted to explain the structure of the SMC by two components, one a dwarf elliptical and the other a contorted gaseous-magnetic arm with no nucleus. The proposed arm ended in the Wing. Undoubtedly, the Wing must be considered a feature of its own. It is important to note that the contorted arm in its brightest part was assumed by JOHNSON to be perpendicular to the plane of the sky.

In the LMC the distribution and motion of the neutral hydrogen is very complex and irregular. MCGEE has attempted to interpret the pattern by two sets of spiral arms, one of which lies in the main plane of the LMC and the other in a plane inclined about 20° to it. Two velocity groups, +300 and +243 $\mathrm{km\,s^{-1}}$, give the normal rotation pattern in the main disk and a +273 $\mathrm{km\,s^{-1}}$ group gives the other rotation curve. This model has been critisized by KERR (1971), who showed that other groupings according to velocity could be formed and quite different results consequently be deducted. A two-plane model should only be accepted when the separation of gas into velocity groupings is complete. For the moment we can only state that the gas in the LMC has a complex distribution with many concentrations, some of which may be well away from the plane of the flattened system. There is no pronounced structure to be seen. Recently an attempt has been made to identify the spiral structure first described by DE VAUCOULEURS (see DE VAUCOULEURS and FREEMAN, 1969) with the aid of extreme Population I objects. CORSO and BUSCOMBE (1970) have studied the distribution of the known Wolf-Rayet stars (60 objects) and the youngest cepheids (86 objects with $P>10$ days). It agrees in all major aspects with the irregular-barred spiral pattern of DE VAUCOULEURS. However, we do not find it necessary for the understanding of the existence of these objects to assume that they have formed in spiral arms. In fact, in systems like the LMC and the SMC with no massive nuclear regions such organized structures may not develop. In both galaxies there is also clear evidence for non-concentrical displacements of the regions of star formation. The Central System of the LMC is undoubtedly the region of the most recent star formation, which probably occurred about 5×10^7 years ago. In this System the Super-Associations (diameter about 1 kpc) form the nuclei of star formation with smaller groups of stars or individual stars forming outside them. All the known WR stars are in the Super-Associations (WESTERLUND and SMITH, 1964), whereas this is not true for the Cepheids.

Significant is in this connection also that DIXON and FORD (1972) found the same ages for the stars at the inner and outer edges of the "major spiral arm". In our description of the Central System the region studied by them is a typical Super-Association.

It is of interest to consider that, if star formation occurs in well separated bursts, the LMC as well as the SMC would in the past have looked like elliptical galaxies. The latter with its rather weak red population would have been rather insignificant, whereas the LMC, with its E 5-like Bar structure (JOHNSON, 1959; WALKER *et al.*, 1969) would have resembled a dwarf galaxy. Both of them would of course have been "peculiar" with a high content of neutral hydrogen.

V. The Rotation of the Clouds

A marked velocity gradient exists across the SMC, but it is not yet completely clear to what extent this is due to rotation or to non-circular motion. The LMC is a flattened system and rotates in a well-defined pattern. Non-circular motions exist in certain areas.

The Small Magellanic Cloud. HINDMAN'S (1967) analysis of the 21-cm data led him to propose that

1) the SMC is seen virtually end-on ($i > 70°$);
2) a marked gradient of velocities exists, indicative of rotation;
3) the major axis has the position angle 55°, and the centre of rotation is at $1^h 03^m$, $-72° 45'$ (1975);
4) the wide range of radial velocities in the core ($0^h 48^m$, $-73° 18'$), its high HI intensity and its richness in blue stars mark the position of an end-on bar;
5) the double-peaked profiles are mainly due to three expanding shells (see above).

The optical observations do not contradict the suggestions in points 1), 2) and 4). They do not clearly show that the SMC is rotating, nor do they at present contribute any proof regarding the existence of the shells. The radial velocities of the supergiant stars are generally associated with the more positive of the double 21-cm peaks; on the other hand the SMC interstellar Ca II gas shows some evidence of being connected with both velocity components, though primarily with the lower velocity one (FEAST *et al.*, 1960; FEAST, 1968).

The planetary nebulae in the SMC show a remarkedly clear division into two velocity groups, 20 $\mathrm{km\,s^{-1}}$ apart (FEAST, 1968; WEBSTER, 1969). This is not a regional effect, nor is it related to excitation class or luminosity. The mean values of their velocities are $-32\ \mathrm{km\,s^{-1}}$ and $+9\ \mathrm{km\,s^{-1}}$, and the corresponding values are $-28\ \mathrm{km\,s^{-1}}$ and $+7\ \mathrm{km\,s^{-1}}$ for the Ca II gas and $-14\ \mathrm{km\,s^{-1}}$, and $+14\ \mathrm{km\,s^{-1}}$ for the HI (over the region of the planetary nebulae observations).

In the core region (see point 4) above) the HI velocities cover a range of about 60 $\mathrm{km\,s^{-1}}$ (-30 to $+30\ \mathrm{km\,s^{-1}}$) and it has been proposed that this is due to gas streaming out of both ends of the end-on bar. The velocities of the planetary nebulae in the core area show similar effects (FEAST, 1968, Fig. 8). A recent study by FEAST (1970) of the kinematics of the HII regions in the SMC shows for these in the core region a much smaller velocity dispersion than for the HI gas; the HII regions appear to be associated with the densest HI concentrations, only. However, the general pattern of gas streaming in the line-of-sight in the core region is confirmed and a high-velocity gradient over the SMC is indicated. The motions in the SMC are still an open question, and they may not be well understood until the velocity pattern of the old red star population is known, and the existence of the Wing with its pure population I (WESTERLUND and GLASPEY, 1971) has been explained.

The Large Magellanic Cloud. There is general agreement that the LMC is a flattened, rotating system. At present the discussion concerns mainly the position of the true mass centre and its relation to the various distribution centroids and to the radio centre of rotation.

The optical centre defined by the symmetry of the Bar ($5^h 24^m - 69°.8$, 1950) falls close to the centroid of the planetary nebulae ($5^h 22^m$, $-69°.5$) and to the centroid of the yellow light isophotes ($5^h 22^m$, $-69°.5$) and is frequently considered as the mass centre. On the other hand the centroids of all Extreme Population I objects are displaced by 1° or more to the north of the optical centre; DE VAUCOULEURS gives as a mean value for the centroid of this component $5^h 34^m$, $-68°.1$ (DE VAUCOULEURS and FREEMAN, 1969).

For the interpretation of the HI rotation curve a "radio centre" at $5^h 20^m$, $-69°$ has been introduced. By this, a rather symmetrical rotation curve was obtained permitting the mass of the LMC to be determined with the aid of existing models. This centre has also been found useful in optical kinematical studies of various Extreme Population I objects (FEAST *et al.*, 1961; FEAST, 1964; SMITH and WEEDMAN, 1971; ANDREWS and EVANS, 1972), and FEAST (1968) has concluded that also the overall rotational pattern of the planetary nebulae agrees with this. WEBSTER (1969) concluded, on the other hand, that the centre of rotation of the planetary nebulae agreed well with the centre of mass of the LMC and considered the displaced radio centre as spurious. DE VAUCOULEURS has also pointed out that due to the basic asymmetric structure of the System, asymmetry of rotation curves with respect to the optical centre are to be expected. There is at present no reason to consider that the centre of mass is displaced from the Bar. It is likely that most of the gas in the LMC is in circular motion (FREEMAN, private communication), but inside the 3° radius of the Central System there are large regions with non-circular motions (cf. WEBSTER, 1969; DIXON and FORD, 1972). It is also likely that the HII regions and supergiant stars share the motions of the HI gas. Considering that they are also mostly in the "disordered" area, a re-analysis of their kinematics appears desirable. Likewise, for the planetary nebulae an investigation with a more extensive material should be carried out. The initial questions to be answered for this class of objects in the LMC are: to what extent should highly elliptical orbits be allowed for, and to what extent is the velocity of a planetary nebula correlated with its brightness.

Bearing all these uncertainties regarding the true mass centre and the motions in mind, we summarize finally an interesting model proposed by FREEMAN (DE VAUCOULEURS and FREEMAN, 1969). It is based on the assumption that the bar in a late-type barred system contains very much less mass than the disk, and that the centre, C_d, of the disk is displaced from the centre, C_b, of the bar. The bar rotates around C_d in such a way that the potential seen by an observer rotating with the bar is time independent. As C_b moves around C_d in a circular orbit, the bar itself spins about C_b in the same sense and with the same period. The effect of the gravitational and centrifugal force fields is that one stable neutral point develops, and that matter is trapped around this point – for LMC the relevant neutral point is at Constellation III (FREEMAN, private communication). The centre of the disk, C_d, is here taken to be the radio centre of rotation, this does not necessarily agree with the centre of the Disk System identified in Table 5.

Acknowledgements

It is a pleasure to thank Drs. P. J. ANDREWS, I. APPENZELLER, A. ARDEBERG, T. LLOYD EVANS, M. FEAST, K. FREEMAN, J. GRAHAM, P. HODGE, TH. SCHMIDT, and D. THACKERAY for providing me with preprints and information about their present research on the Magellanic Clouds.

References

ANDREWS, P. J., EVANS, T. LL.: Monthly Notices Roy. Astron. Soc., in press (1972).
APPENZELLER, I.: (unpublished) (1972).
ARDEBERG, A., BRUNET, J.-P., MAURICE, E., PRÉVOT, L.: Astron. Astrophys. Suppl. Ser. **6**, 249 (1972).
ARDEBERG, A. L., DE GROOT, M.: (unpublished) (1972).
ARP, H.: Astron. J. **64**, 254 (1959).
ARP, H.: Astrophys. J. **149**, 91 (1967).
BERGH, S. VAN DEN: J. Roy. Astron. Soc. Can. **62** (1968) (= Commun. David Dunlap Obs. No 195).
BERGH, S. VAN DEN, HAGEN, L. G.: Astron. J. **73**, 569 (1968).
BIGAY, J. H., BERNARD, A., PATUREL, G., ROUX, S.: Trieste Conference (1972).
BOK, B. J., KIDD, C., ROUTCLIFFE, P.: Publ. Astron. Soc. Pacific **78**, 333 (1966).
BRÜCK, M. T., LAWRENCE, L. C., NANDY, K. N., THACKERAY, A. D., WOOD, R.: Nature **225**, 531 (1970).
BRUNET, J.-P., PRÉVOT, L., MAURICE, E., MURATORIO, G.: Preprint European Southern Observatory (1972).
CAROZZI, N., PEYRIN, Y., ROBIN, A.: Astron. Astrophys. Suppl. Ser. **4**, 231 (1971).
CORSO, G., BUSCOMBE, W.: Observatory **90**, 229 (1970).
DACHS, J.: Astron. Astrophys. **9**, 95 (1970).
DACHS, J.: Astron. Astrophys. **18**, 271 (1972).
DE VAUCOULEURS, G., FREEMAN, K. C.: Univ. of Texas, McDonald Obs. Preprint No 8 (1969) (to appear in Vistas in Astronomy).
DIVAN, L.: IAU Symp. **50** (1972) (in press).
DIXON, M. E., FORD, V. L.: Astrophys. J. **173**, 35 (1972).
FEAST, M. W.: In F. J. KERR and A. W. RODGERS (eds.), The Galaxy and the Magellanic Clouds, Australian Acad., Canberra, p. 330 (1964a).
FEAST, M. W.: Monthly Notices Roy. Astron. Soc. **127**, 195 (1964b).
FEAST, M. W.: Monthly Notices Roy. Astron. Soc. **140**, 345 (1968).
FEAST, M. W.: Monthly Notices Roy. Astron. Soc. **149**, 291 (1970).
FEAST, M. W.: Monthly Notices Roy. Astron. Soc. **159**, 113 (1972).
FEAST, M. W., THACKERAY, A. D., WESSELINK, A. J.: Monthly Notices Roy. Astron. Soc. **121**, 337 (1960).
FEAST, M. W., THACKERAY, A. D., WESSELINK, A. J.: Monthly Notices Roy. Astron. Soc. **122**, 433 (1961).
FEHRENBACH, CH., DUFLOT, M.: Astron. Astrophys. Special Suppl. Ser. **1** (1970).
GAPOSCHKIN, S.: Smithsonian Inst. Astrophys. Obs. Res. Space Sci. Spec. Rep. 310 (1970).
GASPOSCHKIN, S.: Astron. Astrophys. Suppl. Ser. (1972) (in press).
GASCOIGNE, S. C. B.: Monthly Notices Roy. Astron. Soc. **134**, 59 (1966).
GASCOIGNE, S. C. B.: Monthly Notices Roy. Astron. Soc. **146**, 1 (1969).
GENDEREN, A. M. VAN: Bull. Astron. Inst. Netherl. Suppl. **3**, 299 (1969).
GENDEREN, A. M. VAN: Astron. Astrophys. **7**, 49 (1970).
GEORGELIN, Y., MONNET, G.: Astrophys. Letters **5**, 213 (1970).
GIACCONI, R., MURRAY, S., GURSKY, H., KELLOGG, E., SCHREIER, E., TANANBAUM, H.: Astrophys. J. **178**, 281 (1972).
GRAHAM, J. A.: In J. D. FERNIE (ed.), IAU Coll. **21** (in press).
GRAHAM, J. A., ARAYA, G.: Astron. J. **76**, 768 (1971).
HAVLEN, R. J., WEST, R. M., WESTERLUND, B. E.: Astron. Astrophys. **16**, 404 (1972).
HINDMAN, J. V.: Australian J. Phys. **20**, 147 (1967).
HODGE, P. W.: Publ. Astron. Soc. Pacific **72**, 308 (1960).
HODGE, P. W.: Smithsonian Inst. Astrophys. Obs. Res. Space Sci. Spec. Rep. 306 (1969).
HODGE, P. W.: Bull. Am. Astron. Soc. **4**, 223 (1972a).

HODGE, P.W.: Smithsonian Inst. Astrophys. Obs. Res. Space Sci. Spec. Rep. (1972b) (in press).
HODGE, P.W., LUCKE, P.B.: Astron. J. **75**, 933 (1970).
HUTCHINGS, J.B.: Monthly Notices Roy. Astron. Soc. **131**, 299 (1966).
JOHNSON, H.M.: Publ. Astron. Soc. Pacific **71**, 301 (1959).
JOHNSON, H.M.: Publ. Astron. Soc. Pacific **73**, 20 (1961).
KERR, F.J.: In A.B. MULLER (ed.), The Magellanic Clouds, p. 50, Reidel Publ. Co., Dordrecht, Holland (1971).
LE MARNE, A.E.: Monthly Notices Roy. Astron. Soc. **139**, 461 (1968).
LEONG, C., KELLOGG, E., CURSKY, H., TANANBAUM, H., GIACCONI, R.: Astrophys. J. **170**, L 67 (1971).
LINDSAY, E.M.: Monthly Notices Roy. Astron. Soc. **118**, 172 (1958).
LUCKE, P.B.: Bull. Am. Astron. Soc. **4**, 223 (1972).
LYNGÅ, G., WESTERLUND, B.E.: Monthly Notices Roy. Astron. Soc. **127**, 6 (1963).
MATHEWSON, D.D., FORD, V.L.: Astrophys. J. **160**, L 43 (1970).
MCGEE, R.X., MILTON, J.A.: Australian J. Phys. **19**, 343 (1966).
MULLER, A.B. (editor): The Magellanic Clouds, Reidel Publ. Co., Dordrecht, Holland (1971).
OSMER, P.S.: Astrophys. J. **171**, 393 (1972).
PAYNE-GAPOSCHKIN, CECILIA H.: Smithsonian Contr. Astrophys. No **13** (1971).
PAYNE-GAPOSCHKIN, C., GAPOSCHKIN, S.: Smithsonian Contr. Astrophys. **9** (1966).
PHILIP, A.G.D.: Bull. Am. Astron. Soc. **3**, 367 (1971).
PRICE, R.E., GROVES, D.J., RODRIGUEZ, R.M., SEWARD, F.D., SWIFT, C.D., TOOR, A.: Astrophys. J. **168**, L 7 (1971).
PRZYBYLSKI, A.: Monthly Notices Roy. Astron. Soc. **139**, 313 (1968).
PRZYBYLSKI, A.: Monthly Notices Roy. Astron. Soc. **152**, 197 (1971).
PRZYBYLSKI, A.: Monthly Notices Roy. Astron. Soc. **159**, 155 (1972).
ROSENDHAL, J.D., SNOWDEN, M.S.: Astrophys. J. **169**, 281 (1971).
SANDAGE, A., TAMMANN, G.A.: Astrophys. J. **167**, 293 (1971).
SANDULEAK, N.: Astron. J. **73**, 246 (1968).
SANDULEAK, N.: Astron. J. **74**, 47 (1969a).
SANDULEAK, N.: Astron. J. **74**, 877 (1969b).
SANDULEAK, N.: Contr. Cerro Tololo Interam. Obs., No. 89 (1970).
SANDULEAK, N., MACCONNELL, D.J., HOOVER, P.S.: Bull. Am. Astron. Soc. **3**, 239 (1971).
SCHMIDT, TH.: Astron. Astrophys. **6**, 294 (1970).
SMITH, M.G., WEEDMAN, D.W.: Astrophys. J. **169**, 271 (1971).
TIFFT, W.G.: Monthly Notices Roy Astron. Soc. **125**, **199** (1963).
TIFFT, W.G., SNELL, CH.M.: Monthly Notices Roy. Astron. Soc. **151**, 365 (1971).
TURNER, K.C., VARSAVSKY, C., TUVE, M.A.: Ann. Report of Dir. of Dept. of Terr. Magn. 1967–68, Carnegie Inst. Year Book 67, p. 290 (1968).
WALKER, G.A.H., MORRIS, S.C.: Astron. J. **73**, 772 (1968).
WALKER, M.F.: Astrophys. J. **161**, 835 (1970).
WALKER, M.F.: Astrophys. J. **167**, 1 (1971).
WALKER, M.F.: Monthly Notices Roy. Astron. Soc. **156**, 459 (1972a).
WALKER, M.F.: In: S. LAUSTSEN and A. REIZ (eds.), ESO/CERN Conference on Auxiliary Instrumentation for Large Telescopes, p. 399 (1972b).
WALKER, M.F., BLANCO, V.M., KUNKEL, W.E.: Astron. J. **74**, 44 (1969).
WEBSTER, B.L.: Monthly Notices Roy. Astron. Soc. **143**, 97 (1969).
WESSELINK, A.J.: Monthly Notices Roy. Astron. Soc. **122**, 503 (1961).
WESTERLUND, B.E.: Uppsala Astron. Obs. Ann. **4**, No 7 (1960).
WESTERLUND, B.E.: Uppsala Astron. Obs. Ann. **5**, No 1 (1961).
WESTERLUND, B.E.: In K. LODÉN, L.O. LODÉN, U. SINNERSTAD (eds.), Spectral Classification and Multicolour Photometry, IAU Symp. **24**, Academic Press, London, p. 353 (1966).
WESTERLUND, B.E.: In D. OSTERBROCK and C. O'DELL (eds.), Planetary Nebulae, IAU Symp. **34**, Reidel Publ. Co., Dordrecht, Holland, p. 23 (1968).
WESTERLUND, B.E.: In A. BEER (ed.), Vistas in Astronomy **12**, 335, Pergamon Press, Oxford (1970).
WESTERLUND, B.E., GLASPEY, J.: Astron. Astrophys. **10**, 1 (1971).
WESTERLUND, B.E., SMITH, L.F.: Monthly Notices Roy. Astron. Soc. **128**, 311 (1964).
WOLF, B.: Astron. Astrophys. **20**, 275 (1972a).
WOLF, B.: In L. MAURIDIS (ed.), Proceedings of the First European Astronomical Meeting, **2** (1972b).

Discussion

Contopoulos, G.:
1) What is the distance of the center of rotation in Freeman's model with respect to the dimension of the bar, and
2) What are the masses of the bar and the axisymmetric background?

Westerlund, B. E.:
1) The center of rotation is almost 0.8 kpc from the center of the Bar. The Bar is about 2.5×0.8 kpc.
2) The mass of the Bar is probably 0.1 or less of the total mass.

Kippenhahn, R.:
Five or ten years ago Cuno Hoffmeister announced that outside of our Galaxy (and I think near the Magellanic Systems) he has found an absorbing cloud. What is the latest word on that?

Westerlund, B. E.:
Hoffmeister's absorbing cloud is in Microscopium ($20^h 45^m$, $-42°$). He proposed that it is similar to the Magellanic Clouds and a member of the Local Group of galaxies.

Oort, J. H.:
In connection with the question whether there is any indication of a bridge between the Magellanic Clouds and the Galaxy, I call attention to a remarkable cloud of neutral hydrogen near the south-galactic pole. This extends over about 70° on the sky as a rather narrow feature appearing to move at a space velocity of about 400 $km\,s^{-1}$ relative to the local standard of rest. It seems probable that this cloud lies outside the Galaxy. It points rather precisely away from the Large Magellanic Cloud and might possibly be due to gas expelled from it into the general direction of the Galaxy.

Galactic Nuclei

(Invited Lecture)

By G. R. BURBIDGE
University of California, San Diego
Department of Physics
La Jolla, California, U.S.A.

Abstract

Evidence of violent activity in galactic nuclei was reviewed. Some of the more important results given in a recent article (G. BURBIDGE, Annual Reviews of Astronomy and Astrophysics, **8**, 369, 1970) were described. It was pointed out that the radio, microwave, infrared, and X-ray observations of the nucleus of our own Galaxy suggest that two major components giving rise to activity are involved. There is a small non-thermal source which may be a remnant of the event which gave rise to the large-scale violent ejection of matter 10^6–10^7 years ago. Also, it is likely that much of the infrared emission, thermal radio emission, and some of the X-ray sources may arise in different stages of the evolution of stars with masses in the range 30–100 $M_\odot$. A comparatively small cluster of massive stars in the galactic nucleus going through their normal processes of stellar evolution is required.

The problem of accounting for the large amounts of energy and large masses ejected from galactic nuclei was discussed. Processes involving the evolution and collapse of dense star clusters leading to the formation of large gravitationally collapsing masses were described. It was pointed out that an attractive alternative approach is to follow AMBARTSUMIAN and HOYLE and NARLIKAR in arguing that either the nuclear activity is due to the appearance of "prestellar" matter (AMBARTSUMIAN) or that matter is being created in galactic nuclei (HOYLE and NARLIKAR).

Finally, some time was spent in discussion of the problem of the discrepant redshifts. Is there evidence that some extragalactic objects (quasi-stellar objects and peculiar galaxies) with large redshifts are physically associated with galaxies with much smaller redshifts? It seems likely that this is the case, though there is also evidence that some QSOs with comparatively small redshifts are associated with galaxies with similar redshifts.

Discussion

OORT, J. H.:
The problem of the ejection of coherent objects is still in a very debatable state, but a thing that appears quite certain is that very large amounts of gas have

been ejected from nuclei. This has worried me more than the large energy output. What produces all the gas that is being ejected? Is it an "Ambartsumian object" or is it gas falling in from outside?

BURBIDGE, G.:
I agree that this is very difficult to explain but perhaps it indicates that very high density states of matter are present in the nuclear region.

SANDQVIST, A.:
With regards to the very massive stars that you suggest they are being formed in the galactic center region, what order of magnitude do you consider for their masses?

BURBIDGE, G.:
Objects responsible for the discrete infrared sources and some of the X-ray sources etc. could arise in different stages of evolution of stars with masses $\sim 50\, M_\odot$.

SANDQVIST, A.:
So you do not think at all of the order of the mass suggested by BECKLIN-NEUGEBAUER for their infrared point source?

BURBIDGE, G.:
I am only saying that some of these phenomena can be explained in terms of the formation and evolution of stars in the mass range we know about $\leqq 100\, M_\odot$.

MACKAY, C. D.:
Is there an estimate of the mass which is being ejected from NGC 4939?

BURBIDGE, G.:
It is very difficult to estimate the mass. However, masses of order $10^6\, M_\odot$ or greater may be involved.

POTTASCH, S. R.:
What conditions are present in the galaxy NGC 4939 (where the Balmer lines are double) in the system where the O^{++} lines are not seen?

BURBIDGE, G.:
We believe that in situations where the Balmer series is seen and not the O^{++} lines, we are looking at clouds with very high densities.

MCVITTIE, G. C.:
The association of galaxies and quasars appears to be based on statistics which refer to quasars that are radio emitters. What happens to the statistics if radio quiet quasars are included?

BURBIDGE, G.:
If radio quiet objects are considered, it is very difficult to do a statistical study since in very few areas of the sky have radio quiet QSO's been identified. The BAHCALLS and MCGEE have tried to do such a study. Their results do not show a correlation, but their data is very poor.

SHAPIRO, M. M.:
When you discussed certain QSO's that appear close to relatively nearby galaxies and suggested that they are "genetically related" did you mean to imply that the QSO's were ejected from the galaxies? If so, why just redshifts: would we not expect a comparable number of blueshifted objects?

BURBIDGE, G.:
Yes. I did imply that they are ejected. If this is true it also follows that the redshifts are not velocity shifts, otherwise there would be a blueshift problem. As HOYLE and NARLIKAR have pointed out, one possible way of explaining such redshifts would be to argue that the masses of the electrons (or equivalently the values of G) in these objects are different.

Galactic Nuclei

(Invited Lecture)

By L. M. OZERNOY
Lebedev Physical Institute, USSR Academy of Sciences, Moscow, USSR.

With 3 Figures

I. Introduction

The problem of galactic nuclei has stirred theoreticians for most part of this century. During the last decade most significant and exciting facts have been collected (they are summarized in a fundamental review by BURBIDGE (1970); see also VORONTSOV-VELYAMINOV (1965) where a large amount of empirical data is presented on 173 galactic nuclei). This lecture is not intended to be an exhaustive review of all the attempts to interpret the observational results. Instead, I shall try to consider critically the present state of the problem concerning the nature of ultimate energy source for active nuclei and quasars.

This problem is one of the few in astrophysics that may lead to the discovery of new physical laws. This point of view seemed to be most strongly supported by WEBER'S experiments who claimed that the gravitational wave radiation from the center of our Galaxy is responsible for the pulses detected. As is known, if being correct, this interpretation would have had dramatic consequences for physics and astrophysics. However, the recent searches for gravitational radiation pulses in Moscow University (BRAGINSKY *et al.*, 1972) did not confirm WEBER'S results. Moreover, another group (ADAMJANTS *et al.*, 1972) finds that there is a significant correlation of WEBER'S pulses with geomagnetic perturbations and, finally, with solar activity.

In this situation the attempts to explain the nature of active nuclei in the framework of usual physics seem quite natural and even necessary. Just these attempts will be the prime concern of the following. First of all I shall briefly touch upon the main conceptions of the nature of active nuclei (Section II). Then the difficulties encountered by some of these conceptions will be indicated (Sections III and IV). The conception of a supermassive body which seems to me less open to criticism is then considered in more detail (Section V). From the point of view of this conception the treatment of main observational results will be given (Section VI). Further thoughts and questions raised by the theory will be discussed in Section VII.

II. Three Basic Conceptions about the Nature of the Ultimate Energy Source

As is now known, compact sources in the nuclei of normal elliptical (E) and spiral (S) galaxies (including the active center of our own Galaxy), Seyfert galaxies

(SyG), radiogalaxies (RG), blue compact galaxies (BCG), N-galaxies (NG), quasi-stellar galaxies (QSG), and, finally, quasars (QSS) show, together with specific peculiarities, a great similarity in the manifestations of their activity. Apparently, the nature of the activity in quasars and nuclei of galaxies of various kinds is the same and differs mainly in quantitative rather than in qualitative respect.

Let us put the above mentioned populations in a one-dimensional sequence of increasing concentration of the matter to the centre:

$$\mathrm{Ir}-(\mathrm{S}-\mathrm{SyG})-(\mathrm{E}-\mathrm{RG})-(\mathrm{BCG}-\mathrm{NG})-(\mathrm{QSG}-\mathrm{QSS})$$

(as discussed below, populations in brackets are apparently connected genetically, so that the order of priority inside brackets is conventional). It appears that this sequence of galaxies of increasing mean density is the sequence of increasing non-thermal luminosity. Because of this fact, the contribution of stars (which most likely are contained in quasars as well) to the total luminosity of an active galaxy decreases progressively from left to right (e.g. SANDAGE, 1971).

What is then the ultimate source of energy release present in very different populations but leading to very similar results?

Curiously, among recent hypotheses on the nature of the galactic activity there is hardly one which could not be classified with some conception formulated in the very first years after the discovery of quasars. However, this fact is not an evidence of scantiness of human imagination but rather demonstrates objectively the limited number of possibilities to construct a non-contradictory model which may explain the main known observational data in the framework of established physical laws.

All hypotheses on the ultimate source of activity of galactic nuclei and quasars which are the subject of earnest discussions in current literature can be attributed to one of the three main conceptions:

1) Compact star system where star collisions occur or/and explosions of massive stars like supernovae;

2) supermassive collapsed body ("black hole") on which the accretion of surrounding gas takes place;

3) supermassive rotating magnetoplasmic body called magnetoid, spinar, giant pulsar etc.

In these conceptions the gross parameters of the source of activity, if we restrict ourselves for a moment by the rough estimations, are of the same order of values: the mass of a source is $M \leqq 10^8$–$10^9\,M_\odot$, and the radius is $R \leqq 10^{16}$–10^{17} cm.

Of course, the classification presented is oversimplified. As a result of the subsequent consideration I would suggest a more realistic model which seems to be the following hybrid: compact star cluster containing a magnetoid in the centre which is apparently a black hole. However, the principal problem to be solved is the following: which component of this hybrid is most important from the point of view both of energy release and its transformation into observed forms of activity? Now let us turn to the consideration of this problem.

III. Conception of a Compact Star System

A number of authors (e.g. COLGATE, 1967; VON HOERNER, 1968; SPITZER, 1971) postulate that the star concentration in nuclei of galaxies and quasars is so large that non-elastic star collisions become very frequent. According to an other variant of this conception, the star collisions took place mainly in the past. More massive stars which could appear as a result of collisions exploded then like supernovae and produced a cluster of neutron stars (HOYLE and FOWLER, 1967).

As for the first version, two ways of energy release under collisions were analyzed. GOLD *et al.* (1965), SPITZER and SASLAW (1966) consider the situation when stars have such large relative velocities that their collisions lead to the disruption and the release of gas with kinetic energy of about 10^{51} erg/$M_\odot$ per each collision. COLGATE (1967) considers another situation when the relative velocity of stars is small compared with parabolic velocity on the star surface, so that as a result of the collision, the coalescence of two stars into a more massive one may occur. It is assumed that the latter can then explode like a supernova. According to BLANFORD and REES (1971) such explosions leading to the formation of neutron stars may explain the radiovariability of compact sources in nuclei of galaxies and quasars.

The known difficulty of compact star cluster as a model of energy source is that the star density at which the collisions are important is too large (up to $10^{11}\ M_\odot\ pc^{-3}$), and it is not clear whether it can be reached during the normal evolution of galactic nuclei (note that in the nucleus of M31 $\rho_* \lesssim 10^7\ M_\odot\ pc^{-3}$ (OORT, 1971)). The direct arguments against this model follow from numerical calculations of the evolution of the cluster of 10^7–$10^8\ M_\odot$ (SANDERS, 1970). SANDERS finds that the rate of the formation of supernovae from the coalescence is rather small so that the maximal luminosity from supernovae is only 10^{42} erg s^{-1}. Meanwhile the non-thermal luminosities of Seyfert nuclei needed to be explained are at least 10^{44} erg s^{-1}.

Even more strong contradictions arise under the attempt to explain by this model the luminosities of quasars. GUDZENKO, OZERNOY and CHERTOPRUD (1968, 1971) find that the model of independent occasional explosions being confronted with data on optical variability of 3C 273 may be rejected with certainty of 99–98%. This rejection does not depend on the still controversial question whether or not the optical variability of this quasar is quasi-periodic (CHERTOPRUD *et al.*, 1973). By the way, there is now evidence for quasi-periodic character of optical variability for already several quasars and galaxies with active nuclei (see below), and therefore the rejection of the model of independent occasional explosions demonstrated in the case of 3C273 is of great common interest.

The second variant of this conception (the cluster of neutron stars) may be rejected by another method (OZERNOY, 1972a) on the basis of the fact that the chemical abundance of the emission regions of 3C273 and 3C48 studied in detail (BAHCALL and KOZLOVSKY, 1969a, b) is very similar to that of Population I. Meanwhile the chemical compositions of the Supernova of type II in NGC4496 observed near its maximum (MUSTEL, 1971) and bright knots of type I supernova remnant in Cas A (PEIMBERT, 1971) show the great abnormalities. Comparing the abundances in quasars and supernovae, OZERNOY (1972a) has found the strong upper limit of the total number of supernovae explosions in the quasar 3C 273: $N \leqq 5 \times 10^2$. The average rate of supernovae explosions in 3C 273 does not exceed 5×10^{-1}–$5 \times 10^{-4}\ \mathrm{y}^{-1}$, in dependence of the assumed age 10^3–10^6 years of this quasar. In any case this upper limit is much lower than the rate of supernova flares needed to explain the power of optical radiation as well as the level of its variations. A similar strong constraint on the number of supernovae, and therefore on the number of neutron stars, can one obtain for nuclei of Seyfert galaxies where there is no evidence for large abundance anomalies (OSTERBROCK, 1971).

To summarize, although the explanation of the activity of galactic nuclei and quasars by multiple explosions of supernovae or any massive stars seems, at first glance, very natural, the more detailed analysis exposes a number of significant qualitative difficulties.

IV. Conception of an Accreting "Black Hole"

The gravitational collapse was among the first hypotheses on the source of energy. The idea of "gravitational grave" which faded away for several years is now revived under the name of "black hole". This idea is attractive by the following reasons:

1) The formation of a black hole in the centres of galaxies as a result of collapse of gas or a compact star cluster seems to most people inevitable in the framework of general relativity.

2) The energy release in the course of collapse is limited only by the value of $0.42\ \mathrm{Mc}^2 \approx 10^{54}\ (\mathrm{M}/\mathrm{M}_\odot)$ erg which is very great for a large mass.

The main problem is to find the mechanism for the release of such an energy during the large time because the collapse in itself goes on during a very short time of the order of $10^3\ (\mathrm{M}/10^8\ \mathrm{M}_\odot)$ sec.

It was suggested first by SALPETER (1964), ZEL'DOVICH (1964) and then, in more detail, by LYNDEN-BELL (1969) that infall of matter from the exterior might supply a continuous energy source. Such an accreting "black hole" of 10^4 to $10^8\ \mathrm{M}_\odot$ at the center of the Galaxy has been postulated by LYNDEN-BELL and REES (1971) to explain radio and infrared phenomena observed there.

It is expected that the stationary radiation is produced by energy release as a result of spiralling of interstellar gas to the central hole, whereas to explain the explosional energy release, the process of some other kind is needed. PENROSE (1969) and CHRISTODOULOU (1970) assume that any "particle" (e.g. a star) which encounters the black hole splits into two "particles" in such a way that one falls into the hole and the other escapes back to infinity with great energy extracted from the rotational energy of the black hole. However, BARDEEN, *et al.* (1972) showed recently that the extraction of the rotational energy of a Kerr black hole, although possible in principle (e.g. "Penrose-Christodoulou" process), is unlikely in any astrophysically plausible context.

Of course, not all the properties of black holes are yet investigated in detail at present. However, the confrontation of observations with some already known properties of black holes reveals a number of noteworthy difficulties:

1) AMBARTSUMIAN (1958, 1965) repeatedly emphasized that the idea of collapse as the source of energy encounters the fundamental difficulty that we observe only an ejection of matter rather than its infall. Very large rates of outflow of the matter ($\sim 1\ M_\odot$ per year from the nucleus of our Galaxy, up to 10–$10^2\ M_\odot$ per year from nuclei of some Seyfert galaxies) are not explained by the "black hole" conception. Meanwhile the ejection of gas prevents the accretion, and does this more effectively than the radiation (SCHWARTSMAN, 1970). The ejection of matter observable in quasars and nuclei of Seyfert galaxies has the power of kinetic energy (up to 10^{45} erg s^{-1} in quasars) which is much larger than the threshold power ($\sim 10^{43}$ erg s^{-1} for $M = 10^9\ M_\odot$) for which the outflow of gas prevents the accretion (OZERNOY, 1971 a). Of course, the coexistence of infall in some directions and ejection in others is logically possible, but this possibility seems to be rather artificial.

2) The most radiation from a black hole is of thermal origin and produced by gas and dust (LYNDEN-BELL and REES, 1971). However, the spectrum, polarization and other properties of infrared radiation of quasars and galactic nuclei (this radiation gives the most contribution into the total luminosity) suggest definitely that it is of non-thermal origin. The rapid variation in the infra-red region from a number of objects are also in the contradiction with the dust mechanism.

3) If the quasi-periodic behaviour of the optical luminosity demonstrated by some quasars and active nuclei is connected with the rotation of the single body (discussed in the next Section), then its angular momentum

$$J \gtrsim 10^{66} \frac{M}{10^8\ M_\odot} \frac{v}{10^9\ \text{cm s}^{-1}} \frac{R}{10^{16}\ \text{cm}}\ \text{g cm}^2\ \text{s}^{-1}$$

is much greater than the critical value $J_{cr} \approx 3 \times 10^{64} (M/10^8\ M_\odot)^2\ \text{g cm}^2\ \text{s}^{-1}$ which may prevent collapse (BARDEEN and WAGONER, 1971).

4) The radiation from a black hole must fluctuate with a characteristic time of the order of

$$\frac{\text{gravitational radius}}{\text{velocity of light}} \approx 10^3 \frac{M}{10^8\ M_\odot}\ \text{sec}.$$

However, LYUTYJ and CHEREPASHCHUK (1971) who searched such fluctuations from 3C 273, find them only very rare and small, if any at all. The explanation of such fluctuations is most likely not connected with a black hole (OZERNOY and CHERTOPRUD, 1971, 1972).

To summarize, the above mentioned difficulties as well as a number of others (OZERNOY, 1971 a) allow us to conclude that the conception of an accreting black hole, although attractive and exciting, encounters some principal difficulties in the explanation of the essential properties of active galactic nuclei.

V. Conception of a Supermassive Body

a) Observational Evidence

The difficulties of the two preceding conceptions make us turn to the alternative of a single body. But there is, in addition, more direct evidence:

1) The main argument is the quasi-periodic character of optical variability discovered in a number of quasars and active nuclei. These objects are listed in Table 1. Note that we do not know any cases of very strong periodicity. Light variations from 3C 273 which are the subject of many discussions in the literature (see CHERTOPRUD *et al.*, 1973 for references) is an example of the situation when the chaotic flare component is of the same order as the regular one. In some other cases the regular component is larger than the chaotic one. It seems more correct to name the optical variability quasi-periodic, or cyclic, because in some cases the phase does not remain constant but shifts slowly with time. However, in any case the very existence of such a regular component of brightness is in contradiction with the multiple nature of a source or, more precisely, with the model of independent occasional flares.

Table 1. Compact objects with quasi-periodic behaviour of optical brightness

Source	Type	Observed period	References
3C 273	QSS	$9 \pm 1.5\,y$	OZERNOY *et al.* (1969)
3C 345	QSS	$P_1 = 80^{d}.37$; $P_2 = 321^{d}.5$(?)	KINMAN *et al.* (1968)
2135-14	QSS	between 11 and 13 *y*	ANGIONE and SMITH (1972)
0405-12	QSS	between 11 and 13 *y*	ANGIONE and SMITH (1972)
3C 446	QSS	380^{d}	KINMAN (1970)
3C 454.3	QSS	$339^{d}.58$	KINMAN (1970)
			LÜ and HUNTER (1969)
BL Lac ≡	QSO (?)	$369^{d}.6$ (optical)	KINMAN (1970)
≡ VRO 42.22.01		$P_1 \approx 140^{d}$; $P_2 \approx 450^{d}$ (radio)	GORSHKOV and POPOV (1972)
NGC 4151	Seyfert	$5.1\,y$	PACHOLCZYK (1972)
3C 120	N-type or Seyfert	350^{d}	JURKEVICH *et al.* (1971)

2) Among 82 extra-galactic radio sources (MCDONALD *et al.*, 1968) there are at least 5 sources consisting of two pairs of components which are lying along the same axis. This fact is in favour of a single rather than a multiple source producing the "double-double" radiocomponents. The alternative, suggested by REES (1971), namely that interstellar gas concentrating to the galactic plane may determine the direction of ejection, is not operating at least for strong radiogalaxies which are mostly ellipticals.

3) Very informative are the data about the radiovariability. In the case of 3C 279, the small change in polarization position angle relative to the large change in polarized flux density, as well as the fact that this source has a preferred plane of polarization variability which remained the same during the observations, both indicate that the variations were produced by a series of successive outbursts within a single physical region rather than in physically separated regions of the source (ALLER, 1970). The same conclusion was obtained by SHKLOVSKY (1970) from a different analysis of radioflares.

Some additional evidence in favour of a single object has been discussed at the Fowler Symposium on Supermassive Objects (see TRIMBLE, 1971).

b) Variants of the Theory of a Supermassive Body

As is known, any model of the source of activity is restricted by the following conditions: for a mass of the source as large as 10^4–$10^9\ M_\odot$ the energy release must be a not too small share of Mc^2, and the life time of the source must be not less than 10^5 or 10^6 years. True, the fact that Seyfert galaxies amount to about 1% of normal spirals, is sometimes considered as an evidence that the active ("Seyfert") phase is as long as 10^8 years. However, if the active phase is recurrent, from the statistics of nuclei it can only be infered that 10^8 years is the cumulative life-time. This means that we may have e.g. 100 active phases with duration of each about 10^6 years and an interval of repetition of about 10^8 years.

The first and simplest model of a single body as a source of energy was a non-rotating supermassive star supported by radiation pressure (HOYLE and FOWLER, 1963). The theory of such superstars, developed in detail (FOWLER, 1964; CHANDRASEKHAR, 1964; ZEL'DOVICH and NOVIKOV, 1971) shows that because of scanty reserve of stability these superstars must be subject to the fragmentation or/and collapse. Their properties (such as primary thermal luminosity etc.) are in bad agreement with non-thermal activity of nuclei.

However, the supermassive body which possesses rotation and magnetic field differs radically in the stability as well as in observational manifestations. In the most common case, the supermassive body is supported in the equilibrium state by rotation, magnetic field and radiation pressure; such a body I called magnetoid (OZERNOY, 1966).

There are two extreme situations, depending on the amount of thermal radiation: the "hot" (high-entropy) and the "cold" (low-entropy) situation. In the "hot" case the form of the body is nearer to a sphere, and in the "cold" case to a very thin disc. Special models for each of these cases have been studied in detail. For the "hot" case – with rotation but without magnetic field (FOWLER, 1964, 1966), with rotation and toroidal magnetic field (OZERNOY, 1966; OZERNOY and USOV, 1971a), with rotation and poloidal magnetic field (OZERNOY and USOV, 1971a); for the "cold" (disc) case – with rotation without magnetic field (BARDEEN and WAGONER, 1969, 1971; SALPETER and WAGONER, 1971).

The supermassive spheroidal and disc models are not to be confronted, because they may be connected genetically and describe the various stages of evolution of the source of activity in a nucleus. Most likely, the magnetoid, being at its formation more or less spherical, aquires the disc form during final stages of its life (OZERNOY, 1971a). It is interesting that the optical continuum of quasars, after subtracting the (synchrotron) component with power spectrum, has the quasi-planckian form (GREWING and LAMLA, 1968). This form is caused most likely by the thermal radiation of a supermassive body (OZERNOY, 1972b) and is an evidence that the "hot" magnetoid variant is realized in quasars at the observed stage of their evolution. Although the evidence for this interpretation needs further confirmation, especially for active nuclei of galaxies (OZERNOY, 1972c), we shall consider below mostly the results concerning the "hot" variant in more detail.

Due to the rotation and the poloidal magnetic field, the "hot" magnetoid will produce not only thermal radiation L_{th}, but also the magnetodipole one L_{md} (MORRISON, 1969; CAVALIERE *et al.*, 1969, 1971; FOWLER, 1971; PIDDINGTON, 1970; WOLTJER, 1972; OZERNOY and USOV, 1971a, 1973a, b). The possibility of the acceleration of the particles up to relativistic energies by magnetodipole (MD) radiation (GUNN and OSTRIKER, 1969) makes the MD radiation most interesting and important from the observational point of view.

It appears that two conditions: 1) $L_{md} \gtrsim L_{th}$ and 2) $\tau_{evol} \gtrsim 10^5$ y determine unequivocally the character of the rotation and of the magnetic field. Namely, long and powerful enough MD-radiation may be only in the case of uniform rotation and quasi-dipol magnetic field with a magnetic energy of the order of a gravitational one (OZERNOY and USOV, 1971a, 1973a, b). Some results of the latter paper containing the detailed calculations of the electrodynamics and evolution of such a supermassive oblique rotator are presented in the next section.

c) Supermassive Oblique Rotator

The supermassive "hot" oblique rotator has the thermal luminosity

$$L_{th} \approx 10^{46} \frac{M}{10^8\,M_\odot} \text{erg s}^{-1} \tag{1}$$

and the magnetodipole luminosity

$$L_{md} = \frac{2}{3c^3} H_P^2 \,\Omega^4 R^6 \sin^2\chi \approx 10^{48}\,\xi \left(\frac{M}{10^8\,M_\odot}\right)^4 \left(\frac{R}{10^{16}\text{ cm}}\right)^{-4} \sin^2\chi \text{ erg s}^{-1} \tag{2}$$

where $H_P \approx 10^6\,\xi^{\frac{1}{2}}\,(M/10^8\,M_\odot)\,(R/10^{16}\text{ cm})^{-2}$ oe is the magnetic field on the pole; $\xi \equiv E_m/|E_g| \lesssim 1$ is the ratio of magnetic and gravitational energies; Ω is the angular velocity; and χ is the angle between magnetic and rotational axes. Plasma of various origin (such as the interstellar plasma outside of the rotator; plasma ejected electrodynamically; plasma flowing out due to rotational instability) does not violate the vacuum approximation at which formula (2) is valid. However, it appears that outflowing plasma plays a very important role because it transforms the MD-radiation of very low frequencies (as small as 10^{-6}–10^{-8} Hz) into observable non-thermal radiation. This transformation occurs when plasma outflowing as a stream reaches the light cylinder ($r = c/\Omega$) and spreads "embracing" the rotator. Now the low-frequency electromagnetic wave is absorbed by a thin plasma layer and accelerates the particles of the layer up to the energies $\gamma = E/mc^2 \sim 10^3$. The accelerated relativistic electrons radiate by the synchrotron mechanism in the magnetic field (~ 1 oersted near the light cylinder) mainly at the frequency $\nu_m \sim 10^6\,H\gamma^2$ Hz which corresponds to sub-millimeter and infra-red regions. The spectrum of this radiation at high-frequency side may be steep enough due to the cut-off of relativistic electrons. At the low-frequency side the synchrotron reabsorption takes place at the frequency $\nu_r \sim 5 \times 10^{12}$ Hz, weakly depending on the parameters of the rotator. Therefore, a supermassive oblique rotator as the model of the source of energy is in a rather good agreement with the main property of non-thermal radiation from active nuclei and quasars which is localized in infra-red peaks, very similar in different active populations.

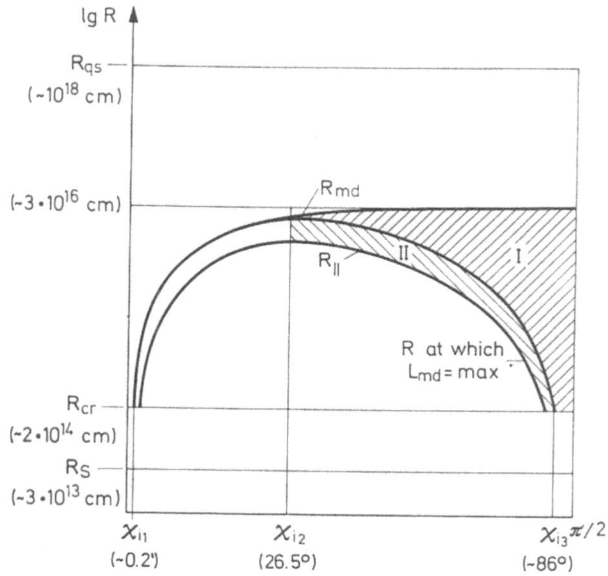

Fig. 1. Characteristic radii of a supermassive oblique rotator when luminosity is changing qualitatively, in dependence of initial angle χ_i between magnetic and rotational axes. The numerical values are given for the case $M=10^8\,M_\odot$ when $z=E_{rot}/|E_g|=\frac{1}{4}$ (equipartition between magnetic, rotational and thermal energies). In the region I $dL_{md}/dt>0$, and in the region II $dL_{md}/dt<0$

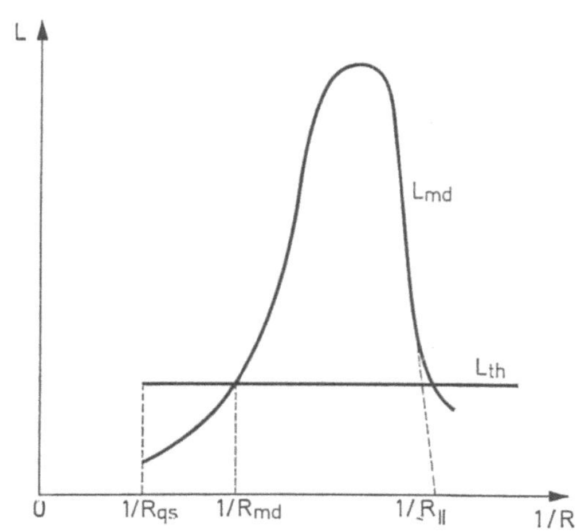

Fig. 2. Sketh of the secular behaviour of the thermal L_{th} and magnetic dipole L_{md} luminosity of a supermassive oblique rotator in dependence of its inverse radius

Now, let us consider how this non-thermal radiation from the rotator changes during its secular contraction due to the loss of the entropy and angular momentum. Fig. 1 shows the radii of the supermassive rotator with the ratio of rotational and gravitational energies $z=\frac{1}{4}$ at characteristic moments of its evolution. The initial radius is $R_{qs} \approx 10^{18} (z/0.1)^{\frac{2}{5}} \ (M/10^8 \mathrm{M}_\odot)^{\frac{3}{5}}$ cm at which the contracting plasma cloud reaches its centrifugal equilibrium and goes over into the phase of quasi-static secular contraction. At the critical radius $R_{cr} \approx 4 \times 10^{14} (z/0.1)^{-1}$ $(M/10^8 \mathrm{M}_\odot)$ cm occurs loss of hydrodynamical stability due to the effects of general relativity. A radius at which the character of radiation is changed qualitatively, depends strongly on the initial angle χ_i between magnetic and rotational axes. At the radius $R_{md} = 3 \times 10^{16} \xi^{\frac{1}{4}} \sin^{\frac{1}{2}} \chi_i (M/10^8 \mathrm{M}_\odot)^{\frac{3}{4}}$ cm the MD radiation is equal to the thermal luminosity (1). At $R < R_{md}$ the value L_{md} becomes greater than L_{th} and increases rapidly (as R^{-4}). If the angle χ would remain constant, it would lead to a very sharp decrease of the life time. However, it appears that MD radiation carries off the angular momentum in such a way that the angle χ decreases. As a result, the change of L_{md} with time is non-monotonous when $26.5° < \chi_i < 86°$ (see Fig. 2). At

$$R_{L_{md}=\max} = \frac{5}{4} R_{md} \cos^2 \chi_i$$

the MD radiation reaches its maximal value

$$L_{md}^{\max} = L_{th} \times (5 \sin^2 \chi_i)^{-1} (\tfrac{5}{4} \cos^2 \chi_i)^{-4}$$

after which it decreases. At the radius $R_{\parallel} = R_{md} \cos^2 \chi_i$ the magnetic axis comes so near to the rotational one that L_{md} becomes less or of the order of L_{th}. Such a behaviour corresponds to the expected secular change of the activity of galactic nuclei and quasars as some "flaring up", reaching the maximum of their activity, and subsequent "dying out".

VI. Interpretation of Main Observational Data

a) "Excited" States of Galaxies

As is known, there is a remarkable similarity in many respects between the populations which form the following pairs: quasars (QSS) – quasi-stellar galaxies (QSG); N-galaxies (NG) – blue compact galaxies (BCG); strong radio galaxies (sRG) – D-galaxies; Seyfert galaxies (SyG) – normal spirals (S). This similarity allows one to regard QSS, NG, sRG and SyG as "overexcited" states of QSG, BCG, D and S, respectively (OZERNOY, 1970a). I call an "excited" state of a galaxy the formation in its nucleus of a magnetoid, and "overexcited" is such a stage of evolution of this body, during which the non-thermal luminosity L_{nth} (being essentially the same as L_{md}) is much larger than the thermal one. All that has been said above may be presented schematically as an extended Hubble sequence by the location of its components, as shown in Fig. 3.[1] BCG and QSG are the

[1] The Zwicky and Markarian galaxies are not included into this sequence because they are non-homogeneous groups of objects. For instance, among Markarian galaxies are SyG, BCG, QSG and, possibly, even NG and QSS.

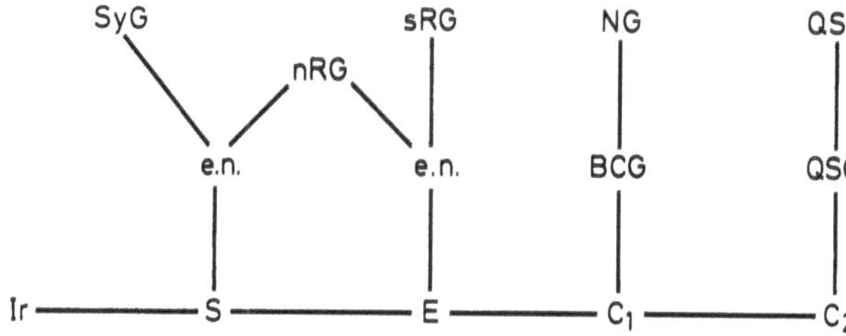

Fig. 3. The extended Hubble sequence of galaxies

intermediate "excited states" analogous to "excited nuclei" (designated as "e.n.") in those normal galaxies where compact sources of non-thermal radiation have been discovered (HEESHEN, 1970; EKERS, 1972; VAN DER KRUIT, 1971a). The galaxies in the nuclei of which the active phenomena are absent, may be regarded as the "ground states" of galaxies. There are S, E and two kinds of compact galaxies designated C_1 and C_2. These compact galaxies which, as may be expected, are among the most massive Zwicky compact galaxies, may be regarded presumably as an initial as well as "candle-end" state of quasars and N-galaxies.

I suppose that the galaxies in the "ground state" are not connected genetically. The transitions between various states of galaxies are expected only in the vertical directions. Under the upward transition we have $\frac{dL_{\text{nth}}}{dt}>0$, then the "ceiling" of the excitement in states SyG, sRG, NG and QSS is reached, after which $\frac{dL_{\text{nth}}}{dt}<0$ and the downward transition into the "ground state" of a galaxy occurs.

The model of a supermassive oblique rotator (Sect. Vc) allows to estimate the ratio of the number of objects with $L_{\text{nth}}>L_{\text{th}}$ to that with $L_{\text{nth}}\lesssim L_{\text{th}}$ as small as 10^{-2}, weakly depending (as $M^{\frac{1}{4}}$) on the mass of the rotator. Such a small percentage of sources in "overexcited" states is explained by the short duration of the stay of galactic nuclei in the state with $L_{\text{nth}}\gg L_{\text{th}}$.

The other consequence of the model, namely, the existence in each of the active populations of two groups of objects having $\dot{L}_{\text{nth}}>0$ or $\dot{L}_{\text{nth}}<0$ is interesting to compare with the same observational result concerning the optical variability of quasars (ANGIONE and SMITH, 1972). The secular trend of the optical variability of 3C 273 and 3C 48 calculated from our model is of the same order as the observational one obtained by GEYER (1964).

The appreciable dispersion of the luminosities in the "over-excited" state may explain why the location of quasars on the $m_v - z$ plot is similar to a scatter diagram.

b) Variability of the Continuum Radiation

As it has been indicated above, the suggested model explains the powerful infra-red radiation of active nuclei and quasars. This radiation is expected to be variable. The cause of variability is the pulsating character of the outflow of matter from a magnetoid because of competition between losses of angular momentum due to centrifugal wind and those due to tension of magnetic force lines (OZERNOY and Usov, 1971b, 1973a, b). The duration of one pulse of outflow is equal to $\tau_{\text{flow}} \approx (2\Omega)^{-1}$, and the interval between successive pulses of the pulsating outflow $\tau_{\text{break}} \approx (c/\Omega R)^2 \tau_{\text{flow}}$.

The variations of the luminosity in radio, optic and X-ray regions may be correlated with infra-red variations, but with some phase difference. LYUTYJ and CHEREPASHCHUK (1972) have found that the variations of H+[N II] in Seyfert galaxies NGC 4151, 3516 and 1068 correlate with continuum variations with some delay (0.5–1 month). The latter can be conditioned by the propagation of ultra-violet radiation or relativistic particles accelerated by low-frequency radiation to the boundary of the gas envelope where they produce the changes in heating and ionization.

c) Ejections of Relativistic Particles and Gas Masses

If the rotation of a magnetoid magnetosphere is differential (this is believed to be at the early stage of the activity), the "magneto-rotational" explosions producing relativistic particles and ejections of plasma are possible (OZERNOY and SOMOV, 1971). At the more later stages when the poloidal magnetic field becomes large enough, the differential rotation is smoothed out, and we deal with the oblique rotator having the essentially uniform rotation. Nevertheless, the pulsating character of outflow of plasma accompanied by readjustment of the magnetic field and acceleration of particles leads to an ejection of clouds of relativistic electrons in two opposite directions. Such ejections can give the radiovariability observed from active nuclei and quasars. When explaining some principal features of this variability (OZERNOY and ULANOVSKY, 1973), we obtain that the sourse of activity must be possessed of a strong regular magnetic field like a magnetoid.

The protons themselves accelerated to high energies, twisting in the magnetic field of gas, accelerate it by their pressure (e.g. WENTZEL, 1971). The small masses ($\lesssim 10^2\ M_\odot$) produced by the pulsating outflow are accelerated as a whole to high speeds, up to the velocity of light, without appreciable increasing of internal velocity dispersion. This process of repeated ejections may be responsible for the appearance of multiple absorption line systems in quasars. The same process may explain the repeated ejections of gas form nuclei of Seyfert galaxies, as in the case of NGC 4151 (ANDERSON and KRAFT, 1969). The efficiency of the acceleration would be less if the cosmic rays act on the external gas clouds surrounding the active nucleus. Nevertheless the massive gas clouds moving in opposite directions from the nucleus of our Galaxy (VAN DER KRUIT, 1970, 1971b) as well as the "ridges" in NGC 4258 (VAN DER KRUIT *et al.*, 1972) could be accelerated by this mechanism. It would be very interesting to compare in detail the dynamics of gas in the near neighbourhood of the center of our Galaxy with the density of cosmic ray energy there (the new approach to the estimate of the latter value using the data on gamma-radiation from the center is developed by GINZBURG and KHAZAN, 1972). The measurements of the magnetic field in the expanding gas by Zeeman-effect in 21-cm may give additional important information about the origin of this gas and its dynamic history.

In summary of the results of this Section, it may be concluded that the model of supermassive oblique rotator (Sect. V c) being one of the concrete realizations of a magnetoid, allows to explain a number of principal observational data. More detailed comparison of the theory with observations as well as a number of observational tests are given in the papers cited above. Further work, both theoretical and observational, is needed for making final conclusions about the validity of the model suggested.

VII. Discussion

Let us turn now to the discussion of some topics arising in the connection with the unique body conception as a source of energy.

a) Single or Recurrent Activity?

It would be very important to have an unquestionable evidence for the fact that the populations which have been labelled as "overexcited" states of galaxies (i.e. SyG, sRG, NG and QSS, see Fig. 3) are really specific stages, rather than specific forms, of galaxies. As for strong radiogalaxies which are always considered as a stage in the life of bright ellipticals, this problem is obviously absent. The same is valid apparently for Seyfert galaxies as well. True, DIBAY (1971) using mainly the measurements of brightness gradients in a number of SyG made by ZASOV and LYUTYJ (1971), concluded recently that Seyfert galaxies differ significantly in their external (non-nuclear) regions from that in normal spirals of the same mass, and therefore supposed that they may be a specific form of galaxies. However ALLEN *et al.* (1971) find that the distribution of H I in Seyfert galaxies does not differ from that in usual spirals. Similarly, LEWIS (1972) discovered that within his radio data the Seyfert galaxies are quite normal in dynamics and gas content outside the nuclei. This is in agreement with the conclusion of MORGAN (see MORGAN *et al.*, 1971) who has shown that a Seyfert galaxy may be considered as a result of acting of some "operator" on a normal spiral after which its nucleus comes to a state of strong activity. The situation with N-galaxies and quasars may be analogous (MORGAN, 1971).

What has been said above, is connected directly with the problem whether the activity of nuclei is repetitive or not. If SyG, sRG, NG and QSS are indeed the active stages of galaxies rather than peculiar "monsters" among normal galaxies, then this activity most likely is of a recurrent character. For radiogalaxies this was suggested ten years ago by SHKLOVSKY (1962). Subsequent observational data have brought new evidence for the recurrency of explosions not only for radiogalaxies but also for Seyfert galaxies (DIBAY and PRONIK, 1967). More recently TIFFT (1970), who investigated the correlation between optical abnormalities and small radio sources in galaxy nuclei, concluded that multiple radio-structure in some galaxies and perhaps optical structure in some spirals suggests that the process is repetitive or cyclical.

To summarize, there is already much evidence that the activity of nuclei is recurrent indeed. The ratio of the durations of an "excited" and a "ground" phase may be as large as about 10^6 y/10^8 y$=10^{-2}$, but neither the dispersion of these figures nor their universality is known at present.

b) Birth and Death of a Magnetoid

There were no attempts to explain the recurrence of active stages of galactic nuclei within the framework of the conceptions of a compact star cluster or accreting black hole. It is of interest to discuss this problem in the light of the magnetoid conception. Two important questions are to be answered: (1) what is the fate of a source of energy in nuclei? and (2) how does the reproduction of a new source after the death of an old one proceeds?

Let us begin with an answer to the second question. All the necessary for such an answer have been ready more than 30 years ago when it has been realized that the gas lost by far evolved stars must accumulate in the galactic nuclei (SPITZER, 1942). The fact that this process is inevitable, was stressed recently once more (ARNY, 1970; MATHEWS and BAKER, 1971). SHKLOVSKY (1972a) underlines especially the role of planetary nebulae, the loss of mass from which is possibly as large as about 1 $M_{\odot}$/year.

The gas lost by stars falls finally into the galactic center. Due to the fact that the spatial density of stars ρ_* is much larger than the initial density of the gas accumulating ρ_g, the star formation from the falling gas is impossible. When the inverse inequality is valid ($\rho_g \gg \rho_*$), the Jeans mass becomes much larger than the mass of stars. As a result, a single object of a very great mass depending on the morphological type of a parent galaxy (mainly on the mass of halo population and angular momentum of the gas) may be formed in the nucleus. The free-fall time of the gas to the center $t_{ff} = (\frac{4}{3}\pi G \rho_*)^{-\frac{1}{2}}$ is as large as 10^8 y at $\rho_* \sim 10^{-25}$g cm^{-3}. During this time the mass of gas up to 10^8 $M_{\odot}$ can be accumulated in the nucleus. The formation from this gas of the active source prevents the further accretion of gas. This is valid also in the case of metagalactic gas which may be accreted by a galaxy and then by its nucleus only at the passive stage of the nucleus (OZERNOY, 1970b).

The transition of the accumulated mass from the infall to the quasi-stationary state must be accompanied by a "bounce" leading to shock waves, heating etc. The details of the formation of the final quasi-stationary configuration such as a magnetoid are not very clear at present. The subsequent evolution may be traced in some detail for the supermassive oblique rotator as a form of magnetoid (Sect. Vc). The necessity to take into account the generation of the magnetic field by the dynamo mechanism makes the consideration of more real models more difficult.

Let us go back to the first question: what is the fate of a magnetoid?

A substantial part of the magnetoid mass is lost due to rotational instability. This gas is rejected outside and can slowly accumulate again in the center only at a "quiescent" stage. However, in such a way all the mass of a magnetoid cannot be lost. There are only three possibilities for the remnant: (1) collapse; (2) non-collapsed rotating disc from neutron stars; (3) nuclear explosion. Although these possibilities are not equal in their rights, it is not yet clear what is more predominant. A great number of neutron stars is hardly compatible with more or less usual chemical abundance of galactic nuclei and quasars, as was suggested above. The nuclear explosion of a disc of very large mass is possible (WAGONER, 1971), but it is not clear whether such an event is compatible with the low abundance of helium in quasars (e.g. MACALPINE, 1972).

From the theoretical point of view, only a black hole does not lead to apparent contradictions with observations if its mass is sufficiently small. A black hole of small mass would not influence appreciably the character and value of the mass of gas accumulating in a nucleus during its passive stage. As a result, a magnetoid of second and further generations with mass M_m having in its center one or few black holes with total mass $M_{bh} \ll M_m$, would not also differ from a first generation magnetoid, i.e. from the magnetoid without a black hole (Sect. V). On the other hand, such a hybrid differs radically from the "standard" black hole of mass M_{bh} surrounded by the accreting mass $M_a \ll M_{bh}$ (Sect. IV). The difference is significant at least in two respects: (i) in the hybrid case the mass outside a black hole mostly outflows due to rotational instability as is also ejected outside rather than accreted inside; (ii) the source of activity of a hybrid is the kinetic energy (and finally the gravitational one) of a supermassive body outside a black hole rather than the energy released by the slow settling of gas into the small mass hole. Further investigations will suggest the observational consequences of the presence of such a small mass hole.

Analogous views may be suggested with respect to the presence of stars in the central parts of an active nucleus or of a quasar. Although supernovae, flare stars etc. apparently do not play any significant role in the activity of nuclei (Sect. III), the stars presumably are present in an active nucleus and a quasar (this statement is supported for quasars by the observable absorption lines of H and K Ca II and G-band in some near quasars with relative weak continuum). These stars, possibly, make some contribution to the gravitational potential of the central part of an active source, but are hardly significant for its luminosity.

To summarize, the astrophysical data give the evidence for the recurrent reproduction of a magnetoid during the "quiescent" (passive) stage of a galactic nucleus from the gas lost by stars or/and accreted from intergalactic medium. Another consideration, more cosmogonical than astrophysical, suggests that a more realistic model of the source of activity for galactic nuclei and quasars is apparently the compact star system in the center of which there is a magnetoid containing possibly a black hole(s) of relatively small mass. The magnetoid plays a leading role in the origin of the observable form of activity.

c) Cosmogonical Activity of a Magnetoid

Let us turn now to the problem, what is the fate of gas ejected from a magnetoid (Sect. VI c).

The gas ejected is expanding and cooling. The cooling promotes both the thermal and gravitational instabilities whereas the expansion acts by an opposite manner. Under some conditions, the fragmentation of the gas and further transformation of it into stars are possible. Therefore the activity of a magnetoid can lead to an eruptional way for the formation of star systems (OZERNOY, 1968).

It is of interest to note some specific properties of a star system which is formed by the plasma ejected from a magnetoid;

1) Not only a spherical (under an explosion), but also a flat (under an outflow of matter) subsystem may be formed.

2) The outflowing plasma possesses some angular momentum so that the flat subsystem being formed would be rotating.

3) The dispersion of ages of stars produced by recurrent explosions in a galactic nucleus would be of the order of the age of the galaxy.

These theoretical possibilities as well as the acceleration by the pressure of cosmic rays of gas clouds, surrounding a magnetoid (Sect. VI c) would be interesting to apply to the small companions which are localized near the large galaxies and are suspected as ejected by some explosional processes in the nuclei of parent galaxies (HOLMBERG, 1969). However, such a physical connection between bright 3CR-quasars and galaxies in some suspicious cases, when the angular distance between the components of such a pair is less than 5′ (BURBIDGE *et al.*, 1971), seems to be impossible. In fact, as was shown by BAHCALL *et al.* (1972), HAZARD and SANITT (1972), such a correlation becomes statistically insignificant when the sample is extended to more weak objects. Moreover, the relation between the angular distance of a quasar from a nearby galaxy and the redshift of this galaxy, which was found by BURBIDGE *et al.* (1972), disappears if we consider a more complete sample of "pairing" systems (OZERNOY, 1972d).

SHKLOVSKY (1972b) suggested that a non-symmetrical explosion in a magnetoid (if it is considered as possible) would give to it an impulse which would lead to a high velocity ejection of the magnetoid from a galactic nucleus. The non-symmetric location of nuclei in some galaxies (BURBIDGE, 1970) is, at first glance, favourable to this view. To check it, it would be important to measure precisely the difference between a redshift of the nucleus and that of the galaxy as a whole. However, in any case the origin of the main part of a normal galaxy, as a result of the activity of its nucleus, seems to be hardly possible. This point of view is supported by the explanation of main parameters of normal galaxies as well as of their clusters without any consideration of the parameters of galaxy nuclei (OZERNOY and CHIBISOV, 1970; OZERNOY, 1971b).

VIII. Concluding Remarks

Quasars and active phenomena in galactic nuclei are considered usually either as the early phase of a new forming stellar system or as the final stage of an ensemble of stars. Most likely neither is true. The activity phenomenon of a galactic nucleus is apparently a secondary one with respect to the galaxy as a whole, and may be repeated during its life alternating with the phases of the more or less "passive" state of the nucleus.

As for the nature of the energy release during the active state of the nucleus, we know mainly what types of models cannot explain satisfactorily the variety of this activity. However it seems now that in this respect astrophysics is nearer to the goal than a decade ago.

Very extensive recent theoretical considerations have shown that there is at least three plausible possibilities to explain the nature of an energy source for the activity of galactic nuclei and quasars. Not all of these possibilities are equal in their rights. A most attractive of them seems to be a supermassive rotating magnetized body (magnetoid) at the hearts of active nuclei and quasars. The explanation, in the framework of this conception, of a number of principal properties of the activity allows to hope that both the further development of this conception and new observations would lead finally to a more complete theory adequate to the richness and complexity of the problem of galactic activity. However we should not overestimate this optimism. The problem of galactic nuclei is so complicated that new discoveries may force us to see this problem from a completely different point of view. Nevertheless there is no other way to solve this problem except to construct new models which would be in more accordance with new observations.

References

ADAMJANTS, R. A., ALEKSEEV, A. D., KOLOSNITSYN, N. J.: Zh. Eksp. Teor. Fiz. Pis'ma **15**, 277 (1972).
ALLEN, R. J., DARCHY, B. F., LAUQUÉ, R.: Astron. Astrophys. **10**, 198 (1971).
ALLER, H. D.: Astrophys. J. **161**, 1 (1970).
AMBARTSUMIAN, V. A.: In Solvay Conference on Structure and Evolution of the Universe (Ed. by R. STOOPS), Brussels, p. 241 (1958).
AMBARTSUMIAN, V. A.: The Structure and Evolution of Galaxies, Proc. 13th Conf. Phys., Univ. Brussels (Interscience: Wiley), p. 1 (1965).
ANDERSON, K. S., KRAFT, R. P.: Astrophys. J. **158**, 159 (1969).
ANGIONE, R. J., SMITH, H. J.: In D. EVANS (ed.), External Galaxies and Quasi-Stellar Objects, IAU Symp. **44**, 171 (1972).
ARNY, T. T.: Monthly Notices Roy. Astron. Soc. **148**, 63 (1970).
BAHCALL, J. N., KOZLOVSKY, B.-Z.: Astrophys. J. **155**, 1077 (1969a).
BAHCALL, J. N., KOZLOVSKY, B.-Z.: Astrophys. J. **158**, 529 (1969b).
BAHCALL, J. N., MCKEE, C. F., BAHCALL, N. A.: Astrophys. Letters **10**, 147 (1972).
BARDEEN, J., WAGONER, R. V.: Astrophys. J. **158**, L 65 (1969).
BARDEEN, J., WAGONER, R. V.: Astrophys. J. **167**, 359 (1971).
BARDEEN, J. M., PRESS, W. H., TEUKOLSKY, S. A.: Preprint OAP-288 (1972).
BLANFORD, R. D., REES, M. J.: Astrophys. Letters **10**, 77 (1972).
BRAGINSKY, V. B., MANUKIN, A. B., POPOV, E. I., RUDENKO, V. N., CHOREV, A. A.: Zh. Eksp. Teor. Fiz. Pis'ma **16**, 157 (1972).
BURBIDGE, E. M., BURBIDGE, G. R., SOLOMON, P. M., STRITTMATTER, P. A.: Astrophys. J. **173**, 233 (1970).
BURBIDGE, G. R.: Ann. Rev. Astron. Astrophys. **8**, 369 (1970).
BURBIDGE, G. R., O'DELL, S. L., STRITTMATTER, P. A.: Astrophys. J. **175**, 601 (1972).
CAVALIERE, A., PACINI, F., SETTI, G.: Astrophys. Letters **4**, 103 (1969).
CHANDRASEKHAR, S.: Phys. Rev. Letters **12**, 114, 437E (1964).
CHERTOPRUD, V. E., GUDZENKO, L. I., OZERNOY, L. M.: Astrophys. J. **182**, L 53 (1973).
CHRISTODOULOU, D.: Phys. Rev. Letters **25**, 1596 (1970).
COLGATE, S. A.: Astrophys. J. **150**, 163 (1967).
DIBAY, E. A.: Astron. Tsirk. No. 660, 1 (1971).
DIBAY, E. A., PRONIK, V. I.: Astron. Zh. **44**, 952 (1967).
EKERS, R. D.: In D. EVANS (ed.), External Galaxies and Quasi-Stellar Objects, IAU Symp. **44**, 222 (1972).
FOWLER, W. A.: Rev. Mod. Phys. **36**, 545, 1104E (1964).
FOWLER, W. A.: Astrophys. J. **144**, 180 (1966).
GEYER, E. H.: Z. Astrophys. **60**, 112 (1964).
GINZBURG, V. L., KHAZAN, YA.: Astrophys. Letters **12**, 155 (1972).
GOLD, T., AXFORD, W. I., RAY, E. C.: Quasi-Stellar Sources and Gravitational Collapse, Univ. of Chicago Press, p. 93 (1965).
GORSHKOV, A. G., POPOV, M. V.: Astron. Zh. **49**, 722 (1972).
GREWING, M., LAMLA, E.: Z. Astrophys. **68**, 473 (1968).
GUDZENKO, L. I., OZERNOY, L. M., CHERTOPRUD, V. E.: Astron. Zh. **45**, 492 (1968).
GUDZENKO, L. I., OZERNOY, L. M., CHERTOPRUD, V. E.: Astron. Zh. **48**, 472 (1971).
GUNN, J., OSTRIKER, J.: Astrophys. J. **165**, 523 (1969).
HAZARD, C., SANITT, N.: Astrophys. Letters **11**, 77 (1972).
HOERNER, S. VON: Bull. Astron., Ser. 3, **3**, 147 (1968).
HOLMBERG, E.: Ark. Astron. **5**, 305 (1969).

HOYLE, F., FOWLER, W.: Monthly Notices Roy. Astron. Soc. **125**, 169 (1963).
HOYLE, F., FOWLER, W.: Nature **213**, 273 (1967).
JURKEVICH, I., USHER, P.D., SHEN, B.S.P.: Astrophys. Space Sci. **10**, 402 (1971).
KINMAN, T.D.: Report at the Symposium IAU No. 44, Uppsala (1970).
KINMAN, T.D., LAMLA, E., CIURLA, T., HARLAN, E., WIRTANEN, C.A.: Astrophys. J. **152**, 357 (1968).
KRUIT, P.C. VAN DER: Astron. Astrophys. **7**, 462 (1970).
KRUIT, P.C. VAN DER: Astron. Astrophys. **13**, 405 (1971 a).
KRUIT, P.C. VAN DER: Astron. Astrophys. **15**, 110 (1971 b).
KRUIT, P.C. VAN DER, OORT, J.H., MATHEWSON, D.S.: Preprint (1972).
LEWIS, B.M.: In D. EVANS (ed.), External Galaxies and Quasi-Stellar Objects, IAU Symp. **44**, 267 (1972).
LÜ, P.K., HUNTER, J.H.: Nature **221**, 755 (1969).
LYNDEN-BELL, D.: Nature **223**, 690 (1969).
LYNDEN-BELL, D., REES, M.J.: Monthly Notices Roy. Astron. Soc. **152**, 461 (1971).
LYUTYJ, V.M., CHEREPASHCHUK, A.M.: Astron. Tsirk. No. 647, 1 (1971).
LYUTYJ, V.M., CHEREPASHCHUK, A.M.: Astron. Tsirk. No. 688, 1 (1972).
MACALPINE, G.M.: Astrophys. J. **175**, 11 (1972).
MATHEWS, W.G., BAKER, J.C.: Astrophys. J. **170**, 241 (1971).
MCDONALD, G.H., KENDERDINE, S., NEVILLE, A.C.: Monthly Notices Roy. Astron. Soc. **138**, 259 (1968).
MORGAN, W.W.: Astron. J. **76**, 1000 (1971).
MORGAN, W.W., WALBORN, N.R., TAPSCOTT, J.W.: Pont. Acad. Sci. Scr. Varia **35**, 27 (1971).
MORRISON, P.: Astrophys. J. **157**, L 73 (1969).
MORRISON, P., CAVALIERE, A.: Pont. Acad. Sci. Scr. Varia **35**, 485 (1971).
MUSTEL, E.R.: Astron. Tsirk. No. 649, 1 (1971).
OORT, J.H.: Pont. Acad. Sci. Scr. Varia **35**, 321 (1971).
OSTERBROCK, D.E.: Pont. Acad. Sci. Scr. Varia **35**, 159 (1971).
OZERNOY, L.M.: Astron. Zh. **43**, 300 (1966).
OZERNOY, L.M.: Highlights of Astronomy **1**, 384 (1968).
OZERNOY, L.M.: Astron. Tsirk. No. 581, 1 (1970 a).
OZERNOY, L.M.: Astron. Zh. **47**, 265 (1970 b).
OZERNOY, L.M.: Astron. Tsirk. No. 661, 4 (1971 a).
OZERNOY, L.M.: Astron. Zh. **48**, 1160 (1971 b).
OZERNOY, L.M.: Astron. Tsirk. No. 712, 1 (1972 a).
OZERNOY, L.M.: Astron. Zh. **49**, 1131 (1972 b).
OZERNOY, L.M.: Astron. Tsirk. No. 718, 1 and 3 (1972 c).
OZERNOY, L.M.: Astron. Zh. **49**, 1148 (1972 d).
OZERNOY, L.M., CHERTOPRUD, V.E.: Astron. Tsirk. No. 635, 1 (1971).
OZERNOY, L.M., CHERTOPRUD, V.E.: Astron. Zh. **49**, 712 (1972).
OZERNOY, L.M., CHERTOPRUD, V.E., CHUVACHIN, S.D.: Astron. Zh. **46**, 1317 (1969).
OZERNOY, L.M., CHIBISOV, G.V.: Astron. Zh. **47**, 769 (1970).
OZERNOY, L.M., SOMOV, B.V.: Astrophys. Space Sci. **11**, 264 (1971).
OZERNOY, L.M., ULANOVSKY, L.E.: Astron. Zh. (in press) (1973).
OZERNOY, L.M., USOV, V.V.: Astrophys. Space Sci. **13**, 3 (1971 a).
OZERNOY, L.M., USOV, V.V.: Astron. Zh. **48**, 240 (1971 b).
OZERNOY, L.M., USOV, V.V.: Astrophys. Letters **13**, 209 (1973 a).
OZERNOY, L.M., USOV, V.V.: Astrophys. Space Sci. (in press) (1973 b).
PACHOLCZYK, A.G.: In D. EVANS (ed.), External Galaxies and Quasi-Stellar Objects, IAU Symp. **44**, 165 (1972).
PEIMBERT, M.: Astrophys. J. **170**, 261 (1971).
PENROSE, R.: Revista de Nuovo Cimento **1**, 252 (1969).
PIDDINGTON, J.H.: Monthly Notices Roy. Astron. Soc. **148**, 131 (1970).
REES, M.J.: Nature **229**, 312 (1971).
SANDAGE, A.: Pont. Acad. Sci. Scr. Varia **35**, 271 (1971).
SANDERS, R.H.: On the Coalescence of Colliding Stars in Dense Stellar Systems, Princeton Univ., Univ. Microfilms Ltd. (Ph. D. Thesis) (1970).
SALPETER, E.E.: Astrophys. J. **140**, 796 (1964).
SALPETER, E.E., WAGONER, R.V.: Astrophys. J. **164**, 557 (1971).
SCHWARTSMAN, V.F.: Astron. Zh. **47**, 760 (1970).
SHKLOVSKY, I.S.: Astron. Zh. **39**, 591 (1962).

SHKLOVSKY, I.S.: Astron. Zh. **47**, 742 (1970).
SHKLOVSKY, I.S.: Astrophys. Letters **10**, 5 (1972a).
SHKLOVSKY, I.S.: In D. EVANS (ed.), External Galaxies and Quasi-Stellar Objects, IAU Symp. **44**, 272 (1972b).
SPITZER, L.: Astrophys. J. **95**, 329 (1942).
SPITZER, L.: Pont. Acad. Sci. Scr. Varia **35**, 443 (1971).
SPITZER, L., SASLAW, W.C.: Astrophys. J. **143**, 400 (1966).
TIFFT, W.G.: Astrophys. Letters **7**, 7 (1970).
TRIMBLE, V.: Nature **232**, 607 (1971).
VORONTSOV-VELYAMINOV, B.A.: Astron. Zh. **42**, 1168 (1965).
WAGONER, R.V.: Highlights in Astronomy **2**, 316 (1971).
WENTZEL, D.G.: Astrophys. J. **163**, 503 (1971).
WOLTJER, L.: In D. EVANS (ed.), External Galaxies and Quasi-Stellar Objects, IAU Symp. **44**, 277 (1972).
ZASOV, A.V., LJUTYJ, V.M.: Astron. Tsirk. No. 658, 1 (1971).
ZEL'DOVICH, YA.B.: Dokl. Akad. Nauk SSSR **155**, 67 (1964).
ZEL'DOVICH, YA.B., NOVIKOV, I.D.: Relativistic Astrophysics. Univ. of Chicago Press, Chicago (1971).

The Chemical Evolution of the Galaxies

(Invited Lecture)

By A. UNSÖLD
Institut für theoretische Physik und Sternwarte der Universität Kiel, Germany

Abstract

The principles of quantitative spectral analysis are reviewed critically. The abundance distributions of the chemical elements are presented in Tables 1 and 2 for the Sun and "normal" stars, for carbonaceous chondrites I, for planetary and galactic nebulae and for galactic cosmic rays (at their origin). Table 3 gives abundances relative to the Sun for normal, metal-poor and metal-rich stars. Apart from stars showing obvious signs of individual nuclear evolution (like helium-, carbon-... stars) cosmic matter consists quite generally of (a) hydrogen and helium in a ratio $\sim 10:1$ by number and (b) the mixture of the heavier elements ($Z \geqq 6$) as observed e.g. in the Sun. The ratio of (b):(a) – relative to the Sun – varies between $\approx 1/500$ and ≈ 1 in metal-poor halo stars; in the disc – and spiral arm-populations it is ≈ 1 (within a factor ≈ 2 up or down); in (super-)metal-rich stars there occur values of up to 3 or 4. Concerning the concept of stellar populations clearer definitions are proposed. The age of our Galaxy is discussed taking into account that practically all the heavy elements must have existed at a very early stage of our Galaxy. The most probable value for its age seems to be indicated by the radioactive elements having no "parents" ≈ 7 to 8×10^9 years. After a brief review of abundances in other galaxies we discuss theories concerning the origin of the heavier elements and their different mixtures with H + He.

The following points are emphasized:

(i) Production of the heavier elements in the observed uniformity by a variety of different processes in individual stars is highly improbable.

(ii) The importance of the empirical abundance-rules detected by SUESS and of pycno-nuclear processes as proposed by AMIET and ZEH is emphasized.

(iii) The luminosity function of old halo stars is shown (Table 4) to be the same as for the corresponding disc stars. It is therefore extremely improbable that the proto-halo had a completely anomalous frequency of massive stars.

Altogether it seems possible to assume that the heavier elements have originated in a gigantic explosion at a very early stage of evolution of the nucleus of our Galaxy which consisted of hydrogen and helium. However it is not sufficiently clear why in quite different types of galaxies the same fraction of the original matter has been transformed into the *same* type of "cosmic mixture" of the heavier elements. We therefore discuss also the other possibility of a return to a modified GAMOW-theory.

I. Introduction

In the twenties the new atomic physics opened the way toward a quantitative interpretation of stellar spectra. It was one of the aims in this new branch of astrophysics to determine in a quantitative way the chemical composition of stellar atmospheres, i.e. of those parts of the stars which are accessible to spectroscopic observation. It was only in the forties that the first reasonably accurate and complete analyses became available.

In the meantime nuclear physics and the theory of the internal constitution of the stars had reached the important conclusion that the main sequence of the Hertzsprung-Russell-diagram is taken up by stars which in their central parts produce energy by the conversion of hydrogen into helium. It was BAADE who in the early fifties stimulated observational and theoretical work concerning the colour-magnitude-diagrams of star clusters and their connection with the nuclear evolution of stars. Obviously, this type of research opened also the way towards much more difficult and far reaching fields of research: the formation of galaxies and the origin of the chemical elements and their abundance distributions.

When we attempt to review the present situation – in its broadest outlines – we soon notice that it is very strange: While atomic and nuclear physics and their applications to astrophysics have reached a high state of perfection, the situation is quite different when it comes to the much older science of hydrodynamics, and this applies even more to the field of magnetohydrodynamics. The nonlinearity of their basic equations presents – even in our age of computers – almost insurmountable difficulties to the interpretation of observed phenomena. On the other hand nature has undoubtedly no particular interest in actually presenting "problems" which have been selected from the viewpoint of mathematical simplicity.

II. Quantitative Analysis of Stellar Spectra

A few remarks may be sufficient here concerning the method of analyzing stellar spectra and the dangers of its pitfalls and limitations (UNSÖLD, 1969).

In high dispersion spectra we may be able to measure the *profiles* of certain Fraunhofer absorption lines, i.e. their actual spectral energy distributions. Normally we can only measure the *equivalent widths* W_λ or the amount of energy "disappearing" in a line. W_λ does not give as much detailed information as the line profile, but it can still be measured in the case of faint lines, where the limited resolving power of a spectrograph may not allow one to obtain true profiles.

As our next step we construct – using what we know about matter at high temperatures and about the energy transfer in stellar atmospheres – theoretical *model atmospheres* which we attempt to fit to the observations largely by trial and error. Such model atmospheres depend upon large – but still finite! – numbers of available constants which cannot simply be determined by straightforward computing programs, but require considerable insight into complicated relations between these constants and observations of lines of different elements, stages of ionization and excitation, and of transition probabilities, broadening mechanisms and so on. The characteristics of stellar atmospheres are

a) their *effective temperature* T_e which is defined in such a way that using Stefan-Boltzmann's law of radiation T_e gives the correct total net flux of radiation leaving the atmosphere. For the Sun, its value is $T_e = 5780$ K;

b) the *surface gravity* g (cm s^{-2}) which determines how strongly the atmosphere is compressed by the gravitational field of the star. For the Sun we have $g = 2.74 \times 10^4$ cm s^{-2};

c) the *abundances* of all the relevant elements. Usually we state (logarithmically) the numbers of atoms $\log \varepsilon$ referred to $\log \varepsilon = 12$ for hydrogen, the most abundant element.

In principle these data ought to be sufficient for the calculation – in connection with observations of a sufficient number of lines distributed over a sufficient range of atomic characteristics – of a model atmosphere appropriate to a particular star.

However, the present unsatisfactory state of hydrodynamics, as applied to astrophysics, necessitates the empirical determination of one more quantity (and possibly of its variations with depth in the atmosphere), the so called *velocity of turbulence* ξ_t. Part of the broadening of spectral lines is due to the Doppler effect of the motions of the atoms or molecules. Such motions are partly thermal. In many cases, however, there is additional broadening due to more or less statistical macroscopic motions in the atmosphere, the so called turbulence. In cooler stars turbulent motions are due to the hydrogen convection zone which keeps e.g. the Sun stirred up down to about one tenth of its radius. In hot supergiants we observe more or less irregular motions which may be related to the oscillations of the Cepheids, but exhibiting instead of a single frequency a rather broad-band frequency spectrum. The velocity of turbulence ξ_t can often be described with sufficient accuracy by a suitable average value; in some cases it may be necessary to consider it explicitly as a function of depth and perhaps even of direction. Such turbulent motions will affect the equivalent widths of the lines only in so far as the range of the Doppler shifts occurs in a volume element which is so small that its optical diameter (= absorption coefficient × geometrical diameter) is <1 even in the line. This part of turbulence is called *microturbulence*. In addition there may be motions affecting larger elements in space. These will only widen the line profile, but will not affect its equivalent width W_λ. In this case we speak of *macroturbulence*. Following the general trend of turbulent flow, the velocity of microturbulence will be smaller than the velocity of macroturbulence. Furthermore the speed of microturbulence will be smaller than the velocity of sound. In the solar atmosphere e.g. recent determinations indicate a microturbulence of 1.0 km s^{-1}, macroturbulence of 1.6 km s^{-1} and a velocity of sound of 8 km s^{-1} (GARZ *et al.*, 1969). As it has been said before, the turbulent velocity should be determined in principle by T_e, g and the chemical composition of the star. Empirically REIMERS (1972) has found that for later type stars ($\sim$F to M) the velocities in their chromospheres and probably also the microturbulence can be represented by interpolation formulae $\xi \sim g^{-0.2} \times T_e$. The mixing-length theory of convection does not seem to be able to explain that result. As to early type stars our knowledge is still more scanty.

In order to understand which general data are important for the interpretation of a particular line or – vice versa – which general data we can hope to obtain from the observation of this or that line we calculate for a particular model atmosphere and for a particular group of lines the equivalent widths W_λ as functions of the number of atoms in the absorbing quantum states or of the abundance ε of the element in question. In this way we obtain a *curve of growth.* For the weakest lines, W_λ is essentially determined by $gf\varepsilon$, where g is the statistical weight of the lower quantum state and f the oscillator strength which is connected according to a well known relation with the transition probability of the line. Somewhat stronger lines become "saturated" and their W_λ depend only slightly upon $gf\varepsilon$ while the Doppler width $\Delta\lambda_D$ which originates from (quadratic) addition of thermal and turbulent Doppler effects becomes most important. In the case of the strongest lines the $gf\varepsilon$ occurs multiplied by the damping constant γ. The latter is due partly to radiation damping, but damping due to collisions with hydrogen atoms or with free electrons is generally more important. Obviously, it is of paramount importance for astrophysics to obtain extensive and accurate knowledge of *oscillator strengths* and *damping constants* for all the relevant lines of astrophysical interest. As to f-values, rising demands for accuracy ($\sim \pm 30\%$) have shifted the emphasis from quantum mechanical calculation to laboratory measurement. Relative measurements of many lines in the electric arc or in the shock tube are reduced to absolute scales by means of beam-foil or Hanle effect measurements for a few selected lines. As to damping constants the accuracy of theoretical work is very limited in most cases. Laboratory work is proceeding fairly well in the case of broadening by electron collisions in plasma. Experiments on broadening by atomic hydrogen are extremely difficult and progress is correspondingly slow. Therefore an obviously very useful programme has been started by HOLWEGER (1971, 1972) who wishes to determine damping constants for suitable lines in the solar spectrum in order to make them available for the interpretation of spectra of other stars.

The turbulence ξ_t should be determined for every star individually with the greatest possible care because an error in the Doppler width $\Delta\lambda_D$ (thermal plus turbulent width) will produce an error about five times as large in the abundance (due to the small inclination of the Doppler branch of the curve of growth). The best procedure is to take an element such as Fe I with well measured f-values and to calculate formally ε for various ξ_t-values until one and the same ε is obtained for lines all along the curve of growth. Since the relative and absolute values of the oscillator strengths for iron, which is by far the most suitable element in such work, had to be revised considerably in recent years it is obvious that many older abundance determinations must be revised before general conclusions can be drawn from them.

In addition to the Fraunhofer absorption lines the *emission lines* of the solar corona and of various types of gaseous nebulae have been used to determine abundances with about the same accuracy. We shall use such determinations too, but we cannot give here an account of the methods used.

Table 1. The abundance distributions of the chemical elements for the Sun and "normal" stars, for carbonaceous chondrites I, for planetary and galactic nebulae and for galactic cosmic rays.

	Planetary nebulae			H II-region	Hot stars main sequence				Supergiants		Middle main sequence		Meteorites	Galactic Cosmic rays
Spectral type Star (HD)	NGC 7027, 2022	NGC 7662	IC 418	Orion-nebula	O 9 V 10 Lac	B 0 V τ Sco	B 3 V ι Her	A 0 V α Lyr	B 1 Ib ζ Per	A 2 Ia α Cyg	G 2 V Sun			
T_{eff}		10^5	35000		37450	32000	20200	9500	27000	9170	5770		Carbona-ceous chondrites I	
log g		5	3.4		4.45	4.1	3.75	4.5	3.6	1.13	4.44			
											phot.	corona		
1 H	12.0	12.0		12.0_0	12.0	12.0	12.0	12.0	12.0	12.0	12.0			10.9
2 He	11.2	11.2		11.0_4	11.2	11.0	10.8		11.3	11.6				9.6
6 C				8.7_1	8.4	8.1	8.1		8.3	8.2	8.55	8.55		8.2
7 N	8.1:	7.7		7.6_3	8.4	8.3	7.7		8.3	9.4	8.00	7.63		7.3
8 O	8.9	8.4		8.7_9	8.8	8.7	8.4	8.8	9.0	9.4	8.83	8.60		8.2
9 F	4.9												4.92	
10 Ne	7.9	7.8		7.8_6	8.7	8.6	8.6	9.3	8.6			7.40		7.5
11 Na	6.6							7.3			6.30		6.36	6.2
12 Mg					8.2	7.5	7.3	7.7	7.8	7.8	7.6	7.33	7.57	7.6
13 Al					7.1	6.2	6.1	5.7	6.8	6.6	6.3	5.95	6.48	6.4
14 Si					7.7	7.6	7.1	8.2	8.0	7.9	7.55	7.55	7.55	7.5
15 P											5.4		5.65	5.5
16 S	7.9			7.5_0		7.2	7.1		7.5		7.25	6.95	7.25	6.7

17 Cl	6.9		5.8_5					5.6		4.79	
18 A	7.0				6.7						5.9
19 K	5.7							5.05		5.13	
20 Ca	6.4:				6.4	6.3	6.5	6.36	6.6	6.42	6.5
21 Sc						3.4	3.2	3.05		3.09	
22 Ti						4.8	5.1	4.6		4.91	
23 V						4.0	3.9	4.0		4.02	
24 Cr						5.6	5.7	5.6		5.63	
25 Mn	(A)	(B)				5.3	5.6	5.0		5.50	
26 Fe				7.3	7.4	(6.5)	7.6	7.60	7.65	7.50	7.6
27 Co							3.7	4.6		4.91	
28 Ni						7.0	4.8	6.25	6.55	6.24	6.2
29 Cu								4.16		4.32	
30 Zn								4.4		4.53	
31 Ga								2.9		3.26	
32 Ge								3.3		3.68	
37 Rb								2.6		2.36	
38 Sr						2.8	3.1	2.8		2.93	
39 Y						2.1		3.2		2.21	
40 Zr						2.9		2.6		3.05	

Literature: Planetary nebulae: ALLER and CZYZAK (1968), FLOWER (1969); OSTERBROCK (1970). H II-regions; Orion: PEIMBERT and COSTERO (1969), MORGAN (1971). 10 Lac: TRAVING (1957). τ Sco: HARDORP and SCHOLZ (1970) (also λ Lep B 0.5 IV). ι Her: KODAIRA and SCHOLZ (1970). (Also η Hya B 3 V and HD 58343 B 3 Ve.) OB-stars: SCHOLZ (1972). (Review article.) α Lyr: HUNGER (1960) and previous papers. ζ Per: CAYREL (1958). α Cyg: GROTH (1961), WOLF (1971). (Also η Leo A 0 Ib.) Sun: MÜLLER (1967), LAMBERT *et al.* (1968, 1969), GARZ *et al.* (1969). Carbonaceous chondrites I: UREY (1967) with extensive bibliography. Galactic cosmic rays: SHAPIRO *et al.* (1971); SHAPIRO (1972) personal communication.

III. Abundances in "Normal" Stars and Related Sources

The vast majority of stars can be classified using *two* parameters. In the *MK-classification* we use the spectral type Sp and the luminosity class L; quantitatively we use the "coordinates" T_e and g. Such stars we term briefly as "normal".

In Table 1 we have made an attempt to compile, what we think to be some of the best abundance determinations. The somewhat ambiguous word "best" comprises: high dispersion measurements extending over a large range of wavelengths; individual determinations of turbulence, careful discussions of blends, but unavoidably also, some personal judgment. In the central part of the table we give data for main-sequence stars extending from quite young objects to the Sun, 4.6×10^9 years of age. In the case of the Sun we give additional data for the corona and for carbonaceous chondrites I which have condensed at quite low temperature and represent solar abundances except for the most volatile elements (compare e.g. ANDERS, 1971).

Two early type supergiants represent stars in an early stage of their evolution away from the main sequence where hydrogen and helium burning in general does not yet reach the surface layers. On the left side we give data for the Orion nebula representing matter from which the youngest stars have originated, and also for some planetary nebulae where we observe matter expelled from the outermost parts of stars which have already reached a fairly advanced state of nuclear evolution. At the right hand side of the table we have put data on the original composition of the galactic cosmic radiation which SHAPIRO *et al.* (1971) have derived by taking into account as well as possible nuclear processes in interstellar space. In Table 2 the abundances of the lightest and of the heavy elements ($Z > 40$) are collected for the Sun and for carbonaceous chondrites I; in stellar spectra their weak lines are mostly inaccessible. It is very remarkable that all these sources have practically the same composition[1]; we may truly speak of "*cosmic abundances*".

The long-time paradox of the iron abundance has been removed in recent years by new laboratory *f*-values (e.g. UNSÖLD, 1971), whereby large systematic errors of the older determinations became evident. At present there is still some uncomfortable feeling about the abundance of neon which in hot stars turns out to be about 8 times higher than in all other sources. In both cases however the abundances derived from several stages of ionization seem to agree with each other. On the other hand any kind of separation of such elements seems quite improbable. We must leave the solution of that puzzle to future work.

For cooler stars of spectral types F to K (and with some reservations even A to M) the quantitative spectral analysis is most suitably made relative to the Sun. This method eliminates errors in the *f*-values and damping constants as long as a line is not transferred from one branch of the curve of growth to another e.g. from the linear to the Doppler- or from the Doppler- to the damping branch. In Table 3 we list in the three first columns the relative abundances $\log \varepsilon_{star} - \log \varepsilon_{Sun}$ for a K3V main sequence star and two red giants. Again there are no significant differences between Sun and stars. However, we have learned

[1] Apart from H and He in the cosmic radiation.

quite recently that this does not hold generally: Several authors (GREENE, 1970; FAWELL, 1971) have shown that in case of Arcturus (α Boo) the isotope ratio $C^{12}:C^{13} \approx 5$ which differs from that obtained for the Sun and for terrestrial and meteoritic carbon ≈ 90 indicating clearly that the CNO-cycle has been at work.

In the right-hand part of Table 3 we deal with a different type of stars, the *metal-poor* high velocity or *halo stars*, represented by the solar type main sequence star HD 140 283, the red giants HD 6833 and HD 122 563 and the horizontal branch *A* star HD 161 817. Detailed analysis shows that in these stars the relative abundances of all the heavier elements ($Z \geqq 6$; of $Z = 3$ to 5 we know practically nothing) are equal. It would indeed be preferable to speak of hydrogen-(and helium-)rich stars. We have reduced the data so that the abundances of the heavier elements can be compared directly with the Sun (average $\log \varepsilon_{\text{star}} - \log \varepsilon_{\text{Sun}} \approx 0$); the first line gives now the overabundance of hydrogen. So far HD 122 563 holds the galactic championship with a hydrogen overabundance given by a factor ~ 500. The relative abundances even of quite heavy elements show no significant differences compared with the Sun. Abundances differing by factors of 2 to 3 refer always to the elements with very few and/or poor lines. The fact that many extremely metal-poor stars have quite normal relative abundances of heavier elements is certainly of great importance in connection with the cosmic evolution of the elements. It does not apply however to all halo stars which, of course, undergo individual nuclear evolution like ordinary disc stars, as we shall see in the following section.

The last column in Table 3 contains data for the (super-)metal-rich star β Virginis. Compared with normal stars the hydrogen abundance is lowered or the metal abundance is increased by a factor ≈ 2. The relative abundances of all the heavier elements are again – within the accuracy of the analysis – the same as e.g. in the Sun.

Table 2. The abundance distributions of the chemical elements for the Sun and carbonaceous chondrites I

Element	Sun	Carbonaceous chondrites I	Element	Sun	Carbonaceous chondrites I
3 Li	0.4	3.25	59 Pr	1.6	0.78
4 Be	1.2		60 Nd	1.8	1.44
5 B	<2.8		62 Sm	1.6	0.91
41 Nb	2.3		63 Eu	0.5	0.51
42 Mo	2.3		64 Gd	1.1	1.15
44 Ru	1.8		66 Dy	1.1	1.11
45 Rh	1.4		68 Er		0.89
46 Pd	1.6	2.17	69 Tm		0.09
47 Ag	0.7	1.53	70 Yb	0.8	0.87
48 Cd	1.5	1.93	71 Lu	0.8	0.09
49 In	1.4	0.78	81 Tl	≦0.2	0.80
50 Sn	2.0	1.81	82 Pb	1.85	1.75
51 Sb	1.9	1.13	83 Bi	≦0.8	0.78
56 Ba	2.1	2.22	90 Th	0.8	~0.6
57 La	1.8	1.11	92 U	≦0.6	~0.0
58 Ce	1.8	1.62			

Table 3. The abundances relative to the Sun for normal, metal – poor and metal – rich stars

Spectral type	G 8 III	F 8 IV–V	K 3 V	~G 0 V	K 1 III	~K 2 III	~sd A 2	F 8 V
Star (HD)	ε Vir	136202	219134	140283	6833	122563	161817	β Vir
T_{eff}	4940	6030	4700	5940	4420	4600	7630	6120
log g	2.7	3.9	4.5	4.6	1.3	1.2	3.0	4.3
1 H	0.00	0.00	0.00	+2.32	+0.96	+2.75	+1.11	−0.30
6 C	−0.12	+0.06		+0.5		−0.39	−0.26	−0.15
11 Na	+0.30	+0.03	+0.04	+0.30	−0.19	+0.33	−0.05	+0.06
12 Mg	+0.04	−0.01	+0.15	−0.01	+0.11	+0.03	+0.08	−0.04
13 Al	+0.14	+0.09	+0.07	+0.26	+0.37	−0.12	−0.26	+0.05
14 Si	+0.13	+0.02	+0.39	−0.07	+0.10	+0.29	−0.19	−0.08
16 S	+0.09							−0.04
20 Ca	+0.10	+0.03	−0.11	+0.03	+0.28	+0.19	+0.09	+0.02
21 Sc	−0.07	−0.07	+0.22	−0.61	−0.15	+0.03	+0.21	+0.10
22 Ti	−0.07	−0.02	−0.10	−0.01	+0.02	+0.08	+0.25	+0.03
23 V	−0.08	−0.14	−0.09		+0.03	−0.06	−0.42	+0.03
24 Cr	+0.01	+0.02	−0.20	+0.09	+0.15	+0.06	−0.27	−0.01
25 Mn	+0.07	−0.08	−0.31	−0.35	−0.17	−0.19	−0.43	−0.06
26 Fe	+0.01	0.00	+0.10	+0.16	+0.11	+0.03	−0.10	−0.04
27 Co	−0.03	−0.27	+0.17	−0.02	−0.10	+0.04	+0.12	−0.07
28 Ni	+0.03	+0.02	+0.10	+0.31	−0.06	+0.16	+0.11	−0.01
29 Cu	+0.06	−0.11						+0.05
30 Zn	+0.05	−0.21	−0.13		+0.14	+0.28		−0.17
38 Sr	+0.02	+0.10	−0.01	0.00		−0.53	+0.24	+0.01
39 Y	−0.16	−0.14	+0.48		−0.46	−0.04	−0.19	+0.03
40 Zr	−0.15	−0.39	−0.30		−0.02	−0.08	−0.14	−0.01
56 Ba	−0.09	−0.05				−0.55	−0.05	0.00
57 La	−0.08				−0.25			+0.11
58 Ce	−0.08							
59 Pr	+0.37							
60 Nd	+0.06							
62 Sm	+0.01							
63 Eu						−0.27		
72 Hf	+0.18							

Literature: ε Vir: CAYREL and CAYREL (1963). HD 136202: ZIELKE (1970) (also HD 105590, 155646, 208776, 208906). HD 219134 and 6833: CAYREL DE STROBEL (1966) (further K-stars: HD 190404, 48781, 94264, 6497, 35620, ε Vir). HD 140283: BASCHEK (1962); ALLER and GREENSTEIN (1960). HD 122563: WOLFFRAM (1972). HD 161817: KODAIRA (1964). β Vir: BASCHEK *et al.* (1967).

IV. Anomalous Abundance Distributions Produced by Stellar Evolution

The earlier stages of stellar evolution are determined by the burning of hydrogen and helium. Hydrogen burning begins with the pp- or fusion process which *only* transforms hydrogen into helium. At higher temperatures the CNO-cycle transforms moreover most of the mass originally present as CNO into N^{14}. The remaining carbon has – under stationary conditions almost independent of temperature – an isotope ratio $C^{12}:C^{13} \approx 4$ while in cosmic matter there is $C^{12}:C^{13} \approx 90$. Helium burning produces essentially C^{12}. Later stages of evolution (white dwarfs, neutron stars, black holes ...), where contraction (gravitational energy) temporarily produces still higher temperatures need not be considered at the moment. However, in order to explain even qualitatively anomalous abundances of heavier elements, neutron processes must be taken into account. For processes involving charged particles their Gamow barrier is obviously much too high.

Turning now to the empirical aspects of our problem we first consider the *helium stars* or—as they are termed sometimes—the hydrogen-poor stars. All H/He ratios between almost normal and practically zero are observed. The abundance ratios in the CNO-group seem to be sometimes normal-pointing to the pp-process only—or changed roughly in the direction of the CNO-cycle. Helium stars are observed at the temperatures, for which hydrogen *and* helium lines can be expected among all stellar populations: in the young spiral arm population I (σ Ori *E*) as well as in the disk and halo populations (high velocity stars). For the same reason no lines of the heavier elements can be observed. Therefore we can know practically nothing about neutron processes in helium stars.

Carbon Stars are late type giants. Their low isotope ratio $C^{12}:C^{13}$ points toward activity of the CNO-cycle. Hydrogen has been largely transformed into helium and partly probably into carbon. In some carbon stars and certainly in all S stars, Ba II stars etc. neutron processes have heavily affected the abundance ratios of the heavy elements. The occurrence of technetium is the most obvious indication of neutron processes.

For some other types, especially the Ap and Am stars the interpretation of their spectra is not so obvious and the dispute between adherents of nuclear explanation or interpretations in terms of diffusion processes is still going on.

Considering all the different types of stars whose spectra show signs of evolution, we should remember what we said quite generally about stellar hydrodynamics. There exists no case in which it has been possible to follow up definitely a series of nuclear *and* hydrodynamical events leading quantitatively to the observed composition of the stellar atmosphere. Strong nuclear events must have taken place in some cases in early states of evolution near the main sequence: as in a helium star in the Orion association, or perhaps the Ap and Am stars and the recently discovered Pu^{244}-production during the formation of our solar system. Not all of these events can be ascribed to previous close binary systems nor can they have any place in the conventional theory of stellar evolution. Also the "partly burnt" matter in the atmospheres of some, though not all red giants of all populations (α Boo e.g. being a member of the halo population II) is not explained by the current theories of stellar evolution. In this connection we should remember that the present theory of the internal constitution of the Sun does not explain its small observed neutrino production. On the theoretical side it is often forgotten that the theory of the "mixing length" as applied to convection zones represents an extremely poor approximation and has no proper hydrodynamical justification. The contribution which non-stationary "flash"—phenomena may make towards the mixing of burnt matter into the atmosphere is also unknown.

V. Abundance Distributions of the Elements in Our Galaxy. Stellar Populations

In discussing the relations between abundance distributions and galactic structure we must first form a clear cut frame of concepts.

a) We use the term "*stellar population*" primarily in connection with the *distribution* of the stars: halo, disc, spiral arms or—emphasizing more *dynamical aspects*—with the velocity distributions and galactic orbits;

b) Stellar groups of different *age* can be distinguished according to their colour-magnitude diagrams;

c) As to *abundance* we can clearly separate out those stars, whose atmospheres show features of individual evolution connected with known nuclear processes: helium-, carbon-, Ba II-, S-Stars and so on. The hydrodynamical problems concerning the mixing of burnt matter into the atmospheres must be mostly left to future research. Apart from nuclear processes diffusion seems to affect some types of stellar atmospheres. The physics of Ap- and Am-stars have been discussed extensively but may still need further clarification. In the white dwarfs the presence of hydrogen in the atmospheres of the DA's and the very small abundances of the few accessible heavy elements point strongly toward diffusive separation, the detailed theory of which however is beset with difficulties. Also the helium-poor early type stars are now mostly agreed not to be connected with differences in composition.

After these somewhat lengthy explanations and limitations we attempt to summarize our present knowledge. Of course, it will not be possible to give complete references or to discuss controversial observations in detail.

(i) The abundance ratio of hydrogen to helium is the same everywhere; about 10:1 by number. For the justification of this statement it may be sufficient to refer to several extensive summaries (TAYLER, 1967; UNSÖLD, 1969);

(ii) the relative mixture of the heavier elements[2] $Z \geqq 6$ is the same everywhere.

(iii) The ratio of the heavier elements ($Z \geqq 6$) to hydrogen (and He), which in the literature has received the somewhat misleading name "*metal-abundance*", varies in the following way:

α) In the *halo stars* the metal abundance, relative to the Sun, varies between about 1 and $\approx 1/500$. Earlier work, based mainly on proper motion surveys, had indicated a strong correlation between metal abundance and galactic orbits (eccentricity e, inclination i, space velocity component W etc.). More recent work e.g. by BOND (1971), based on spectral classification using Schmidt telescopes, confirmed such a strong correlation only for the "pronounced halo stars" having large eccentricities, inclinations ..., but indicated quite *weak* correlations for intermediate stars having $e \leqq 0.5$ etc.;

β) the *disc and spiral arm stars* have metal abundances roughly equal to that of the Sun. More detailed work, especially by GUSTAFSSON and NISSEN (1972), using the STRÖMGREN narrow band photometry indicate that different

[2] The elements Li, Be and B require special consideration. In the cooler stars (only there can their lines be expected) they are completely destroyed by proton bombardment in the interior and partly already in the deeper parts of the convection zone from where matter is rapidly mixed into the photosphere. The Li abundance of old stars like the Sun is therefore low, while many meteorites have preserved the Li abundance of the proto-Sun.

galactic clusters have metal abundances differing by factors of ≈ 3. The Hyades e.g. have a metal abundance about 3 times larger than the Sun. There are also similar "(super-) metal-rich" field stars, such as β Vir. We should however emphasize that (relative to the Sun) overabundances reach at most a factor of 3 to 4 while underabundances go down to $\approx 1/500$.

BASCHEK *et al.* (1967) have analyzed β Vir (Table 3) with great care (the determination of the turbulence is particularly critical) and obtained an overabundance of *all* the heavier elements by a factor of ≈ 2 relative to hydrogen and compared with the Sun. So far we do not know whether there exist early type (super-) metal-rich stars nor has it been possible to determine the helium abundance spectroscopically in any case. Since the overabundance of the metals e.g. in the Hyades cannot have resulted from the evolution of the individual stars and since similar abundances occur in stars of various ages, we must conclude that such stars were made of matter having their present composition. Variations in the ratio of hydrogen to metals might possibly be a relict from the earliest times of the Galaxy. However, we might also adopt a kind of "restricted B^2FH-hypothesis" and assume that in the early stages of the Galaxy hydrogen was transformed into helium somewhat faster in some places than in others so that e.g. the Hyades were made of a cloud of hydrogen-poor (and helium-rich) matter. Such an interpretation of the spectra would—compared with the standard analysis—involve some change in g which however would be of the same order of magnitude as the present errors of spectroscopical determinations of g. The mass luminosity relation of the Hyades may indeed indicate some helium-enrichment but quantitatively the fraction of hydrogen transformed into helium as obtained from the spectra would still be higher than that obtained from considerations of the integral parameters of the stars (KOESTER and WEIDEMANN, 1973).

VI. Some Considerations Concerning the Age of the Universe, of Our Galaxy and of the Heavier Elements

Without going into details we summarize briefly a few "historical data" which might be relevant concerning the origin of the elements (Also: SANDAGE, 1968).

The "age of the universe" as obtained from relativistic models based on our present knowledge of the Hubble constant, an estimate of the deceleration parameter and complete ignorance of the cosmological constant is probably between 7.5 and 20×10^9 years.

Concerning the ages of clusters we should first remark that their determination depends on model calculations for the internal constitution of their stars. Here however the basic theory of convection zones is beset with unsolved hydrodynamical problems. Also the puzzle of the solar neutrino flux still seems to be unsolved. However the very accurate colour magnitude diagrams obtained by SANDAGE (1970) prove in any case that all the *globular clusters* and the *halo stars* must have originated within a very short time near the beginnings of our Galaxy. On an absolute scale SANDAGE obtains—within considerably larger limits of error—9 to 12×10^9 years. For the *oldest galactic cluster* NGC 188 (EGGEN and SANDAGE, 1969; SANDAGE and EGGEN, 1969) an age of 8 to 10×10^9 years is obtained. Since NGC 188 has more or less normal metal abundance we must conclude that metal poor stars were produced only during a brief time near the very beginnings of the galaxy and that part of the disk stars having normal or nearly normal metal abundance were also produced so early that their age cannot be distinguished at present with certainty from that of the halo.

The age of the heavier elements therefore establishes – within fairly narrow limits – an upper limit to the age of our Galaxy. Since we have found that the abundance distribution of the heavier elements is quite universal, such elements found on earth or in meteorites should have the same age. In this context the well known observed ratio of the uranium isotopes $U^{235}:U^{238}$ is of great interest; similar calculations can be made with Th^{232}. Assuming sudden origin (as we must do, if we want to avoid quite phantastic hypotheses) and beginning with a reasonable, though somewhat uncertain estimate of the original isotope abundances, we obtain (e.g. SANDAGE, 1968) for these elements an age of 7×10^9 years (with an estimated limit of error of $\pm 1 \times 10^9$ years). Taking into account the systematic errors of the absolute ages obtained for globular and galactic clusters – which however can only be estimated roughly – we think that they are not incompatible with the above value. Considering further the rather narrower limits for the radioactive age determinations we are inclined to consider 7 to 8×10^9 years as the most probable age of our Galaxy.

For metal-overabundant objects (galactic clusters and field stars) a clearcut relation between metal abundance and age cannot be established. It seems that such stars have been made at rather different times.

If we believe that the disk originated by collapse of a proto-halo, the duration of that event can have been only of the order of galactic rotation times, $\sim 10^8$ years (EGGEN *et al.*, 1962).

In discussing the genetic relation between the halo and the disc of our Galaxy it will be important to know the axial component of their angular momentum per mass unit P_z, since this quantity must have remained constant in time without regard to a (possible) collapse of the proto-halo.

For nearby disc stars with almost circular orbits there is $P_z \sim 250 \times 10$ km s^{-1} kpc. For the extremely metal-poor halo population II there is certainly $P_z \lesssim 50 \times 10$ km s^{-1} kpc. This is only an upper limit due to selection effect working against the discovery of retrograde stars; the actual value of P_z might be just as well almost zero (UNSÖLD, 1967; SANDAGE, 1969). Our upper limit is near the (rather uncertain) average value for our whole Galaxy.

VII. Abundance of the Elements in Other Galaxies

Before dealing with problems of theoretical interpretation we should briefly summarize also what we know about abundances in other galaxies.

In the Magellanic clouds comparatively detailed information is available from observations of gaseous nebulae as well as of individual spectra of the brightest early type supergiants. Quite generally we can summarize these very diverse observations by saying that relative to the Sun there are no abundance differences larger than the usual errors of measurement and evaluation amounting to a factor of ≈ 2. Even in the S Doradus complex which probably originated in an explosion similar to those in galactic nuclei nothing peculiar can be found. Particular emphasis should be laid on the recent detailed analysis of the A 3 IaO LMC-supergiant HD 33 579 by WOLF (1972) which for twelve well determined elements shows no differences larger than $\Delta \log \varepsilon = 0.23$ relative to the Sun.

Observations of gaseous nebulae are available for other bright and mostly late type galaxies, e.g. M 33 (Sc); the general agreement with solar abundances is also within the errors of observation + theory with one possible exception: PEIMBERT (1968), SEARLE (1971) and others obtained for some galaxies in the inner parts of the disks an overabundance of nitrogen. The possibility that this might be an excitation effect can perhaps not be completely excluded. On the other hand N^{14} might be a product of the CNO-cycle. Compare also a paper by PEIMBERT and TORRES-PEIMBERT (1971) on similar abundance anomalies in planetary nebulae.

About stars in other galaxies we are informed only from composite spectra. The general similarity with what one might expect from the superposition of normal spectra is certainly most striking. Going into a very detailed numerical "decomposition" SPINRAD (1970) and others have laid emphasis on the finding of super-metal-rich stars. Comparing the information available for one galaxy with the number of data to be derived and the amount of rather problematic theory of stellar atmospheres in such attempts we do not feel too confident about drawing further conclusions from them.

As to metal-deficient stars in other galaxies we have very little information. VAN DEN BERGH (1968) showed that the globular clusters near M 31 are comparable to those in our Galaxy, probably somewhat metal-richer. But—as already BAADE (1963) emphasized—there is one clear cut proof that all galaxies (which could be investigated in sufficient detail) contain metal-poor stars, namely the brightest red giants which in a metal-poor star population are $\approx 2^m$ brighter than in metal-rich population. Metal-poor stars—and that appears to be very important—are not restricted to galaxies built according to the halo-disc-scheme, but occur just as well in elliptical and in irregular galaxies.

Only quite recently SEARLE and SARGENT (1972) have described two "dwarf blue galaxies" in which oxygen and neon show significant underabundance. These may be the first metal-poor galaxies having a numerous population I.

Finally we ought to say a few words about the nuclei of galaxies and especially their "excited" states. However little can be added to the reports given by SARGENT, BURBIDGE and OSTERBROCK at the Vatican Study Week in 1970 [O'CONNELL (ed.), 1971]. The accuracy of element abundances obtained for such sources is still low, of course, and no one knows what strange objects may yet be detected.

Concerning the *Seyfert* galaxies OSTERBROCK finds after an extended discussion especially of NGC 4151 that "there is no evidence for large abundance anomalies".

In the case of the much more violent *quasars* we have little detailed information about the mechanisms producing their emission spectra. But just recording only which elements are represented by any lines at all, it is surprising to meet all the elements and only those whose abundance in the Sun is higher or equal to that of iron. So the composition of the matter which we "see" is probably not far from normal.

Cosmic rays are most probably produced[3] in the galactic nucleus and partly perhaps in supernovae and their remnants. It is most remarkable that the abundance destribution (Table 1) at least of the heavier elements in cosmic rays at their origin is the *same* as in the ubiquitous "cosmic matter". We cannot escape the conclusion that cosmic rays are just "ordinary" matter, which has been accelerated without further nucleosynthesis.

VIII. Origin of the Elements

In discussing the origin of the elements and their abundance distributions we must put aside all those stars whose atmospheres have somehow been contaminated by matter from their own nuclear reactor. Then we have to deal with two basic types of matter:

a) Hydrogen and helium in a ratio $\approx 10:1$ and (omitting the special problems of Li, Be, B),

b) the cosmic abundance distribution of the elements $Z \geqq 6$ or "metals" as presented in Tables 1 to 3.

Refering to the solar "metal abundance" as unity, we have in the halo of our Galaxy ratios from $\approx 1/500$ to ≈ 1; in the disc and the spiral arms a ratio ≈ 1 prevails, but mixtures up to about 3 or 4 seem to be present.

Before going into details we want to emphasize that our problem is not only one of *nuclear physics* but still more so one of *astronomy*. The problems which do not belong to nuclear physics are in principle mostly hydrodynamical. But we simply must realize that our knowledge of astronomical hydrodynamics is so poor that the only hopeful way of achieving at least modest progress is through "intelligent observations"!

The two main hypotheses which have been proposed so far concerning the abundance distributions of the elements are well known:

(i) The *Lemaître-Gamow* hypothesis of the "big bang" producing roughly solar-type matter. It was largely abandoned owing to the difficulty of jumping the gap near $A=5$ and to the discovery of metal-poor subdwarfs. Later on the Lemaître-Gamow hypothesis was revived in the modified form of the "big fireball" hypothesis explaining only the production of the universal hydrogen-helium mixture 10:1 (with an insignificant admixture of very light elements) and of the 2.7 K radiation.

(ii) In 1957 E. M. and G. R. BURBIDGE, W. A. FOWLER and F. HOYLE proposed the "synthesis of the elements in stars", their B^2FH-hypothesis. According to them, the galaxy began as a big proto-halo of hydrogen (plus helium in more recent versions) and the heavier elements were produced in several generations of stars forming, evolving and spreading part of the evolved matter into the interstellar medium. As to the nuclear processes involved there can be no doubt

[3] The most energetic component whose composition is unknown and which may be of extragalactic origin, must here be left out of consideration.

about reactions between the lighter nuclei and charged particles. Already GAMOW had provided evidence for reactions between heavier nuclei and neutrons of ~25 keV corresponding to $\approx 2.5 \times 10^8$ K. The reactions were supposed to proceed so slowly (*s*-process) that all intervening β-decays could take place. Fast (explosive) processes – perhaps connected with supernova explosions – were proposed by B^2FH to explain the neutron rich nuclei (*r*-process) and by TRURAN (1971), ARNETT (1971) and others to explain the nuclei up to the iron peak of the abundance curve. Such processes require a considerable number of different sources if they are to explain *all* the elements. As a result in such theories the number of available constants is quite considerable. In the "explosive nucleosynthesis" the number of abundances to be explained is only about twice as large as the number of constants which (so far) must be chosen ad hoc.

Concerning the production of the neutron-rich nuclei we should mention the important papers by AMIET and ZEH (1968). They suggest that the *neutron-rich nuclei* were produced *first* under extremely high pressures (densities) and temperatures $\rho \approx 2 \times 10^{10}$ g cm^{-3} and $T \approx 5 \times 10^9$ K. Under such conditions electrons are pressed into the nuclei and when the pressure is released one gets a quite reasonable abundance curve. The *slow* neutron process (which empirically is the best established one) is thought to have only just "retouched" the abundance distribution produced by the primary pycnonuclear process. One of the main arguments for these ideas is, that the abundance maxima for the heavier nuclei coincide with the magic neutron numbers (closed shells) not for the present matter but for the neutron-richer protomatter under the already stated extreme temperatures and densities (We can, however, not even guess how such conditions might have been realized).

This proposition originated partly in connection with the purely empirical abundance rules detected long before by SUESS (1964, 1966 and with ZEH, 1968). We shall not present these in detail but emphasize SUESS' finding that *s* and *r* nuclei follow the *same* rules without any differences. That shows that the production of neutron-rich and neutron-poor nuclei must have been more closely interwoven than is assumed usually. (The attitude which some astronomers take concerning these empirical relations and their proposed interpretation, reminds me somehow of GALILEI's colleagues who refused to look through the telescope in order not to see the "impossible" satellites of Jupiter!)

After these remarks we return to our astronomical issues and discuss first:

α) The hypothesis of a *galactic origin of the heavier elements* ($Z \geqq 6$). If this hypothesis is true, we must assume that practically all the heavier elements have been produced between the beginnings of the halo and the formation of the galactic disc, that is within one or at most a few hundred million years. Within such a period however only stars with more than about 5 $M_\odot$ can undergo complete evolution. The assumption that star formation might have been much more violent in the earlier halo than it is now in our Galaxy would not be sufficient because we do not observe the required number of metal-poor dwarfs having $\lesssim 1$ $M_\odot$. TRURAN and CAMERON (1971) have therefore suggested the additional hypothesis that in the early halo almost *only* stars $\gtrsim 5$ $M_\odot$ were produced which could evolve fast enough. However, more recently it has become possible to determine from observations the *luminosity function of the unevolved halo sub-*

dwarfs. The main difficulty in such an attempt was, of course, the danger of observational selection. However Sir R. WOOLLEY *et al.* (1971) using stars within 25 pc (thus giving equal chances to all stars of the same magnitude) have published colour diagrams for stars within different ranges of galactic orbits, described by their (generalized) eccentricity e and inclination i. In Table 4 we compare for the most extreme disc- and halo-group the percentage of stars between absolute magnitude 4 (just below the "knee") and 8 and between 8 and 12. These fairly large intervals have been chosen because the halo stars would otherwise show too large statistical fluctuations. It turns out (and the intermediate groups of WOOLLEY *et al.* cornfirm that conclusion) that the luminosity functions in the range of unevolved stars are the *same* for disc and halo stars. It appears therefore highly improbable that the halo-function had originally a very large peak for stars having 5 to 10 times larger masses.

The hypothesis mentioned above makes it equally difficult to explain why the mixture of the heavier elements with $Z \geqq 6$ is the same in normal disc stars as in extremely metal-poor ($\sim 1/500$) halo stars. It appears entirely incredible that *all* the different nuclear processes producing the elements from carbon to barium should have worked in the same way not only under astronomically so different conditions as in the early halo and the almost complete disc, but even in such quite different galaxies as our own system and the Magellanic clouds! The only way out of these difficulties seems to be the assumption of some sort of "mass production" of the heavier elements. Let us examine different possibilities! We remark in passing that the origin of the heavier elements has obviously nothing to do with the disc-halo-structure of our Galaxy because quite different galaxies also have metal-poor and metall-rich stars. If we tentatively retain the assumption that the galaxies originally consisted of hydrogen (plus 10% helium) we might assume that the heavier elements resulted from an explosion in the *galactic nucleus*. Whether, following AMBARTSUMIAN (1971), we consider that nucleus as just something "originally given" or produced by collapse of a diffuse cloud may remain open. The explosion would produce a huge shock front, at the outside of which halo stars with different metal abundance could be produced. The rest would collapse and form the disc stars with fairly uniform metal abundances but some fluctuations explaining the intermediate halo stars with their characteristically weak correlation between metal abundances and galactic orbits and perhaps some super-metal-rich stars. It is in line with this idea that various kinds of "active" galactic nuclei and even cosmic rays seem to show more or

Table 4. Comparison of the luminosity functions for disc- and halo-stars within 25 pc

Absolute magnitude		Number of stars (%)	
		$4 \leqq M_v \leqq 8$	$8 \leqq M_v \leqq 12$
Disk stars	$\begin{cases} e < 0.050 \\ i < 0.050 \end{cases}$	20 (62%)	12 (38%)
Halo stars	$\begin{cases} e \geqq 0.300 \\ i \geqq 0.050 \end{cases}$	10 (67%)	5 (33%)

less "cosmical" abundances of the heavier elements. There remain, however, two very serious unsolved questions, namely: Why have galaxies differing widely according to Hubble type and mass produced so very nearly the same mixture of the heavier elements? And why is the percentage of mass transformed into these heavier elements the same for the Magellanic Clouds or very nearly the same in the case of many other galaxies even in remote clusters of galaxies? The situation is not quite as difficult as it was with the B^2FH-hypothesis, but it is still bad enough! This problem seems to me sufficiently important that we should discuss at least also other possibilities.

β) Realizing that all the world models which have been worked out in connection with the GAMOW "big bang" idea are highly schematical and have been preferred to more general ones only for mathematical reasons, we might consider a woeful return to the idea of the primeval origin of an abundance distribution similar to that of the majority of the stars or – practically speaking – of the Sun. In this case all the galaxies might have been formed and eventually got their discs with about the same "normal" composition. What then? The metal-deficient stars might now be due to a kind of explosion in the galactic nucleus as has been proposed in many recent *theoretical* papers: A large mass exploding and producing almost only hydrogen and helium. Since this mass would come out of the galactic center where the angular momentum per mass unit is small, the same property of the halo stars would be easily explained, while for hypothesis (*a*) its explanation is not obvious. The main difficulty concerning the present hypothesis is that in the actual nuclei of active galaxies we never observe hydrogen plus helium only but always more or less solar type matter.

The super-metal-rich stars would have to be explained – following what we called the restricted B^2FH-hypothesis – as having originated out of clouds of matter in which a moderate fraction of the hydrogen had already been transformed into helium in the course of stellar evolution processes. While under theory (*a*) this explanation is also probable we have there the additional possibility that the mixing of the original H + He with the matter bringing heavier elements from the galactic nucleus might have been not quite uniform.

Coming to an end let us summarize briefly:

We have realized that the origin of the elements is not only a nuclear problem but that in recent years a lot of astronomical observations have been made which may help us to fill the many gaps due to our hydrodynamical ignorance. We have emphasized the two basic element mixtures: hydrogen plus helium on the one hand and the "cosmical" mixture of the heavier elements $Z \geqq 6$ on the other hand. As to the galactic halo we emphasized its low angular momentum per mass unit and the luminosity function of its old stars. The comparison between our Galaxy, the Magellanic clouds and – in a necessarily more restricted way – other galaxies showed that there must exist some mechanism which stops production of heavier elements when a certain fraction of the original matter has been transformed into heavier nuclei. Many astrophysicists will not feel inclined to return to a kind of modified GAMOW theory. However, considering the unsolved problems in any theory of a galactic origin of the heavier elements we should probably at least keep other hypotheses in our mind. In any case it appears almost impossible to maintain the idea that the heavier elements were produced in individual stars. Assuming a production of the heavier elements in explosions of galactic nuclei many objections disappear though there remain others which should not be taken lightly. We should think again therefore also about the possibility of a primeval origin of the heavier elements and (correspondingly) of a possible origin of metal-poor stars through big hydrogen-helium-explosions early in the evolution of galactic nuclei. Certainly the best and probably the only road leading us toward a solution of these problems of chemical cosmology is: "More and better observations".

References

For most items we quote only one or two important (preferably review-) papers containing extensive further references.

Aller, L. H., Czyzak, S. J.: In D. E. Osterbrock and C. R. O'Dell (eds.), Planetary Nebulae, IAU Symp. **34**, 209 (1968).
Aller, L. H., Greenstein, J. L.: Astrophys. J. Suppl. **5**, 139 (1960).
Ambartsumian, V. A.: In D. J. K. O'Connell (ed.), Nuclei of Galaxies, Pont. Acad. Sci. Scripta Varia 35, North Holland Publ. Co., Amsterdam (1971).
Amiet, J. H., Zeh, H. D.: Z. Phys. **217**, 485 (1968).
Anders, E.: Geochim. Cosmochim. Acta **35**, 516 (1971).
Arnett, W. D.: Astrophys. J. **166**, 153 (1971).
Baade, W.: Evolution of stars and galaxies (ed. C. Payne-Gaposchkin), Harvard Univ. Press (1963).
Baschek, B.: Z. Astrophys. **48**, 95 (1959).
Baschek, B.: Z. Astrophys. **56**, 207 (1962).
Baschek, B., Holweger, H., Namba, O., Traving, G.: Z. Astrophys. **65**, 418 (1967).
Bergh, S. van den: Commun. David Dunlop Obs. No. 195 (1968). (With extensive further references.)
Bond, H. E.: Astrophys. J. Suppl. **22**, 117 (1971).
Burbidge, E. M., Burbidge, G. R., Fowler, W. A., Hoyle, F.: Rev. Mod. Phys. **29**, 547 (1957).
Cayrel, G., Cayrel, R.: Astrophys. J. **137**, 431 (1963).
Cayrel, R.: Suppl. Ann. Astrophys. No. 6 (1958).
Cayrel de Strobel, G.: Ann. Astrophys. **29**, 413 (1966).
Eggen, O. J., Lynden-Bell, D., Sandage, A. R.: Astrophys. J. **136**, 748 (1962).
Eggen, O. J., Sandage, A.: Astrophys. J. **158**, 669 (1969).
Fawell, D. R.: Observatory **91**, 182 (1971).
Flower, D. R.: Monthly Notices Roy. Astron. Soc. **146**, 243 (1969).
Garz, T., Holweger, H., Kock, M., Richter, J.: Astron. Astrophys. **2**, 446 (1969).
Greene, T. F.: Astrophys. J. **161**, 365 (1970).
Groth, H.-G.: Z. Astrophys. **51**, 206 (1961).
Gustafsson, B., Nissen, P. E.: Astron. Astrophys. **19**, 261 (1972). (Containing references to earlier papers.)
Hardorp, J., Scholz, M.: Astrophys. J. Suppl. **19**, 193 (1970).
Holweger, H.: Astron. Astrophys. **10**, 128 (1971).
Holweger, H.: Solar Phys. **25**, 14 (1972).
Hunger, K.: Z. Astrophys. **49**, 129 (1960) (and previous papers).
Kodaira, K.: Z. Astrophys. **59**, 139 (1964).
Kodaira, K., Scholz, M.: Astron. Astrophys. **6**, 93 (1970).
Koester, D., Weidemann, V.: Astron. Astrophys. **25**, 437 (1973).
Lambert, D. L., Mallia, E. A., *et al.*: Monthly Notices Roy. Astron. Soc. **138–142** (1968). (Many papers.)
Morgan, L. A.: Monthly Notices Roy. Astron. Soc. **153**, 393 (1971).
Müller, E. A.: Ass. Int. de Geochimie et Cosmochimie, Paris (1967).
O'Connell, D. J. K. (ed.): Nuclei of galaxies, Pont. Acad. Sci. Scipta Varia 35, North Holland Publ. Co., Amsterdam (1971).

OSTERBROCK, D. E.: Quart. J. Roy. Astron. Soc. **11**, 199 (1970).
PEIMBERT, M.: Astrophys. J. **154**, 33 (1968).
PEIMBERT, M., COSTERO, R.: Bol. Obs. Tonantzintla y Tacubaya **5**, 3 (1969).
PEIMBERT, M., TORRES-PEIMBERT, S.: Astrophys. J. **168**, 413 (1971).
REIMERS, D.: Astron. Astrophys. (1973).
SANDAGE, A.: In L. WOLTJER (ed.), Galaxies and the Universe, Columbia Univ. Press, New York, London (1968).
SANDAGE, A.: Astrophys. J. **158**, 1115 (1969).
SANDAGE, A.: Astrophys. J. **162**, 841 (1970).
SANDAGE, A., EGGEN, O. J.: Astrophys. J. **158**, 685 (1969).
SCHOLZ, M.: Vistas Astron. **14**, 53 (1972).
SEARLE, L.: Astrophys. J. **168**, 327 (1971).
SEARLE, L., SARGENT, W. L. W.: Astrophys. J. **173**, 25 (1972).
SHAPIRO, M. M., SILBERBERG, R., TSAO, C. H.: Twelfth Int. Conf. Cosmic Rays, Hobart, Tasmania (1971).
SPINRAD, H.: Quart. J. Roy. Astron. Soc. **11**, 188 (1970).
SUESS, H. E.: Proc. Nat. Acad. Sci. **52**, 387 (1964).
SUESS, H. E.: Z. Naturforsch. **21a**, 90 (1966).
SUESS, H. E., ZEH, H. D.: Naturwissenschaften **55**, 477 (1968).
TAYLER, R. J.: Quart. J. Roy. Astron. Soc. **8**, 313 (1967).
TRAVING, G.: Z. Astrophys. **41**, 215 (1957).
TRURAN, J. W., CAMERON, A. G. W.: Astrophys. Space Sci. **14**, 179 (1971).
UNSÖLD, A.: Z. Phys. **202**, 13 (1967).
UNSÖLD, A.: Verh. Schweiz. Naturf. Ges. – Vers. in Einsiedeln, p. 35 (1968).
UNSÖLD, A.: Science **163**, 1015 (1969).
UNSÖLD, A.: Proc. Roy. Soc. London (A) **270**, 23 (1971).
UREY, H. C.: Quart. J. Roy. Astron. Soc. **8**, 23 (1967).
WOLF, B.: Astron. Astrophys. **10**, 383 (1971).
WOLF, B.: Astron. Astrophys. **20**, 275 (1972).
WOLFFRAM, W.: Astron. Astrophys. **17**, 17 (1972).
WOOLEY, Sir R., POCOCK, S. B., EPPS, E., FLINN, R.: Roy. Observ. Bull. No. 166 (1971).
ZIELKE, G.: Astron. Astrophys. **6**, 206 (1970).

Note added in proof (4. VIII. 1973):

[1] R. D. MCCLURE and S. VAN DEN BERGH (Astron. J. **73**, 313, 1968) and others have shown that dwarf galaxies are metal poor while giant galaxies have normal or even slightly higher metal abundances.

[2] According to M. GRENON (IAU-Symposium Paris 1972: Ages of the stars) the super-metal-rich stars belong to the inner part of the galactic disc, the metallicity increasing generally toward the center. Other galaxies seem to show a similar metallicity gradient which may again point towards an origin of the heavier elements in the central parts of the galaxies.

The Theory of Spiral Structure. Resonances

(Invited Lecture)

By G. CONTOPOULOS
University of Thessaloniki, Greece

With 8 Figures

I. Introduction

The main ideas of the density wave theory of spiral structure have been developed by B. LINDBLAD since 1940 and in recent years by LIN and his associates, and independently by KALNAJS.

Many expositions of this theory have been made already (see, e.g., LIN, 1966, 1970, 1971; CONTOPOULOS, 1970a). A recent review of the present state of the theory can be found in my Lecture Notes on "The Dynamics of Spiral Structure" issued by the University of Maryland (CONTOPOULOS, 1972).

A special subject, which is of particular importance in the theory of spiral structure is the subject of resonances. I will report mainly about work done recently on this subject by our group in Greece.

A resonance occurs whenever the ratio of the mean angular velocity of stars, $\Omega-\Omega_s$, in a frame of reference rotating with the angular velocity of the spiral pattern, to the "epicyclic frequency", κ, is a rational number. Thus there is an infinity of resonances in the Galaxy. However, the only important resonances appear if the ratio $\nu=\frac{2(\Omega_s-\Omega)}{\kappa}$ is 0 (particle resonance, where we have corotation, i.e. $\Omega=\Omega_s$), or ± 1 (Lindblad resonances, in particular the inner Lindblad resonance, where $\nu=-1$).

We will study separately the effects at the particle resonance and at the inner Lindblad resonance.

II. Particle Resonance

According to the theory of LIN and his associates (LIN and SHU, 1966; LIN, 1970, 1971) there are two spiral waves going through the particle resonance. LIN's dispersion relation has two branches that cross each other at corotation (Fig. 1, solid lines).

However, this dispersion relation does not apply near the particle resonance. In fact, SHU (1970) found, in his second order theory, by using LIN's approximate dispersion relation as first order theory, that the amplitude of the wave becomes infinite at corotation. This is, of course, not realistic. Therefore, a more complete linear theory is needed near the particle resonance.

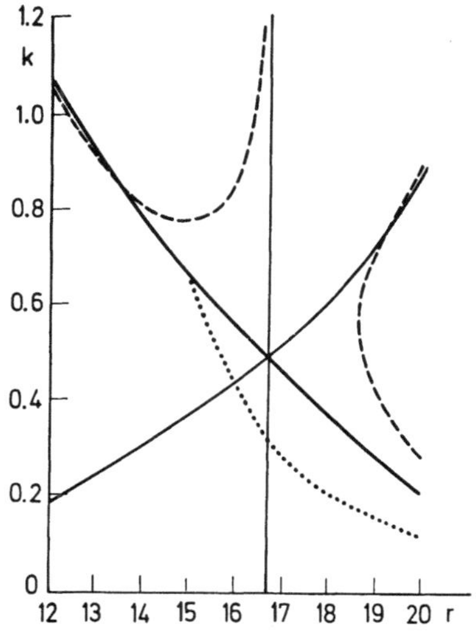

Fig. 1. Dispersion relations near the particle resonance. Solid lines: The two branches of the dispersion relation of LIN and his associates. Dashed line: The change due to the inclusion of the "resonant" term. Dotted line: An improved dispersion relation, assuming k to be small and complex

If we change slightly LIN'S procedure in deriving the dispersion relation, to take into account a small "resonant" term, which is usually omitted, but which is important near corotation, we find the dispersion relation shown as dashed line in Fig. 1 (CONTOPOULOS, 1972).

This dispersion relation is quite different from LIN'S near the particle resonance. The wavelength tends to zero and the corresponding spiral arms approach asymptotically the corotation circle.

However, the above treatment is not complete. In fact, following LIN, we have assumed, up to now, that the wavenumber k is absolutely large, and we have disregarded any variation of the amplitude of the wave.

An improved dispersion relation can be found by allowing k to be small. We have also taken into account the amplitude variations by using a method applied by MARK (1971) in the inner Lindblad resonance. Our preliminary results (CONTOPOULOS, 1973b) are shown in Fig. 1 (dotted line).

Our calculations indicate that there is only one spiral wave going through the particle resonance. Its amplitude is variable but the variations are small. The results are similar in trailing and leading spirals.

Finally, similar results are found if the dispersion of velocities is larger than assumed by LIN, although in such cases the waves derived by using LIN'S dispersion relation do not approach the particle resonance (TOOMRE, 1969).

Up to this point we have not considered non-linear effects. The first indication of the importance of non-linear effects came from BARBANIS' (1970) calculations, who found that stars near corotation are trapped near two points 90° away from the spiral arms.

We can consider here two kinds of problems:

a) The problem of orbits, and

b) The self-consistent problem.

a) We have recently developed a theory to explain the trapping of stars near the particle resonance. The situation is similar to that of the Lagrangian points in the restricted three-body problem. At corotation we have two unstable equilibrium points (L_1, L_2) at the minima of the potential and two stable points (L_4, L_5) at the maxima of the potential. Many particles librate around L_4 or L_5, instead of circulating around the origin 0. Using an integral of motion similar to the "third integral" (CONTOPOULOS, 1960; DEPRIT, 1966) near L_4 and L_5 we could find that the librating orbits near L_4 and L_5 are either rings (Fig. 2) or bananas (Fig. 3). A more complete description of these orbits is given in a forthcoming paper (CONTOPOULOS, 1973 a).

We have made a rough estimate of the total mass of trapped orbits near L_4, or L_5. If the spiral force is about 4% of the axisymmetric force, then the amount of trapped matter is about 30% of the total matter contained in a ring 2 kpc wide, which includes the resonances.

b) If we come now to the self-consistent problem we have to take into account the fact the trapped matter itself produces appreciable perturbations to the orbits.

As a first step in finding a self-consistent model we imposed two mass concentrations near L_4 and L_5, and we calculated the mean position ($\pm x_c$, $\pm y_c$) and amount (m_c) of the trapped mass, as functions of the mean position ($\pm x_i$, $\pm y_i$) and amount (m_i) of the imposed mass. Such mass concentrations are tidally stable. We are now studying the distribution of the trapped mass and try to make it identical to that of the imposed mass.

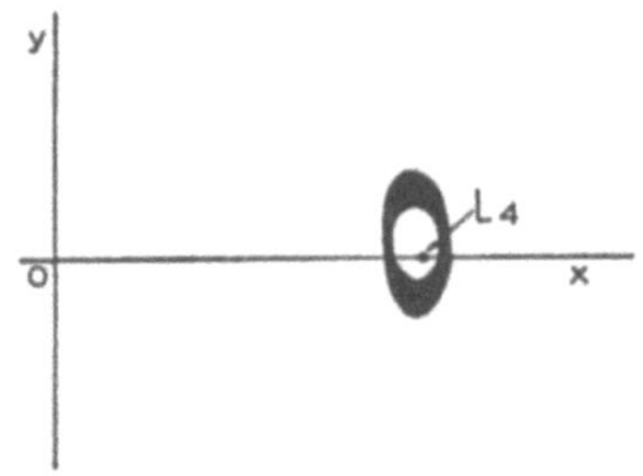

Fig. 2. A ring type orbit near the particle resonance

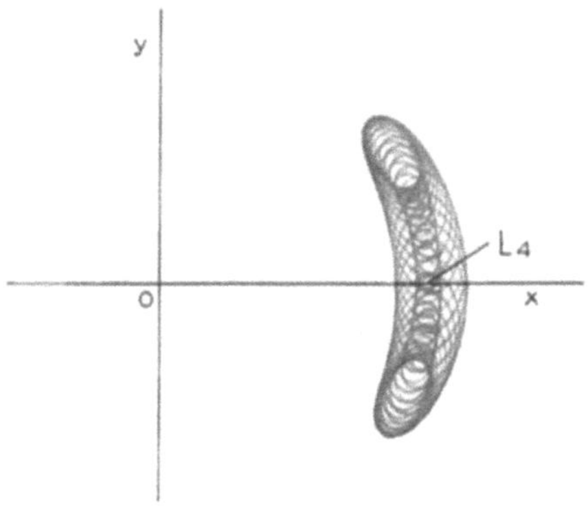

Fig. 3. A banana type orbit near the particle resonance

III. Inner Lindblad Resonance

We come now to the inner Lindblad resonance. The first important effect due to the inner Lindblad resonance is that it discriminates between trailing and leading waves.

It is well known that away from resonances the theory does not give any preference to the trailing waves. Both leading and trailing waves have their maxima of density near the minima of the spiral potential. The original hopes of LIN and SHU (1966; LIN, 1967) that a second order theory would provide a preference of trailing waves were not realized (TOOMRE, 1969; SHU, 1970).

However near the inner Lindblad resonance the situation is different, as it was first indicated by KALNAJS. We have calculated (CONTOPOULOS, 1971) the response of the stars to a slightly growing wave of constant amplitude and we found that near resonance the maxima of density form a trailing spiral both in the case of a leading and of a trailing potential (Fig. 4).

The imposed spiral potential of a two-armed galaxy can be written

$$V_1 = A\,e^{i(\Phi+\omega t-2\vartheta)}, \tag{1}$$

where $\omega=\omega_R+i\,\omega_I$. We have $\omega_R=2\Omega_s$, where Ω_s is the angular velocity of the spiral pattern, and ω_I is negative, zero, or positive, if the spiral field is growing, neutral, or damping, respectively. The spiral is trailing if the wave number $k=\dfrac{d\Phi}{dr}<0$ and leading if $k>0$.

The resulting density response is

$$\sigma_1 = -A_1\,e^{i(\Phi+2\phi_1+\omega t-2\vartheta)}. \tag{2}$$

The quantity ϕ_1 is called the azimuth advance and is always positive. Well outside the resonance $\phi_1 \to 0$, i.e. the minima of potential and maxima of density

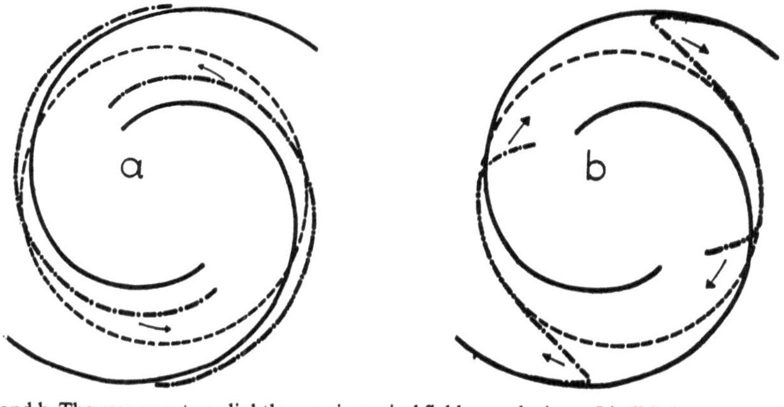

Fig. 4a and b. The response to a slightly growing spiral field near the inner Lindblad resonance (dashed circle). Solid lines represent the minima of potential which is trailing in (a) and leading in (b), and dash-dotted lines represent the maxima of the response density

are in phase. However, if ω_I is small negative (slightly growing field), as we move inwards ϕ_1 increases, tending to 90° in the trailing case, and to 270° in the leading case. In the leading case it overcomes the leading character of $\Phi\left(\frac{d\Phi}{dr}>0\right)$ and changes it into trailing, i.e. $\frac{d(\Phi+2\phi_1)}{dr}<0$.

Recently we have made similar calculations of the azimuth advance in the case of the particle resonance. There, however, the quantity ϕ_1 and its derivative are small and cannot change the character of the wave from leading to trailing. Thus the only place where there is a preference of trailing waves is the inner Lindblad resonance.

In our calculations above we used the usual assumption of LIN that the variation of A is small. But if we allow an arbitrary change of A near resonance, we can construct both trailing and leading self consistent models.

If we write the amplitude A in the form

$$A=e^{i(-i\ln A)}, \tag{3}$$

Eq. (1) is written

$$V_1=e^{i(\Phi+\omega t-2\vartheta)}, \tag{4}$$

where now Φ is complex and

$$k=\frac{d\Phi}{dr}=k_R+i\,k_I, \tag{5}$$

with

$$k_I=-\frac{d\ln A}{dr}. \tag{6}$$

MARK (1971) used a complex k and found a trailing self-consistent model near resonance, in the case that ω is real (neutral waves). His result is that the wave is evanescent inside the inner Lindblad resonance, i.e. $k_I<0$.

Recently GEORGALA and I made several calculations of self-consistent models, for various values of ω_I, both for trailing and leading waves.

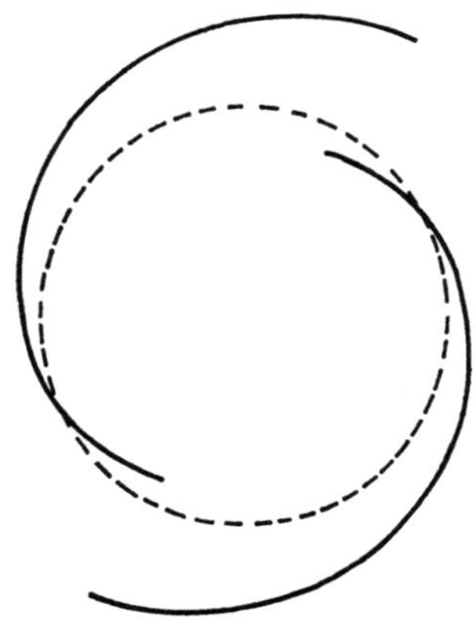

Fig. 5. A self-consistent trailing spiral galaxy

A self-constistent trailing spiral for $\omega_I=0$ is given in Fig. 5; this is not very different from the response spiral of Fig. 4a.

On the other hand, in the leading case the wave is amplified inwards. Furthermore, the greatest contribution to the wave at resonance is from stars coming from the central region of the Galaxy. Therefore, a study of spiral waves near the galactic center is important. This is the subject of the thesis of GEORGALA.

In the above discussion we have again considered only linear effects. Thus the theory is applicable only during the early phases of the formation of spiral waves, when their amplitude is very small (infinitesimal). However, as soon as the amplitude is appreciable non linear effects become important.

We have studied (CONTOPOULOS, 1970a, b) the non-linear response of stars near the inner Lindblad resonance, both analytically, using the theory of the "third integral" applied to resonance cases, and with the help of numerical calculations.

Again here we can separate the problem of the orbits from the self-consistent problem.

a) The orbits near the inner Lindblad resonance are different from the usual epicyclic orbits away from resonance. While far from resonance there is a stable periodic orbit which is a slightly perturbed circle, near resonance there are two stable periodic orbits which are like two ellipses perpendicular to each other. These orbits can be thought of as special cases of LINDBLAD's (1955) dispersion orbits, with orientations which are not arbitrary, but are fixed with respect to the spiral potential. Stars in the neighborhood of these periodic orbits oscillate around them and fill elliptic rings as shown in Fig. 6.

b) Now we come to the problem of the distribution of resonant orbits. If we make a reasonable assumption about the initial distribution of velocities

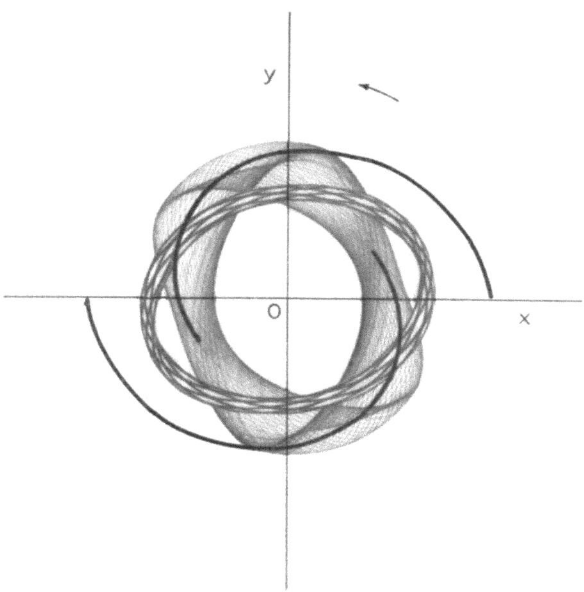

Fig. 6. Resonant orbits near the inner Lindblad resonance

(say, ellipsoidal) we form two bars by superposing several orbits of this type. One bar is continuation of the spiral arms and we call it the main bar (Fig. 6).

As we move outwards from the resonance there are fewer orbits corresponding to the secondary bar. The outer parts of the main bar become more and more circular, until we reach the regular epicyclic orbits away from resonances.

On the other hand as we move inwards from the inner Lindblad resonance the other set of orbits, perpendicular to the above, prevails. Thus we have a secondary bar perpendicular to the first.

Of course these bars rotate with the spiral arms, having a fixed azimuth in the rotating plane. Further, the relative importance of the two bars depends also on the initial distribution of matter. In particular the gas would move mainly along the dominant bar, as we cannot expect to have streams of gas to go through each other.

These resonant phenomena are very important. We have found that if the spiral force is about 5% of the axisymmetric force the resonant orbits cover an important part of the Galaxy, between 2 and 5.5 kpc.

All the models considered so far are not self consistent. We are at present considering the effects of these barred contributions on the orbits. Although we have not yet any final results, we expect that self-consistent bars should not be quite different from the response bars described above.

Lynden-Bell and Kalnajs (1972) have recently emphasized the importance of such resonant bars in barred spirals. It is our opinion that the bars are associated with the orbits trapped near the inner Lindblad resonances. In our picture the location of the inner Lindblad resonance is related to the size of the bar. If the inner Lindblad resonance is in the outer parts of a galaxy then the galaxy would look more barred-like.

A systematic study of orbits in barred spirals has started in Thessaloniki by Mihalodimitrakis, extending the work of De Vaucouleurs and Freeman (1970, 1972) to other models. He found that resonance phenomena are very important in barred spirals and cover wide regions on both sides of the bar (Fig. 7).

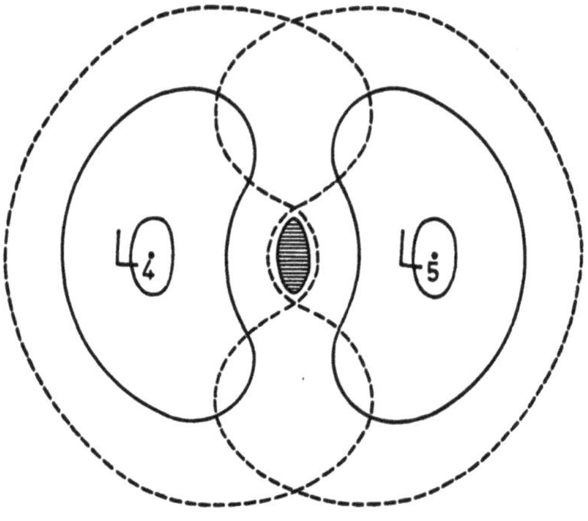

Fig. 7. Periodic orbits around the Lagrangian points L_4, L_5 of a barred galaxy

An interesting application of the non-linear theory near the inner Lindblad resonance concerns the observed motions of gas. If the gas moves along elliptical orbits then its velocity has an appreciable outward component along certain directions. Thus the observed radial motions in the central parts of our Galaxy may not mean an expansion as proposed by OORT and others (ROUGOOR and OORT, 1959; OORT, 1971) but just eccentric motions along resonant orbits. This explanation was proposed first by SHANE (1972). A more detailed study was made recently by SIMONSON and MADER (1972). It seems that by placing the resonant orbits at a convenient angle with respect to the direction of the Sun, most of the observed features in the 21-cm profiles in the central region of the plane of our Galaxy can be explained.

More recently TULLY (1972) has made an interesting study of the kinematics of M51, having at his disposal a large number of velocities at various points of this galaxy.

His conclusions are summarized in Fig. 8a and b. The outer parts of the galaxy are best explained by TOOMRE's theory (TOOMRE, 1970; TOOMRE, A. and J., 1972) of tidal action due to the companion galaxy. The intermediate parts are explained by the usual LIN theory. In the central parts it seems that the theory of the resonant dispersion orbits explains the observations better than an expansion model.

Thus TULLY's thesis makes happy many theoreticians at the same time, TOOMRE for the outer parts, LIN for the intermediate, and CONTOPOULOS for the inner parts. But this is not unnatural. After all galaxies are such rich objects that one cannot expect that one single theory will explain away everything we observe in them.

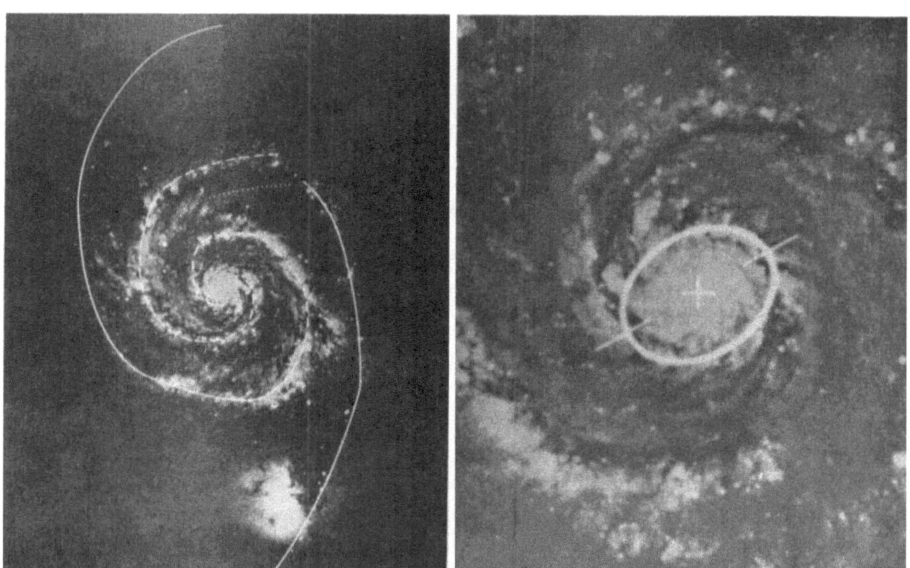

Fig. 8a and b. The spiral structure of the galaxy M51 (TULLY): (a) Solid lines represent TOOMRE's encounter theory and dashed lines LIN's theory. (b) The dispersion ring model near the inner Lindblad resonance

References

BARBANIS, B.: In W. BECKER and G. CONTOPOULOS (eds.), The Spiral Structure of our Galaxy, IAU Symp. **38**, 343 (1970).
CONTOPOULOS, G.: Z. Astrophys. **49**, 273 (1960).
CONTOPOULOS, G.: In W. BECKER and G. CONTOPOULOS (eds.), The Spiral Structure of our Galaxy, IAU Symp. **38**, 303 (1970a).
CONTOPOULOS, G.: Astrophys. J. **160**, 113 (1970b).
CONTOPOULOS, G.: Astrophys. J. **163**, 181 (1971).
CONTOPOULOS, G.: The Dynamics of Spiral Structure, Astronomy Program, University of Maryland (1972).
CONTOPOULOS, G.: Astrophys. J. **181**, 657 (1973a).
CONTOPOULOS, G.: (In preparation) (1973b).
DEPRIT, A.: In G. CONTOPOULOS (ed.), The Theory of Orbits in the Solar System and in Stellar Systems, IAU Symp. **25**, 170 (1966).
DE VAUCOULEURS, G., FREEMAN, K.C.: In W. BECKER and G. CONTOPOULOS (eds.), The Spiral Structure of our Galaxy, IAU Symp. **38**, 356 (1970a).
DE VAUCOULEURS, G., FREEMAN, K.C.: Vistas in Astronomy **14**, 163 (1972).
KALNAJS, A.: In W. BECKER and G. CONTOPOULOS (eds.), The Spiral Structure of our Galaxy, IAU Symp. **38**, 318 (1970).
LIN, C.C.: SIAM J. Appl. Math. **14**, 876 (1966).
LIN, C.C.: In J. EHLERS (ed.), Relativity Theory and Astrophysics **2**, American Math. Society, Providence (1967).
LIN, C.C.: In W. BECKER and G. CONTOPOULOS (eds.), The Spiral Structure of our Galaxy, IAU Symp. **38**, 377 (1970).
LIN, C.C.: In C. DE JAGER (ed.), Highlights of Astronomy **2**, D. Reidel Publ. Co., Dordrecht (1971).
LIN, C.C., SHU, F.H.: Proc. Nat. Acad. Sci. **55**, 229 (1966).
LIN, C.C., SHU, F.H.: In H. VAN WOERDEN (ed.), Radio Astronomy and the Galactic System, IAU Symp. **31**, 313 (1969).
LINDBLAD, B.: Stockholm Obs. Ann. **18**, No. 6 (1955).
LYNDEN-BELL, D., KALNAJS, A.: Monthly Notices Roy. Astron. Soc. **157**, 1 (1972).
MARK, J.W.K.: Proc. Nat. Acad. Sci. **68**, 2095 (1971).
OORT, J.H.: 135th Meeting of the Amer. Astron. Soc., Amherst (1971).
ROUGOOR, C.W., OORT, J.H.: In R.N. BRACEWELL (ed.), Paris Symposium on Radio Astronomy, IAU Symp. **9**, 416 (1959).
SHANE, W.W.: Astron. Astrophys. **16**, 118 (1972).
SIMONSON, S.C., MADER, G.L.: Preprint, University of Maryland (1972).
SHU, F.H.: Astrophys. J. **160**, 99 (1970).
TOOMRE, A.: Astrophys. J. **158**, 899 (1969).
TOOMRE, A.: In W. BECKER and G. CONTOPOULOS (eds.), The Spiral Structure of our Galaxy, IAU Symp. **38**, 334 (1970).
TOOMRE, A., TOOMRE, J.: Astrophys. J. **178**, 623 (1972).
TULLY, R.B.: University of Maryland (Ph. D. Thesis) (1972).

Discussions

KIPPENHAHN, R.:
In the last slides you showed how different solutions give agreement with different parts of the spirals in M 51. But does not LIN's theory only hold for spirals which are tightly wound? And did I understand right that some of your solutions are not consistent, in the sense that the density distribution which causes the assumed potential does not agree with the density distribution which you obtain from the computed orbits?

CONTOPOULOS, G.:
LIN's original theory is valid for tightly wound spirals away from the main resonances of a galaxy, namely the Lindblad resonances and the particle resonance. Extensions of this theory to take into account linear effects near resonances are under study now. However the resonance theory that is applied to the inner parts of M 51 is non-linear, therefore it goes beyond LIN's theory. TOOMRE's tidal theory gives open spirals.
As regards the second question, I made a clear distinction between two kinds of problems:
a) The problem of orbits, where the spiral potential is assumed as *given*, and
b) The self-consistent problem, where the potential is derived from the distribution of the stars, which move under the influence of this potential.
It is obvious that the second problem presupposes the solution of the first.

On Spiral Generating

(Invited Lecture)

By D. LYNDEN-BELL
Institute of Astronomy, Cambridge, U.K.

Consider two particles in circular orbits in an axially symmetrical galaxy-like potential. Of all the different pairs of orbits having the same total angular momentum, the pair in which one particle is at the nucleus and the other has all the angular momentum, has the least energy. Energy can be freed from any other pair of orbits, if a mechanism can be found to transfer angular momentum from the particle in orbit closer to the centre to the particle further out. Such freed energy can be used to increase the random energies of vibration about circular motion and thus to increase the (coarse-grained) entropy (LYNDEN-BELL, 1962, 1967) of the galaxy. Anthropomorphically, we can say that just as entropy wants to increase, so does a galaxy want to transfer its angular momentum outwards. The energy store that could be tapped if a mechanism were formed, is the total kinetic energy of galactic rotation.

In axially symmetrical perturbations of a galaxy the different elements preserve their angular momenta, so there is no possibility of moving the angular momentum from particle to particle. Thus our attention is called to non-axially symmetrical disturbances in which the gravitational couples can transfer angular momentum from one part of a galaxy to another. To avoid the difficulty that material spirals would wind up too quickly to agree with observation, we must look for disturbances that may last a long time. BERTIL LINDBLAD (1963) gave an ingenious explanation as to why even in the absence of self gravity in the disturbance bar-type perturbations would last a long time. Since this is probably the reason why bars and two-armed normal spirals are dominant, I would like to give brief account of the idea.

In an axially symmetrical gravity field, a typical star oscillates in towards the galactic centre and out again as it moves around the galaxy. Viewed from fixed axes the orbit describes a rosette. Viewed from axes with the same average rotation rate about the galactic centre as the star, it describes a small orbit about its epicentre. This orbit does not enclose the galactic centre. Thus the same orbit looks very different as viewed in different rotating axes. Particularly interesting are those rotating axes in which the orbit closes exactly. There are many of these for any one orbit since for example we may take axes in which the star apparently goes around the galactic centre once for m oscillations in and out. The angular velocity of such axes is

$$\Omega_A = \Omega - \frac{\kappa}{m},$$

where Ω, the angular rotation of the star around the galaxy and κ, the angular frequency of radial oscillation, are evaluated for the orbit in question. The value of Ω_A will vary from star to star so there are no axes in which all the orbits can be seen as closed m-lobed figures. However the particular case $m=2$ is remarkable in that over a wide range of most galactic rotation curves $\Omega-\kappa/2$ varies only a little. Thus if we analyse stellar motions into 2-lobe orbits and look at them in axes rotating with some average $\overline{\Omega_A}$, we shall see those 2-lobe orbits in the middle as fixed. Those closer to the centre do not close, but we can analyse them into closed orbits whose lines of apsides rotate slowly forwards. Those further from the centre we can likewise analyse into closed orbits whose lines of apsides rotate slowly backwards.

If we now imagine that gravity has bunched together the lines of apsides of the orbits rotating with $\overline{\Omega_A}$, it is plausible that the weak gravity of such a disturbance could overcome the weak shearing of Ω_A and so hold such a disturbance in a coherent shape for a long time. LINDBLAD sought the origin of barred spirals this way and KALNAJS points out that any shape made up of complete 2-lobe orbits may be expected to be long lived. He believes open 2-armed spiral fields could be dependent on such effects. When analyses of self-gravitating bar-type perturbations are made, it is found that the self gravity leads to the disturbance rotating somewhat faster than purely kinematic theory gives.

Angular Momentum Transport

My aim in discussing these disturbances was to find a non-axially symmetrical gravity field capable of transferring angular momentum between different parts of the galaxy. With KALNAJS I recently analysed the mechanisms of angular momentum transport in galaxies (LYNDEN-BELL and KALNAJS, 1972). The basic mechanism is by the gravitational couple of one part of the galaxy on another. This may be prettily described if, following Maxwell's lead in electromagnetic theory, we introduce the gravitational stress tensor

$$\boldsymbol{T}=\left(\frac{1}{4\pi G}\right)(\boldsymbol{g}\,\boldsymbol{g}-\tfrac{1}{2}g^2\,\boldsymbol{I}).$$

Here $\boldsymbol{g}$ is the gravitational acceleration field and $\boldsymbol{I}$ the unit tensor. The gravitational torque on that part of the galaxy outside a centred cylinder due to that part within is

$$\boldsymbol{C}=\int \boldsymbol{R}\times\boldsymbol{T}\cdot d\boldsymbol{S}$$

where the integration is over the surface of the cylinder and $\boldsymbol{R}$ is the vector distance from the galactic axis. Using the expression for $\boldsymbol{T}$

$$C_z=(4\pi G)^{-1}\int R\, g_\phi\, g_R\, dS=(4\pi G)^{-1}\int(\partial\psi/\partial\phi)(\partial\psi/\partial R)\, dS.$$

To transfer angular momentum outwards one needs $\partial\psi/\partial\phi$ and $\partial\psi/\partial R$ to be positively correlated over the cylinder. Thus ψ, the gravitational potential, must have a trailing spirality. I believe this to be the basic reason why spirals trail.

KALNAJS earlier pointed out that the action density of TOOMRE (1969) and SHU (1970) was in reality angular momentum density. Their expressions for transport by tightly wrapped waves do not agree with the gravitational contribution described above, because of a further mechanism which we analysed quite generally following enlightenment from TOOMRE. Consider a system of many local buses each of which, on average, picks up tourists whenever its excursions lead it nearer to Athens and these disembark while the excursion of the bus has taken it farther from Athens. Although on average tourists may not accumulate on the bus and although any individual local bus may not go far from its home nevertheless the system of buses will transport tourists from Athens to the borders of Greece. Now each star in its orbit picks up angular momentum when further from the galactic centre and deposits it when closer in. Thus the stars convect angular momentum just as buses convect tourists. This convective angular momentum flow can be in either direction but typically it offsets some of the gravitational couple without destroying the net sense of transport.

I have described buses that convect tourists without accumulating them on the bus. However occasionally there is a persistent resonance such as when a local bus ride is particularly scenic. Then more and more tourists wish to sit in the bus and there may be a steady rate of accumulation there. Typically, stars feeling a gravitational disturbance in the galaxy merely convect angular momentum without accumulating it. However we concur with KATO'S conclusion (1971) that resonant or near resonant orbits, those which close or almost close in the rotating axes of the spiral structure, are sources and sinks of angular momentum. The physical mechanisms by which the orbits give up or absorb angular momentum at each of the principal resonances are analysed in our paper. At the inner Lindblad resonance the resonant gravity field forces an eccentricity onto the star's orbit. This becomes two-lobed with the apogalactica lying in the most bound regions of the perturbing potential. We analyse the differences in the perturbation forces felt by the star as it was in its old circular orbit and in its new eccentric one. The azimuthal components of these extra forces is everywhere of the same sign and it removes angular momentum from the star's orbit. Both by mathematics and by physical explanation we find that

(i) The inner Lindblad resonance is an angular momentum emitter.
(ii) The outer Lindblad resonance is an angular momentum absorber.
(iii) The co-rotation resonance is a (rather weak) angular momentum absorber.

The absorbers and emitters are the right way around to transfer angular momentum outwards. Thus Lin density waves which carry negative angular momentum might be emitted in the outer parts[1] and absorbed in the central regions.

KALNAJS and I have recently analysed the details of orbits close to resonances of spiral structure using BORN'S (1927) method for resonant perturbations in Hamilton Jacobi theory. We find we can recover with a neat analysis the orbits discovered by the numerical computations of CONTOPOULOS (1970) and BARBANIS (1970). Whereas in plasma physics it is the trapping of particles by a wave that

[1] The wave velocity is actually outwards but the group velocity is inwards and this carries the (negative)angular momentum.

leads to Landau's damping and Landau excitation, we find that it is the more general concept of the trapping of orbits such as the catching of the lines of apsides near a given orientation that leads to energy and angular momentum absorption or emission in stellar dynamics.

Even in Greece I imagine that there may be occasions when local buses are full and tourists must be warned against assuming they can all accumulate in the buses on the most scenic routes. Likewise we find that it is only orbits whose apses are currently being trapped by a spiral wave that contribute to the angular momentum emission. When we calculate how many may have been trapped, we find that except for barred spirals this mechanism cannot have provided sources (or sinks) of angular momentum that have persistently emitted for all galactic history. Worse still, such a theory is dependent on stars alone whereas BAADE pointed out long ago that observationally spiral structure only occurs when gas and young stars are present. It is likely that these have a greater part to play than the mere lighting up of a spiral density wave that exists even in their absence, for no *spiral* structure has been detected in the purely stellar SO galaxies. HOHL's (1970) beautiful numerical experiments do show purely stellar transient spirals that wind up, but this is beside the point, for in reality such young galaxies would still have gas and dust. Some tidally deformed galaxies do show spiral features, but this is not the general phenomenon whose origin we are seeking.

Shock Waves

FUJIMOTO (1968), followed by LIN and ROBERTS (1969), recognised that gaseous motions generated by a tightly wrapped density wave would be dominated by the appearance of tightly wrapped shock waves. Later work (SHU *et al.*, 1972) has fulfilled LIN's belief (LIN, 1967) that the density wave itself might trigger star formation and it is the shock that seems to be the trigger. At present I think it is natural to consider that organisation of H II regions described by BAADE as "beads in a string" is precisely demarcated because there is a sharp shock front in the interstellar gas with the H II regions of star formation sitting very close by. In fact, what we see is not the density wave but the shock, and the density wave is something the theories postulate in order to make the shock. KALNAJS has been studying model shock waves in his more open spirals and he finds that the shock waves cross over the maxima and minima of the potential wave that generates them. The shock waves are wrapped as tightly as galaxies even when the disturbance of the gravitational potential is a very open spiral.

It is relevant to remark here that QUIRK (1971) in his numerical studies with MILLER and PRENDERGAST remarked that it was the somewhat unrealistic "gas" particles that seemed to make the spirals while the star particles merely provided a non-axially symmetrical gravity field.

It seems to me that there is a great deal to be said for KALNAJS's very open or even bar-like perturbation as the dominant stellar perturbation. Bars are common among galaxies with no gas, so the objection of BAADE no longer holds. Shock waves driven by the gravity fields of such perturbations acting on the gas would then make the spirals of similar form in both the barred spirals, where the gravity field of the bar is strong, and in normal spirals where any bar-like disturbance is submerged under the light from the shock-generated spiral.

Shock Waves with no Gravity Disturbance

There is one other possibility that I consider worth mentioning although, as far as I know, it has not been worked out at all. In theories of disc accretion onto black holes to form either quasars or X-ray sources, one meets the same basic problem as one does in the theory of spiral structure. How does the material find a way to pass its angular momentum outwards? Here the material being accreted may not have sufficient mass to give any non-axially symmetrical gravity field worth mentioning and yet the viscosity may be too weak to transfer angular momentum in the usual way. Could it be that there are non-axially symmetrical cold gas flows involving spiral shock waves in which viscosity, acting on a tangential discontinuity of the velocities across the shock, carries the angular momentum outwards. This would eventually be deposited on material in the outer parts while the inner material would gradually orbit its way down towards the central objects. Might it be that we can after all neglect the gravity of the spiral structure and leave only the shock waves? While I consider this an unlikely possibility, is it not one worth thinking about? If it were true, then a strong kick from outside might take the gas from the steady swirling in circles around a galaxy into a mode in which some systematically swirled into the centre. TOOMRE has remarked to me that a number of the galaxies with excited nuclei have companions that could be influencing them tidally. Such a picture is a remote but interesting possibility as it would tie together theories that promote nuclear activity by the feeding of black holes with other aspects of the spiral structure of galaxies.

References

BARBANIS, B.: In BECKER and CONTOPOULOS (eds.), Spiral Structure of our Galaxy, IAU Symp. **38**, 343 (1970).
BORN, M.: The Mechanics of the Atom, Blackie (1927), also Geneva winter school 1973.
CONTOPOULOS, G.: Astrophys. J. **160**, 113 (1970).
FUJIMOTO, M.: Astrophys. J. **152**, 391 (1968).
HOHL, F.H.: NASA TR R-343 (1970).
KATO, S.: Publ. Astron. Soc. Japan **23**, 467 (1971).
LIN, C.C.: An. Rev. Astron. Astrophys. **5**, 453 (1967).
LINDBLAD, B.: Stock. Obs. Ann., 22, No. 5 (1963).
LYNDEN-BELL, D.: Monthly Notices Roy. Astron. Soc. **124**, 279 (1962).
LYNDEN-BELL, D.: Monthly Notices Roy. Astron. Soc. **136**, 101 (1967).
LYNDEN-BELL, D., KALNAJS, A.: Monthly Notices Roy. Astron. Soc. **157**, 1 (1972).
QUIRK, W.J.: Astrophys. J. **167**, 7 (1971).
ROBERTS, W.W.: Astrophys. J. **158**, 123 (1969).
SHU, F.H.: Astrophys. J. **160**, 89 and 99 (1970).
SHU, F.H., MILIONE, V., GEBEL, W., YUAN, C., GOLDSMITH, D.W., ROBERTS, W.W.: Astrophys. J. **173**, 557 (1972).
TOOMRE, A.: Astrophys. J. **158**, 899 (1969).

Discussion

COURTÈS, G.:
The new data about velocity fields in nearest galaxies are now of excellent quality owing to the interference method. Several thousand points have been obtained (4000 on M 33) on NGC6946, M 31, M 51. It is an observing fact that there exists a discontinuity of about 10 km s^{-1} in the rotational velocity (faster than the disk) along the spiral features, this systematic effect found in 1968 on M 33 [Annals d'Astrophysique **31**, 63 (1968)] is now comfirmed on these several galaxies and with a spatial resolution of few seconds of arc owing to the use of the 200″ Hale telescope–(See Dr. MONNET's lecture in this symposium).

LYNDEN-BELL, D.:
Yes. This clearly indicates that the shock picture is correct but we can't tell whether these observations discriminate between gravity-field-forced shocks and the no gravity shock idea I mentioned until we have worked out whether such an idea is actually a possibility.

KIPPENHAHN, R.:
In your picture a galaxy wants to condense but it cannot because of its angular momentum; therefore spirals are being formed in order to transport momentum outwards by gravitational fields. How is this related to LIN's theory or to the results presented here by Prof. CONTOPOULOS?

LYNDEN-BELL, D.:
The waves that propagate to move the angular momentum from place to place are rather long-wave examples of the waves discussed by LIN. I do not hold with his suggestions that the waves go out and back again to reinforce themselves. CONTOPOULOS is investigating the same theory as we are but we are laying particular stress on how the angular momentum goes.

CONTOPOULOS, G.:
Is there any angular momentum transfer from the central region of the galaxy if there is no inner Lindblad resonance?

LYNDEN-BELL, D.:
Not for a steady wave. For a slowly growing wave there would be a little.

On the Possible Role of the Stellar Drift Motions for the Dynamics and Structure of Differentially Rotating Stellar Systems

By M. N. MAKSUMOV
Astrophysical Institute of the Academy of Sciences of the Tadjik SSR, Dushanbe, USSR

Abstract

The differential rotation of stellar systems leads to the appearance of specific drift motions of stars. These motions may cause the excitation of spiral waves. This possibility is investigated locally in the present paper. We conclude that drift motions may be of importance for generating the spiral structure of galaxies, especially of galaxies of the Sc type.

I. Introduction

The density wave theory has been rapidly developed recently in Stellar Dynamics. To a great extent, this development was stimulated by the resemblance between a rarefied plasma and collisionless self-gravitating systems. For this reason the methods used in plasma physics are widely used in stellar systems also.

Initially, the density wave theory was applied to purely dynamical problems – stability and collisionless evolution problems (SIMON, 1961; LYNDEN-BELL, 1962; SWEET, 1963; TOOMRE, 1964; LEBEDEV *et al.*, 1965; MAKSUMOV and MAROCHNIK, 1965 a, b; LEE, 1967; MAROCHNIK and PTITSINA, 1968; WU, 1968; SUCHKOV, 1969; BISNOVATY-KOGAN *et al.*, 1969). LIN (LIN and SHU, 1964, 1966; LIN *et al.*, 1969) and many others (BECKER and CONTOPOULOS, 1970) applied the density wave theory to the problems of spiral structure of galaxies.

But the wave theory of spiral structure encounters difficulties due to the antispiral theorem (LYNDEN-BELL and OSTRIKER, 1967) and the need of an energy source in order to maintain the spiral pattern (TOOMRE, 1968). Both these difficulties can be overcome, probably, if there is an effective instability (overstability).

One attempt to avoid the above difficulties, is based on the interaction between the flat and spherical subsystems of spiral galaxies (MAROCHNIK and SUCHKOV, 1969). This interaction is supposed to be responsible for a beam-type overstability, similar to that of plasma physics. This mechanism implies that a considerable part of the rotation energy of the flat subsystem can be transformed into density wave energy. But in fact this is not quite correct, because in such a case the flat subsystem would be scattered by the spherical one very fast. The energy of density fluctuations arising from the flat system cannot exceed the energy of its random stellar motions. This energy is two orders of magnitude less than the rotational energy. Perhaps the interaction between spherical and flat subsystems can maintain the tightly wound spiral structure of Sa and Sb galaxies.

But this is impossible in the case of Sc galaxies, where the spherical subsystem is rather weak.

However there is another energy source which is comparable with the energy of the random motions, and even more powerful. This is the differential rotation. The differential rotation energy of the whole galactic disk is only one order of magnitude less than the total rotation energy.

As it will be shown below, some dynamical effects are connected with the kinematic and spatial inhomogeneity of stellar systems. These effects are due to the drift motions of stars. As in the case of phenomena studied in plasma physics (TCERKOVNIKOV, 1957; RUDAKOV and SAGDEEV, 1959, 1961; VEDENOV *et al.*, 1961; ROSENBLUTH *et al.*, 1962; KADOMTSEV and TIMOFEEV, 1962; GALEEV *et al.*, 1963a; MIKHAILOVSKY, 1963; MIKHAILOVSKAYA and MIKHAILOVSKY, 1963; RUKHADZE and SILIN, 1964), these effects are expressed as a specific drift overstability of the density waves. This overstability arises because the drift motions of stars excite density waves coherently. Drift overstabilities are of great importance for the structure and dynamics of stellar systems, in particular for the formation and maintenance of spiral waves in galaxies. They are considered in detail elsewhere (MAKSUMOV, 1973a, b, c; MAKSUMOV and MISHUROV, 1973), but their results are summarized here. A similar opinion was expressed by LYNDEN-BELL (1970), without a quantitative analysis.

According to rough quasilinear estimates (MAKSUMOV, 1973b, c) almost all the energy of differential rotation may be transformed into energy of density waves.

A simple axisymmetric model will be considered, namely a thin stellar disk. Boundary conditions are not taken into account because the problem is a local one. The treatment is based on the collisionless kinetic equation because we deal with either a collisionless gas of "clouds" or a collisionless gas of stars. Moreover, drift instabilities are a specific "kinetic" effect.

II. Differentially Rotating Disk Galaxy

Consider a differentially rotating infinitely thin inhomogeneous stellar disk. In configuration space introduce a cylindrical coordinate system (ρ, φ, z), z being the rotation axis. Separate the circular rotational velocity of stars and introduce residual velocity variables as follows

$$v_\perp = (v_\rho^2 + v_\varphi^2)^{\frac{1}{2}}, \qquad \alpha = \tan^{-1}(v_\varphi / v_\rho),$$

where v_ρ and v_φ are projections of the residual velocity on the radial coordinate and perpendicularly to it. The angular velocity of rotation $\Omega = \Omega(\rho)$ is a function of the ρ variable only. In these variables, the kinetic equation takes the form

$$\begin{aligned} &\frac{\partial f}{\partial t} + v_\perp \cos\alpha \frac{\partial f}{\partial \rho} + \left(\frac{v_\perp \sin\alpha}{\rho} + \Omega\right) \frac{\partial f}{\partial \varphi} \\ &+ \left(\left[\frac{\partial \psi}{\partial \rho} + \Omega^2 \rho\right] \cos\alpha + \frac{\sin\alpha}{\rho} \frac{\partial \psi}{\partial \varphi} - \rho \frac{d\Omega}{d\rho} v_\perp \cos\alpha \sin\alpha\right) \frac{\partial f}{\partial v_\perp} \qquad (1) \\ &+ \left(-\left[\frac{\partial \psi}{\partial \rho} + \Omega^2 \rho\right] \frac{\sin\alpha}{v_\perp} + \frac{\cos\alpha}{v_\perp \rho} \frac{\partial \psi}{\partial \varphi} - \frac{v_\perp \sin\alpha}{\rho} - \rho \frac{d\Omega}{d\rho} \cos^2\alpha\right) \frac{\partial f}{\partial \alpha} - 2\Omega \frac{\partial f}{\partial \alpha} = 0. \end{aligned}$$

The distribution function $f(\rho, \varphi, v_\perp, \alpha, t)$ is normalized to the surface number density of stars. $\psi(\rho, \varphi, t)$ is the gravitational potential of the system.

In the theory of tightly wound spiral arms, the inclination angle i is assumed to be extremely small $\left(\tan i = \frac{m}{k_\rho \rho}\right.$, where m/ρ and k_ρ are the azimuthal and radial wave numbers of a spiral mode). If i is small, the drift effects may be ignored. In fact, the ratio of the third term of the kinetic equation (1)

$$\left(\frac{v_\varphi}{\rho}\frac{\partial f}{\partial \varphi} = \frac{v_\perp \sin\alpha}{\rho}\frac{\partial f}{\partial \varphi}\right),$$

which describes the drift, to the second one

$$\left(v_\rho \frac{\partial f}{\partial \rho} = v_\perp \cos\alpha \frac{\partial f}{\partial \rho}\right)$$

is just of the order of $\tan i$. But it is clear that the magnitude of the inclination angle of real spirals is finite. Generally speaking, this fact cannot be ignored.

The kinetic equation which describes the slow, in comparison with the epicyclic motion, processes, may be obtained by averaging of Eq. (1) over the angle α. Approximately, this angle may be considered as varying uniformly in time.

Writing out the characteristics of the Eq. (1) and omitting the part of the gravitational potential which is balanced by rotation, we find by expansion in inverse powers of (2Ω) and subsequent averaging over the angle α, the following averaged characteristics (BOGOLYUBOV and ZUBAREV, 1955; BOGOLYUBOV and MITROPOLSKY, 1963; RUDAKOV and SAGDEEV, 1958):

$$\begin{aligned} &\dot{\bar{\rho}} = \frac{1}{\kappa \rho}\frac{\partial \psi}{\partial \bar{\varphi}}, \qquad \dot{\bar{\varphi}} = \Omega + \frac{1}{4}\frac{\bar{v}_\perp^2}{\kappa^2 \bar{\rho}}\frac{d\Omega}{d\bar{\rho}} - \frac{1}{\kappa \bar{\rho}}\frac{\partial \psi}{\partial \bar{\rho}}, \\ &\dot{\bar{v}}_\perp = \frac{1}{4}\frac{\bar{v}_\perp}{\kappa^2 \bar{\rho}}\frac{d\Omega}{d\bar{\rho}}\frac{\partial \psi}{\partial \bar{\varphi}}, \qquad \kappa = 2\Omega + \frac{1}{2}\bar{\rho}\frac{d\Omega}{d\bar{\rho}}. \end{aligned} \tag{2}$$

The bar here denotes averaging, and it will be suppressed below. It is seen that there are two types of drift, one due to differential rotation and another one due to the orthogonality between the angular velocity and the gravitational force.

In averaged variables $(\rho, \varphi, v_\perp)$ the kinetic equation for $f(\rho, \varphi, v_\perp, t)$, as a continuity equation, has the form

$$\frac{df}{\partial t} + \frac{1}{\rho}\frac{\partial}{\partial \rho}(\rho\, \dot{\rho}\, f) + \frac{1}{\rho}\frac{\partial}{\partial \varphi}(\rho\, \dot{\varphi}\, f) + \frac{1}{v_\perp}\frac{\partial}{\partial v_\perp}(v_\perp\, \dot{v}_\perp\, f) = 0. \tag{3}$$

where $\dot{\rho}$, $\dot{\varphi}$ and $\dot{v}_\perp$ are defined by Eq. (2).

Note that the kinetic equation may be written in the form

$$\frac{\partial f}{\partial t} + \dot{\rho}\frac{\partial f}{\partial \rho} + \dot{\varphi}\frac{\partial f}{\partial \varphi} + \dot{v}_\perp \frac{\partial f}{\partial v_\perp} = 0. \tag{4}$$

Of course, (3) and (4) are equivalent. This requires

$$\frac{1}{\rho}\frac{\partial}{\partial \rho}(\rho\, \dot{\rho}) + \frac{1}{\rho}\frac{\partial}{\partial \varphi}(\rho\, \dot{\varphi}) + \frac{1}{v_\perp}\frac{\partial}{\partial v_\perp}(v_\perp\, \dot{v}_\perp) = 0. \tag{5}$$

Thus Eq. (2) must satisfy the relation (5). Substituting (2) into (5) we see that it is satisfied in our approximation in which by differentiation with respect to ρ, (i) only $\left(1+\frac{1}{4}\rho\,\Omega^{-1}\frac{d\Omega}{d\rho}\right)$ in the expression for κ must be differentiated, because (2Ω) is assumed to be a fixed parameter, in obtaining the relations (2), and (ii) the second derivative of Ω with respect to ρ and the second power of the derivative of Ω with respect to ρ must be neglected.

After this remark substitute the relations (2) into Eq. (3). Then we find the kinetic equation needed

$$\begin{aligned}\frac{\partial f}{\partial t}+\frac{1}{\rho}\frac{\partial}{\partial\rho}\left(\frac{1}{\kappa}\frac{\partial\psi}{\partial\varphi}f\right)+\frac{1}{\rho}\frac{\partial}{\partial\varphi}\left[\rho\left(\Omega+\frac{1}{4}\frac{v_\perp^2}{\kappa^2\rho}\frac{d\Omega}{d\rho}-\frac{1}{\kappa\rho}\frac{\partial\psi}{\partial\rho}\right)f\right]&\\ +\frac{1}{v_\perp}\frac{\partial}{\partial v_\perp}\left(\frac{v_\perp^2}{4\kappa^2\rho}\frac{d\Omega}{d\rho}\frac{\partial\psi}{\partial\varphi}f\right)=0.&\end{aligned} \tag{6}$$

Now we linearize Eq. (6) assuming that the stationary state is axisymmetric and the gravitational force is balanced by the centrifugal force. Then

$$\frac{\partial f_0}{\partial t}=\frac{\partial f_0}{\partial\varphi}=\frac{\partial\psi_0}{\partial\varphi}=\frac{\partial\psi_0}{\partial\rho}=0,$$

where f_0 and ψ_0 are the distribution function and the gravitational potential in the undisturbed state respectively. Since we have changed the definition of the gravitational potential before the relations (2), the equation $\partial\psi_0/\partial\rho=0$ means that the gravitational force is balanced by the centrifugal force. Recall that for this reason the centrifugal force falls out of the characteristic equations (2).

For the disturbed part of the distribution function, denoting it as f, we derive

$$\frac{\partial f}{\partial t}+\frac{1}{\kappa\rho}\frac{\partial\psi}{\partial\varphi}\frac{\partial f_0}{\partial\rho}+\left[\Omega+\frac{v_\perp^2}{4k^2\rho}\frac{d\Omega}{d\rho}\right]\frac{\partial f}{\partial\varphi}+\frac{v_\perp}{4k^2\rho}\frac{d\Omega}{d\rho}\frac{\partial\psi}{\partial\varphi}\frac{\partial f_0}{\partial v_\perp}=0. \tag{7}$$

The Poisson equation for the gravitational potential disturbance ψ is

$$\frac{\partial^2\psi}{\partial\rho^2}+\frac{1}{\rho}\frac{\partial\psi}{\partial\rho}+\frac{1}{\rho^2}\frac{\partial^2\psi}{\partial\varphi^2}+\frac{\partial^2\psi}{\partial z^2}=-8\pi^2 GM\int_0^\infty f\,v_\perp\,dv_\perp\,\delta(z), \tag{8}$$

where $\delta(z)$ is Dirac's function, G the Newtonian gravitational constant, and M the mass of the individual star. The expression of the gravitational potential at $z=0$ must be substituted in (7).

The solution of Eq. (8) may easily be found making use (i) of suitable integral transforms of ψ and f and (ii) of the plane wave expansion of Dirac's function. By substituting the expression of the gravitational potential ψ into the solution of Eq. (7) for f, one may reduce the problem to an integral equation, which is a singular homogeneous Fredholm equation of the second kind. But its analysis is rather difficult. For this reason, it is more appropriate to look for local solutions of our problem (MIKHAILOVSKY, 1963; VEDENOV *et al.*, 1962; GALEEV, 1963). In the case of local solutions, boundary conditions do not enter and k_ρ is assumed equal to zero $\left(k_\rho\simeq 0,\text{ i.e. }\frac{m}{k_\rho\rho}\gg 1\right)$. The local dispersion relation to be found in this way, permits us to consider the stability problem and to find order of magnitude estimates for the frequencies and the growth rate.

We look for local solutions of Eqs. (7)–(8) in the form

$$f=f(\rho, v_\perp)\,e^{i(-\omega t+m\varphi)}$$
$$\psi=\psi(\rho, z)\,e^{i(-\omega t+m'\varphi)}. \tag{9}$$

Substituting (9) and (10) into (8) and solving it with the aid of Fourier-transforms, we obtain

$$\psi(\rho, z=0)=\frac{4\pi^2 GM}{\dfrac{|m|}{\rho}}\int_0^\infty f\,v_\perp\,dv_\perp. \tag{10}$$

Then from (7) we derive the dispersion relation

$$1=\frac{4\pi^2 GM}{\kappa}\,\mathrm{sgn}(m)\int_0^\infty v_\perp\,dv_\perp\,\frac{\left(\dfrac{\partial}{\partial\rho}+\dfrac{v_\perp}{4\kappa}\dfrac{d\Omega}{d\rho}\dfrac{\partial}{\partial v_\perp}\right)f_0}{\omega-m\,\Omega-\dfrac{m}{4\kappa^2\rho}\dfrac{d\Omega}{d\rho}v_\perp^2}. \tag{11}$$

Eq. (11) is of the type

$$F(\omega_r-m\,\Omega+i\,\omega_i)=0.$$

Assuming the overstability to be weak ($d\Omega/d\rho$ to be small), consider ω_i as a small quantity in comparison with the quantity $(\omega_r-m\,\Omega)$. The calculations below give results which are qualitatively correct only for small ω_i.

Expanding the dispersion relation in powers of ω_i we find

$$F(\omega_r-m\,\Omega+i\,0)+\frac{\partial F(\omega_r-m\,\Omega+i\,0)}{\partial(\omega_r-m\,\Omega)}\,i\,\omega_i+\cdots=0.$$

If we write

$$F(\omega_r-m\,\Omega+i\,0)=F_r(\omega_r-m\,\Omega+i\,0)+i\,F_i(\omega_r-m\,\Omega+i\,0),$$

where F_r and F_i are both real functions, we find that

$$F_r(\omega_r-m\,\Omega+i\,0)=0,$$

and

$$\omega_i=-\frac{F_i(\omega_r-m\,\Omega+i\,0)}{\dfrac{\partial F_r(\omega_r-m\,\Omega+i\,0)}{\partial(\omega_r-m\,\Omega)}}. \tag{12}$$

If f_0 is of Maxwellian type, i.e.

$$f_0=\frac{n(\rho)}{\pi\,\overline{v^2}}\exp\left(-\frac{v_\perp^2}{\overline{v^2}}\right),$$

where n is the projected number density, we find

$$F_r=1-\chi\int_0^\infty\frac{\dfrac{dn}{d\rho}-2\beta\,n\,\xi}{\tilde\omega-\gamma\,\xi}\,e^{-\xi}\,d\xi, \tag{13}$$

$$F_i=\frac{\pi\,\chi}{\gamma}\int_0^\infty\left(\frac{dn}{d\rho}-2\beta\,n\,\xi\right)e^{-\xi}\,\delta\left(\xi-\frac{\tilde\omega}{\gamma}\right)d\xi, \tag{14}$$

where

$$\xi=\frac{v_\perp^2}{\overline{v^2}},\quad \tilde{\omega}=\omega_r-m\Omega,\quad \beta=\frac{1}{4\kappa}\frac{d\Omega}{d\rho},\quad \gamma=\frac{m\beta\overline{v^2}}{\kappa\rho},\quad \chi=\frac{2\pi GM}{\kappa}\operatorname{sgn}(m).$$

It is seen from (13) that the integral is valid if the zero of the denominator is eliminated by a zero numerator, i.e. when $\xi=\frac{\tilde{\omega}\kappa\rho}{m\beta\overline{v^2}}$ we must have $\frac{dn}{d\rho}=2n\frac{\tilde{\omega}\kappa\rho}{m\overline{v^2}}$. If $dn/d\rho$ is assumed to be negative, it follows from this fact that $\tilde{\omega}/m<0$. This means that those waves will be overstable whose relative angular velocity has the same direction as the drift motion of the stars.

Since under the integral in (13) the function decreases rapidly as ξ increases only the contribution of small ξ is of importance. Then

$$\begin{aligned}F_r&=1-\frac{\chi}{\tilde{\omega}}\int_0^\infty\left(\frac{dn}{d\rho}-2\beta n\xi\right)\left(1+\frac{\gamma}{\tilde{\omega}}\xi\right)e^{-\xi}d\xi\\&\simeq1-\frac{\chi}{\tilde{\omega}}\left(\frac{dn}{d\rho}-2\beta n+\frac{\gamma}{\tilde{\omega}}\frac{dn}{d\rho}\right)\\&=1-\frac{A}{\tilde{\omega}}-\frac{B}{\tilde{\omega}^2},\end{aligned}$$

where

$$A=\chi\left(\frac{dn}{d\rho}-2\beta n\right),\qquad B=\chi\gamma\frac{dn}{d\rho}.$$

From the equation $F_r=0$ we have

$$\tilde{\omega}=\frac{A}{2}\pm\left(\frac{A^2}{4}+B\right)^{\frac{1}{2}}.$$

Those $\tilde{\omega}$ must be taken into account here, for which the condition $\tilde{\omega}/m<0$ holds.

Finally

$$\omega_i=-\frac{\pi\tilde{\omega}^3}{\gamma}\,\frac{\dfrac{dn}{d\rho}-2\beta n\dfrac{\tilde{\omega}}{\gamma}}{(2\gamma+\tilde{\omega})\dfrac{dn}{d\rho}-2\beta n\tilde{\omega}}\,e^{-\frac{\tilde{\omega}}{\gamma}}. \tag{15}$$

As it follows from (15), $\omega_i\to0$ at $\gamma\to0$; ω_i will be positive for $m>0$ in the following cases:

$$\left|\frac{dn}{d\rho}\right|<\left|2\beta n\frac{\tilde{\omega}}{\gamma}\right|,\qquad \left|\frac{dn}{d\rho}\right|>\left|\frac{2\beta n\tilde{\omega}}{2\gamma+\tilde{\omega}}\right|.$$

The first inequality corresponds to a positive numerator and a negative denominator and the second one to the reverse situation. The remaining gap in the negative $dn/d\rho$ region is connected with the approximation used, as it is well known in plasma physics (Galeev *et al.*, 1963b). Thus the magnitude of the density gradient conditions the strength of the overstability. Finally we note once more that both $\tilde{\omega}$ and ω_i vanish when the n and Ω gradients vanish. Hence, in fact, this type of waves is specific for inhomogeneous systems.

The expression for $\tilde{\omega}$ shows that drift waves are hydrodynamically stable, in the degree of accuracy we have used.

The order of magnitude of $\tilde{\omega}$ may be estimated as

$$\tilde{\omega} \simeq \frac{m}{4k^2\rho}\frac{d\Omega}{d\rho}\overline{v^2} \sim \frac{\overline{v^2}}{4\Omega^2}\frac{\Omega}{4\rho^2} = \frac{1}{4}\frac{r^2}{\rho^2}\Omega,$$

where $r=(\overline{v^2})^{\frac{1}{2}}/2\Omega$. If ρ is of the order of few r, then $\tilde{\omega}\simeq 0.1\,\Omega$, or one order less. The corresponding growth rate is one or two orders less than the frequency, i.e. $\omega_i \simeq (10^{-2}\text{–}10^{-3})\,\Omega$. These estimates characterize the situation for weak gradients only.

III. Concluding Remarks

Drift effects are of great importance for piral density waves because the inclination angle is finite. Indeed, in the theory of tightly wound spiral arms, the tangent of the inclination angle "i" is regarded as small quantity $\left(\tan i = \frac{m}{k_\rho \rho}\right)$. If "$i$" is small, one may, indeed, neglect drift effects since the ratio of the third term in kinetic equation (1) $\left(\frac{v_\varphi}{\rho}\frac{\partial f}{\partial \varphi}\right)$, which presents the drift motion, to the second one $\left(v_\rho \frac{\partial f}{\partial \rho}\right)$ is just of the order of $\tan i$. It is clear, however, that the inclination angle is finite for real spiral arms. As for those in galaxies of Sc type, this angle is even wide.

A limiting case, opposite to the one considered by LIN and other authors, was discussed above, where $\frac{m}{k_\rho \rho} \gg 1$, or $k_\rho \to 0$. This assumption simplifies the local analysis though it limits the size of the vicinity of any fixed point, say ρ_0, within which the results of the local analysis are valid (MIKHAILOVSKY, 1963; MIKHAILOVSKAYA and MIKHAILOVSKY, 1963; RUKHADZE and SILIN, 1964). It does not present a difficulty, of course, to adopt a step-by-step method in order to take into account the finiteness of the ratio $\frac{m}{k_\rho \rho}$.

References

BECKER, W., CONTOPOULOS, G.: The Spiral Structure of our Galaxy, IAU Symp. **38**, D. Reidel Publ. Co., Dordrecht (1970).

BISNOVATY-KOGAN, G. S., ZEL'DOVICH, YA. B., SAGDEEV, R. Z., FRIDMAN, A. M.: Zh. Prikladnoj Mekhaniki i Tekhnich. Fiziki, No. 3,3 (1969).

BOGOLYUBOV, N. N., MITROPOLSKY, YU. A.: Asimptoticheskie metody teorii nelineinykh kolebany, Moscow (1963).

BOGOLYUBOV, N. N., ZUBAREV, N. D.: Ukr. Matem. Zh. 7, 5 (1955).

GALEEV, A. A.: Zh. eksp. teor. Fiz. **44**, 1920 (1963).

GALEEV, A. A., MOISEEV, S. S., SAGDEEV, R. Z.: Atomnaya Energiya **15**, 451 (1963a).

GALEEV, A. A., ORAEVSKY, V. N., SAGDEEV, R. Z.: Zh. eksp. teor. Fiz. **44**, 903 (1963b).

KADOMTSEV, B.B., TIMOFEEV, A.V.: Dokl. Akad. Nauk **146**, 581 (1962).
LEE, E.P.: Astrophys. J. **148**, 185 (1967).
LEBEDEV, V.I., MAKSUMOV, M.N., MAROCHNIK, L.S.: Astron. Zh. **42**, 709 (1965).
LIN, C.C., SHU, F.H.: Astrophys. J. **140**, 646 (1964).
LIN, C.C., SHU, F.H.: Proc. Nat. Acad. Sci. **55**, 229 (1966).
LIN, C.C., YUAN, C., SHU, F.H.: Astrophys. J. **155**, 721 (1969).
LYNDEN-BELL, D.: Monthly Notices Roy. Astron. Soc. **124**, 279 (1962).
LYNDEN-BELL, D.: In W. BECKER and G. CONTOPOULOS (eds.), IAU Symp. **38**, 331 (1970).
LYNDEN-BELL, D., OSTRIKER, J.P.: Monthly Notices Roy. Astron. Soc. **136**, 293 (1967).
MAKSUMOV, M.N.: Bull. Inst. Astrofiz. Akad. Nauk Tadj. SSR, No. 64, 3 (1973a).
MAKSUMOV, M.N.: Bull. Inst. Astrofiz. Akad. Nauk Tadj. SSR, No. 64, 37 (1973b).
MAKSUMOV, M.N.: Bull. Inst. Astrofiz. Akad. Nauk Tadj. SSR, No. 64, 47 (1973c).
MAKSUMOV, M.N.: MAROCHNIK, L.S.: Astron. Zh. **42**, 1261 (1965a).
MAKSUMOV, M.N., MAROCHNIK, L.S.: Dokl. Akad. Nauk **164**, 1019 (1965b).
MAKSUMOV, M.N., MISHUROV, YU, N.: Bull. Inst. Astrofiz. Akad. Nauk Tadj. SSR, No. 64, 18 (1973).
MAROCHNIK, L.S., PTITSINA, N.G.: Astron. Zh. **45**, 516 (1968).
MAROCHNIK, L.S., SUCHKOV, A.A.: Astron. Zh. **46**, 524 (1969).
MIKHAILOVSKAYA, L.V., MIKHAILOVSKY, A.B.: Zh. eksp. teor. Fiz. **45**, 1566 (1963).
MIKHAILOVSKY, A.B.: Voprosy teorii plasmy **3**, Moscow (1963).
ROSENBLUTH, M.N., KRALL, N.A., ROSTOKER, N.: Nuclear Fusion, Suppl. Part 1, 143 (1962).
RUDAKOV, L.I., SAGDEEV, R.Z.: Fizika plasmy i problemy upravlyaemikh termoyadernykh reaktsy, Moscow **3**, 268 (1958).
RUDAKOV, L.I., SAGDEEV, R.Z.: Zh. eksp. teor. Fiz. **37**, 1337 (1959).
RUDAKOV, L.I., SAGDEEV, R.Z.: Dokl. Akad. Nauk. **138**, 581 (1961).
RUKHADZE, A.A., SILIN, V.P.: Usp. Fiz. Nauk, **82**, 499 (1964).
SIMON, R.: Bull. Cl. Sci. Acad. Roy. Belg. Ser. 5, **7**, 731 (1961).
SUCHKOV, A.A.: Astron. Zh. **46**, 534 (1969).
SWEET, P.A.: Monthly Notices Roy. Astron. Soc. **125**, 285 (1963).
TCERKOVNIKOV, YU.A.: Zh. eksp. teor. Fiz. **32**, 67 (1957).
TOOMRE, A.: Astrophys. J. **139**, 1217 (1964).
TOOMRE, A.: Astrophys. J. **158**, 899 (1968).
VEDENOV, A.A., VELIKHOV, E.P., SAGDEEV, R.Z.: Usp. Fiz. Nauk, **73**, 701 (1961).
VEDENOV, A.A., VELIKHOV, E.P., SAGDEEV, R.Z.: Nuclear Fusion. **1**, 82 (1962).
WU, C.S.: Phys. Fluids **11**, 545 (1968).

The Dynamics of Nearby Galaxies

By P. J. WARNER
Mullard Radio Astronomy Observatory, Cambridge, U.K.

Abstract

Maps of the radial velocity field of M 33, derived from 21-cm observations made with the Cambridge Half-Mile radio telescope, are presented at two angular resolutions. At the lower resolution of $4.5' \times 9'$ the dynamics of the HI within 5 kpc of the nucleus is found to be well described by pure circular motions alone: beyond this distance non-circular motions of more than 10 km s^{-1} are observed. These non-circular motions are attributed either to motions of the HI through the plane of the galaxy, or more probably to radial motions associated with mass asymmetries in the plane.

The lower resolution velocity field is examined for the effects of a density wave as proposed by LIN and SHU. The radial velocity perturbations due to a logarithmic spiral density wave which coincides with the optical arms, are not found. If present, the peak value of these perturbations must be smaller than 5 km s^{-1}.

At the higher resolution of $1.5' \times 3'$, corresponding to an approximately circular beam of 600 pc in the plane of M 33, the HI dynamics shows considerable small scale variation. Differential rotation extends to within 600 pc of the nucleus. A rotation curve derived from all measured velocities, is used to obtain a mass model for the galaxy. When projected on to the plane of the sky this model shows a nearly exponential decrease of surface density. The variation of mass to light ratio, with distance from the nucleus is shown to be small. The existence of near resonant orbits, over most of the galaxy, for a pattern speed of 5 km s^{-1} kpc^{-1} is noted.

This work, and a similar analysis of M 101 and M 31 will be published in the Monthly Notices of the Royal Astronomical Society.

Structural Changes in Globular Galaxies due to Collisions

By S. M. ALLADIN, K. S. SASTRY and G. M. BALLABH
Centre of Advanced Study in Astronomy, Osmania University, Hyderabad, India

Abstract

The effects of a collision between two globular galaxies on their internal structure are considered. Numerical estimates are made for the change in the internal energy, the mass of escaping matter and the change in the average radius of a galaxy due to different collisions, to determine how these factors depend on the distance and velocity at closest approach of the two galaxies and on their density distributions. A collision between two galaxies may lead to the formation of a double galaxy by tidal capture or to the tidal disruption of one of them. One particular collision, in which the galaxies pass through each other, is studied in greater detail to determine how the mass distribution and the shape of a galaxy are altered by tidal effects.

During a collision of two galaxies, the stars are accelerated on account of the tidal forces, and as a result the total internal energy of the galaxies (i. e. the energy due to the distribution and motion of the stars) increases at the expense of the total external energy (i. e. the energy due to the relative orbital motion of the galaxies). If $|U|$ is the magnitude of the initial internal energy of a galaxy and ΔU is the increase in the internal energy of the galaxy during the encounter, the ratio $\Delta U/|U|$ provides a measure of the change in the structure of the galaxy due to the encounter.

ΔU can be obtained by selecting a number of stars as representative of the test galaxy and calculating at each instant the tidal force exerted on them by the field galaxy. The tidal force, integrated over all time, will yield the change in velocity of the representative stars during the collision, from which the change in internal energy of the entire galaxy during the collision can be inferred. It is convenient to represent the galaxies as polytropes and to derive the relevant tidal forces from the polytrope theory as explained in ALLADIN (1965) and SASTRY and ALLADIN (1970). The polytrope of index 4 closely represents the density distribution of a typical globular galaxy (ROOD, 1965).

The following simplifying assumptions often considerably reduce the computational labour without causing much loss of accuracy:

(i) The galaxies may be treated as spherically symmetric configurations whose density distribution does not change during the encounter in the first approximation.

(ii) The relative motion of the two galaxies may be considered as uniform rectilinear motion with the distance and velocity at closest approach taken as those for the actual relative orbit of the galaxies.

(iii) The motion of the stars in the galaxies may be neglected in comparison with the orbital motion of the galaxies during the encounter. This is the so-called "impulsive approximation".

Making the above simplifying assumptions, SASTRY (1972) made numerical estimates for $\Delta U/|U|$ for collisions between galaxies having different density distributions for various distances of closest approach. The mass of escaping matter, $M^{(e)}$, from a galaxy in units of the initial mass, M, of the galaxy, is derived for each collision from the number of escaping stars. From the law of conservation of energy and the virial theorem, the increase in the average radius ΔR of the galaxy is estimated, where R is defined as in CHANDRASEKHAR (1942) for a cluster of stars. Some of the results are given in Tables 1 and 2. The initial masses and the outer radii of the two galaxies are taken as $10^{11}\,M_{\odot}$ and 10 kpc respectively for all the collisions.

Table 1 gives the dependence of $\Delta U/|U|$ on the distance, p, and velocity, V, at closest approach for a collision between two typical globular galaxies considered as polytropes of index 4. For this model of the galaxy, the root mean square radius is 1.88 kpc, and the magnitude of the internal energy, $|U|$, which by the virial theorem is equal to half the total potential energy, is 1.3×10^{59} c.g.s. units. It may be noted from Table 1 that a head-on collision between two typical globular galaxies with a relative velocity of 1000 km s^{-1} leads to the formation of a double galaxy by tidal capture, since the velocity of escape exceeds 1000 km s^{-1} at the centre of the galaxy for the models chosen.

Table 2 gives the dependence of $\Delta U/|U|$ on density distribution of the galaxies for a collision defined by $p=2$ kpc and $V=1000$ km s^{-1}. $\Delta U/|U|$ scales as $1/V^2$ in the impulsive approximation. Therefore collisions with smaller values of V may lead to the tidal disruption of a galaxy. For example, a collision between a homogeneous spherical galaxy and a typical globular galaxy having a mass distribution that of polytrope $n=4$, with $V=700$ km s^{-1} and $p=2$ kpc results in the tidal disruption of the homogeneous galaxy since $\Delta U/|U|$ becomes greater than unity in this case.

Table 1. Changes in the structure of a colliding galaxy as a function of the distance and velocity at closest approach

p in kpc	$V=1000$ km s^{-1}			$V=2000$ km s^{-1}		
	$\frac{\Delta U}{\lvert U\rvert}$	$\frac{M^{(e)}}{M}$	$\frac{\Delta R}{R}$	$\frac{\Delta U}{\lvert U\rvert}$	$\frac{M^{(e)}}{M}$	$\frac{\Delta R}{R}$
0	Tidal capture			2.0×10^{-1}	Nil	2.5×10^{-1}
2	2.3×10^{-1}	4×10^{-2}	1.6×10^{-1}	5.7×10^{-2}	1×10^{-3}	6×10^{-2}
6	1.3×10^{-2}	4×10^{-5}	1.3×10^{-2}	3×10^{-3}	Nil	3×10^{-3}
10	2×10^{-3}	Nil	2×10^{-3}	5×10^{-4}	Nil	5×10^{-4}
20	9×10^{-5}	Nil	1×10^{-4}	2×10^{-5}	Nil	2×10^{-5}

Table 2. Changes in the structure of a colliding galaxy as a function of the density distribution of the two galaxies

Field galaxy	Test galaxy polytrope $n=4$			Test galaxy homogeneous		
	$\frac{\Delta U}{\|U\|}$	$\frac{M^{(e)}}{M}$	$\frac{\Delta R}{R}$	$\frac{\Delta U}{\|U\|}$	$\frac{M^{(e)}}{M}$	$\frac{\Delta R}{R}$
Homogeneous	0.003	Nil	0.003	0.14	Nil	0.16
Polytrope $n=4$	0.23	0.04	0.16	0.52	0.01	0.96
Mass point	0.67	0.05	0.56	0.67	0.01	0.89

The changes in the orbits of the representative stars have been studied in detail for a collision between two typical globular galaxies. It is assumed that initially when the galaxies are 50 kpc apart (at which distance the tidal effects are negligible) the velocity distribution of the stars in the test galaxy is spherically symmetric and all the stars move in circular orbits. Taking the relative motion of the two galaxies as a uniform rectilinear motion with a speed of 1000 km s^{-1} and the distance of closest approach to be 2 kpc, the final positions and velocities of the stars when the galaxies have receded to a distance of 50 kpc after the collision, are derived, making use of the polytrope theory to obtain the forces between galaxies and the forces between the representative star and each of the two galaxies. From the results, the final mass distribution of the galaxy is derived. This is compared in Table 3 with the initial mass distribution of a polytrope of index 4 given by LIMBER (1961). This more detailed treatment indicates that the mass of escaping matter from the test galaxy is about 6% of its total mass. The average velocities of the non-escaping stars after the collision at various distances from the centre, are also tabulated in Table 3. These are compared with the corresponding initial values (SASTRY and ALLADIN, 1970). The root mean square velocity of all the stars in the galaxy decreases from 360 km s^{-1} before the collision to 330 km s^{-1} after the collision.

Table 3. Mass distribution and average velocities before and after collision

r in kpc	$M(r)/M$		V in km s^{-1}	
	initial	final	initial	final
1	0.30	0.29	361	361
2	0.74	0.70	400	376
3	0.92	0,77	363	306
4	0.97	0.83	324	296
5	0.99	0.87	292	258
6	1.00	0.89	267	198
7	1.00	0.89	248	252
8	1.00	0.89	232	170
9	1.00	0.90	219	143
10	1.00	0.91	207	197
13	1.00	0.94	–	192
20	1.00	0.94	–	–

To get an idea of the change in the shape of the galaxy due to the collision, let us choose the origin of the coordinate system at the centre of the test galaxy, the x-axis along the direction of the field galaxy when it is at its closest approach, the y-axis along the direction of motion of the field galaxy, and the z-axis perpendicular to the orbital plane of the two galaxies. The root mean square coordinates of the stars with respect to the centre of the test galaxy before the collision were

$$\langle x^2\rangle^{\frac{1}{2}}=\langle y^2\rangle^{\frac{1}{2}}=\langle z^2\rangle^{\frac{1}{2}}=1.1 \text{ kpc}$$

and the corresponding values for the non-escaping stars after the collision become:

$$\langle x^2\rangle^{\frac{1}{2}}=1.8 \text{ kpc}; \quad \langle y^2\rangle^{\frac{1}{2}}=2.2 \text{ kpc}; \quad \langle z^2\rangle^{\frac{1}{2}}=1.4 \text{ kpc}.$$

Thus the final root mean square radius, r_c, of the galaxy is 3.2 kpc, which implies that r_c has increased by about 70%. The increase in r_c is chiefly due to the expansion of the outer shells of the galaxy.

The escaping stars do not follow the perturbing galaxy, as indicated by CONTOPOULOS and BOZIS (1964) in the case of a faster and more distant collision. When the galaxies have receded to a distance of 50 kpc after the collision, 50% of the escaping matter, which amounts to 3% of the initial total mass of the galaxy, lies within a distance of 2 kpc from the plane defined by the z-axis and the negative x-axis. It will appear like a tail if viewed from the direction of the z-axis. Recent investigations by TOOMRE and TOOMRE (1972) on slow collisions between two disk galaxies, indicate that an encounter between two galaxies of equal mass results typically in a long and curving 'tail' of escaping debris from the far side of the victim disk, and in an avalanche of near-side particles, most of which are captured by the perturbing galaxy.

References

ALLADIN, S. M.: Astrophys. J. **141**, 768 (1965).

CHANDRASEKHAR, S.: Principles of Stellar Dynamics, University of Chicago Press, Chicago, Ch. 5, p. 200 (1942).

CONTOPOULOS, G., BOZIS, G.: Astrophys. J. **139**, 1239 (1964).

LIMBER, D. N.: Astrophys. J. **134**, 537 (1961).

ROOD, H. J.: The Dynamics of the Coma Cluster of Galaxies, University of Michigan, p. 124. (Ph. D. thesis) (1965).

SASTRY, K. S.: Astrophys. Space Sci. **16**, 284 (1972).

SASTRY, K. S., ALLADIN, S. M.: Astrophys. Space Sci. **7**, 261 (1970).

TOOMRE, A., TOOMRE, J.: Astrophys. J. **178**, 623 (1972).

Waves in Rotation Curves of Galaxies as Population Effects

By P. Pişmiş
Instituto de Astronomía, Universidad de México, Mexico

Abstract

A brief survey is given of two earlier papers by the author where it is argued that the waves in the rotation curves of galaxies are physically significant phenomena and that variations of density in a galaxy are insufficient to produce these. It is proposed that the deviations from a smooth rotation curve may be due to kinematical differences of populations. In the present report new data showing waves in rotation curves are listed. It is shown that a crucial test to the proposed interpretation will be afforded by the dispersion of velocities in the rotation curve; namely at the minima of the rotation curve the dispersion of velocities will be higher than at the maxima.

I. Introduction

Several years ago I called attention to the circumstance that rotation curves of galaxies tended to show deviations from a smooth curve and that in many cases these deviations appeared sufficiently pronounced as to be considered physically significant features. An interpretation of these "waves" in the rotation curves was then proposed (Pişmiş, 1965, 1966). It was argued that the waves or dips would arise as a consequence of the coexistence of the different population groups (subsystems) in our Galaxy and presumably in other spiral galaxies.

In the intervening years data on the rotation of galaxies have increased considerably and it now seems unnecessary to defend the reality of waves in the rotation curves. But although such property is generally recognized, it is currently ascribed to "streaming motion" along spiral arms, in the frame of the density wave theory. I shall continue using the term "waves" without attaching a preconceived interpretation to the phenomenon. Observationally one cannot distinguish the effects of streaming motions from real variation from an orderly rotation curve.

The present report is intended to renew the emphasis on my interpretation of the wavy form of rotation curves, to present new observational data and to call to attention a crucial test.

II. Observational Data; Review of Earlier Work

In the first paper (Pişmiş, 1965) I listed the galaxies with rotational curves determined by the Burbidge group (Astrophys. J. 1959 through 1964), to secure the homogeneity of the data. Spectra are obtained by the long-slit technique and most of the observed velocities are determined based on emission lines. Of the 22 galaxies listed in my first paper[1], 16 exhibit waves with significantly large amplitude. It should be noted that the observed amplitude of the wave is usually smaller than the real one due to the projection effect.

Regarding the Galaxy, I pointed out that the rotation curve from Cepheids (Kraft and Schmidt, 1963) shows a detectable dip which occurs between 10.5 kpc and 12.5 kpc from the galactic center ($R_0=10$ kpc is assumed). At $R=11.5$ the deviation from uniform rotation reaches a minimum and the rotational velocity there is lower by some 20 km s^{-1} than on the smooth curve. Quantitative discussion of that rotation curve will not be given here. Suffice it to say that galactic Cepheids do seem to show an appreciably wavy rotation curve.

Although the form of the rotation curve from 21-cm hydrogen observations by Kerr is wavy (1964) the interpretation of it is not unique; as is well known it may be also explained by the absence of neutral hydrogen in the interarm regions.

That the waves (the maxima) of the rotation curve cannot be produced by concentration of mass in spiral arms was shown in two ways:

(i) By considering the concrete case of NGC 4258, the spiral galaxy which showed three minima with the data of Burbidge *et al.* (1963). The density distribution was computed by the method of Brandt (1960). The resulting computed ratio of the density of arm to interarm regions turned out to be so very high as to be incompatible with the observed light ratio and hence sensibly with that of mass in those regions (Pişmiş, 1965).

(ii) In a later estimation (for details see Pişmiş, 1966) a model galaxy was constructed where the mass distribution of the old population was assumed to be similar to that in an elliptical galaxy and the extreme population I was assumed to be concentrated in circular rings with density on the average 4 times that of the underyling old population. The computed circular velocity curve of this model showed extremely shallow waves, not exceeding 5 km s^{-1}, well within the observational uncertainties even if such a galaxy were observed edge-on.

III. Observational Data; Recent Work

Galaxies with optical rotation curves published since 1965 are listed in Table 1. The first two columns give the designation and type respectively. The third column gives the observed radius within which velocities are measured, the fourth gives the number of minima, and the fifth the inclination of the galaxy; column six gives the references.

[1] NGC 139 should be deleted as its inclusion in the list is due to an error.

Table 1. Spiral galaxies with rotation curves

Name	Type	$\tilde{\omega}$ pc	No of maxima	Inclination[a]	References
NGC 7331	Sb	18 000	3	69.1	RUBIN *et al.* (1965a)
4826	Sb	2 000	2	59.9	RUBIN *et al.* (1965b)
681	Sa	5 250	2	81.0	BURBIDGE *et al.* (1965a)
6181	Sc	3 510	2	60	BURBIDGE *et al.* (1965b)
972	Sb	3 860	?	66.3	BURBIDGE *et al.* (1965c)
4736	Sb	2 500	3	35.1	CHINCARINI and WALKER (1967)
3310	Sb	1 375	1	45.6	WALKER and CHINCARINI (1967)
1808	Sb	4 850	3	66.5	BURBIDGE, E., BURBIDGE, G. (1968)
6574	Sc	2 720	1	50	DEMOULIN and TUNG CHAN (1969a)
7625	So	3 500	0	53	DEMOULIN (1969b)
3593	So_p	1 150	1+	≈90	DEMOULIN (1969c)

[a] Angle between the plane of galaxy and plane of sky.

Of the 11 galaxies listed nine show at least one wave in the rotation curve.

Another significant result should be mentioned. An extensive study of M 33 (3 000 points) by the Fabry-Pérot technique (MONNET, 1970) yields lower velocities in the interarm region as compared to the spiral arms. The difference is of the order of 15 km s^{-1}.

Wavy rotation curves are also observed in the radio region. Aperture synthesis studies of neutral hydrogen in NGC 2403, M 101 and NGC 4946 have shown significant deviations from circular motion (ROGSTAD and SHOSTAK, 1972). These deviations "typically of the order of 10 km s^{-1} tend to be systematic along arcs coinciding with optical spiral arms". ROGSTAD and SHOSTAK consider further that their observations provide qualitative confirmation of the density wave theory of spiral structure. However, the same observations can equally well be interpreted as a population effect.

In our Galaxy the 21-cm neutral hydrogen line provides a rather significant evidence for a wavy form of the rotation curve. A study by SHANE and BIEGER-SMITH (1966) covering galactic longitudes 22° to 70° yields a rotation curve with waves which these authors attribute to streaming. The curve is based on an adopted model for the galaxy which they consider the most likely among the seven models they discuss.

The observations of the 21-cm line by VARSAVSKY and QUIROGA (1970) in the interval 281° to 345° of galactic longitude yield a decidedly wavy rotation curve, between 4 and 9 kpc from the center ($R_0 = 10$ kpc), with two minima falling roughly at 4.5 and 6.7 kpc respectively. This is the region for which KERR (1964) has also derived a rotation curve showing two minima; the location of the minima agrees rather well with those of VARSAVSKY and QUIROGA.

While the interpretation of the radio data to obtain the velocity field depends on the choice of a model for our own Galaxy, no such restriction exists for external galaxies. The radio data of the latter yield the radial component of the field unambiguously although the resolution required is much higher than that for the Galaxy.

In discussing the kinematical behavior of young objects in the Galaxy, such as Cepheids, B stars, young clusters and neutral hydrogen, FEAST (1967) finds that between $R=9$ and 11 kpc and $r=2$ to 6 kpc these objects tend to show negative residuals in the southern section of the galactic plane and positive residuals in the northern section, the residuals being the differences between the observed radial velocities and those calculated, assuming $A=14.3$. He concludes, therefore, that his results lend support to the model proposed by SHANE and BIEGER-SMITH showing streaming motions. But as we mentioned earlier, streaming is another way of describing that which we have called waves in the rotation curve.

In concluding this section we may state that the existence of waves in the rotation curves of galaxies seems to be rather the rule than the exception and that waves are shown by the gaseous as well as the stellar component.

IV. The Cause of the Waves

That our Galaxy is an assembly of subsystems – or populations – is at present firmly established. The subsystems differ from one another in spatial distribution and kinematical properties which constitute the best criteria of population types in the Galaxy.

We assume, what appears plausible, that other spiral galaxies are also composed of subsystems much as in our Galaxy, although the relative masses of the subsystems may differ from galaxy to galaxy. We go a step further and suggest that gas clouds in a galaxy may also exhibit population characteristics similar to those of the stellar populations. We further believe that all subsystems show a smooth mass distribution, with rotational symmetry, except that in extreme population I the mass is distributed in a spiral form.

If we make these assumptions we can ascribe all known properties of the population types in our Galaxy to the stellar and gaseous components of other spiral galaxies. These properties may be briefly described as follows: large radial gradient of density (high concentration of mass towards the center), low average rotational velocity and large random motions (dispersion of velocities) correspond to the extreme population II or the spheroidal subsystem. These characteristics grade into the flattest subsystem, the extreme population I (spiral structure) where the average density gradient vanishes, the average rotational velocity is almost equal to the circular velocity and the random motions are very small. In a general way these properties – excepting the density gradient – were observationally found by STRÖMBERG (1924, 1925).

The properties summarized above can be described theoretically if one makes the plausible assumption that a galaxy is at present in steady state and that its mass distribution has rotational symmetry. Of the hydrodynamical equations of flow, the component in the direction $\tilde{\omega}$ is the equation relevant to our discussion. This may be written as follows:

$$\theta_c^2-\theta_r^2=-\tilde{\omega}\left[\frac{1}{\nu}\frac{\partial(\nu\langle\pi^2\rangle)}{\partial\tilde{\omega}}+\frac{\langle\pi^2\rangle-\langle\theta^2\rangle}{\tilde{\omega}}\right]. \tag{1}$$

Here θ_c is the circular velocity at point $\tilde{\omega}$ in the galaxy, θ_r the average rotational velocity of an element of volume around that point, $\langle\pi^2\rangle$ and $\langle\theta^2\rangle$, mean square random velocities (a measure of the dispersion) in cylindrical coordinates and ν, the density of particles (stars or gas clouds) of the subsystem in question.

It should be mentioned that it is the quantity $\theta_c - \theta_r$ which—following STRÖMBERG—is usually discussed and referred to as the asymmetrical drift. However, for the sake of simplicity of argument, $\theta_c^2 - \theta_r^2$ in formula (1) will be retained as such, since it gives a measure of the asymmetrical drift, θ_c and θ_r having always the same sign.

Eq. (1) can also be written in the following way:

$$\theta_c^2 - \theta_r^2 = -\tilde{\omega}\langle\pi^2\rangle \frac{\partial \ln(\lambda\,\nu)}{\partial\tilde{\omega}} \tag{2}$$

where λ satisfies the relation $\lambda^2 = \langle\pi^2\rangle/\langle\theta^2\rangle$.

In the solar neighborhood $\lambda \simeq$ constant appears to be satisfied by all subsystems judging from the values of $\langle\pi^2\rangle$ and $\langle\theta^2\rangle$ estimated by OORT (1967, Table 5). I shall assume that the constancy of λ is also true in other spirals at a given distance, $\tilde{\omega}$, from the center. At point $\tilde{\omega}$, in the plane of a galaxy, θ_c^2 will be the same for all subsystems, as $\theta_c^2/\tilde{\omega}$ is equal to the gravitational force acting at the point. However, θ_r^2 will be different for different subsystems and so will $\theta_c^2 - \theta_r^2$. The latter will vary with $\langle\pi^2\rangle$ and ν according to (2). Thus, inspection of (2) for λ = constant shows that for vanishing density gradient (for which $\langle\pi^2\rangle$ is also small) $\theta_c^2 - \theta_r^2$ nearly vanishes. These are the properties of extreme population I and further, when the density gradient is negative and large in absolute value for which $\langle\pi^2\rangle$ is also large, $\theta_c^2 - \theta_r^2$ is large. It is worth noting that the three components of the velocity dispersion all increase going from population I towards population II while θ_r decreases and the density concentration increases. That $\langle\pi^2\rangle$ [$\langle\theta^2\rangle$ and $\langle Z^2\rangle$] and the density gradient should vary together and in the same sense seems to be a vestigial property inherent in the evolution of the subsystems. This point will be treated in detail in a forthcoming paper.

We shall now give a rough estimate of the expected amplitude of a wave in a rotation curve of a galaxy like ours, at the position of the Sun.

We assume the density in a spiral arm (extreme Pop. I only) to be four times that of the underlying population II (the latter term we use to indicate all populations except the extreme Pop. I). At the spiral arm the rotational velocity is the average over population I and the underlying population II. While at mid-interarm the velocity is entirely that of population II. Table 2 gives the relevant assumed numerical values and the estimated average rotational velocities as well as the dispersion of velocities for arm and mid-interarm.

Table 2. Estimated values for rotational velocity and dispersion at arm and interarm regions

	Pop. I extreme	Pop. II	Arm	Interarm
θ_r	260 km s^{-1}	220 km s^{-1}	250 km s^{-1}	220 km s^{-1}
Relative density	4	1	5	1
$\langle\pi^2\rangle^{\frac{1}{2}}$	10 km s^{-1}	40 km s^{-1}	20 km s^{-1}	40 km s^{-1}

$\langle\pi^2\rangle^{\frac{1}{2}}$ is the dispersion in the radial direction.

With the very rough values appearing in Table 2 we find a wave with amplitude of 30 km s^{-1}. In an external galaxy seen edge-on one would observe the total amplitude, which would decrease as the inclination of the galaxy decreased. For the Andromeda nebula (inclination 77°) this amplitude could reduce to 28 km s^{-1}. It might be interesting to make a model of a galaxy and estimate the amplitude of the waves all along a radius. This requires knowledge of the mass distribution of the different subsystems, variation of θ_r and the dispersion along the radius of the model. The scanty data do not warrant such elaborate work at this time.

Especial attention should be called to the values of the dispersion of the arm and interarm regions. At the interarm region, the dispersion in the radial direction, $\langle\pi^2\rangle^{\frac{1}{2}}$, is about twice that at the arm. It is precisely this property that provides a crucial test of our interpretation: If indeed the waves in a rotation curve are due to population effects at the minima one should observe a higher velocity dispersion as compared to the maxima.

The dispersion of velocity is not usually measured; it is difficult to obtain it through slit spectra, as the necessary elimination of the instrumental profile of the spectral line from the observed profile cannot easily be performed. In the optical region the use of the Fabry-Pérot technique appears promising for the determination of dispersion of velocities as well as the average velocity.

In conclusion we emphasize again that the waves in rotation curves may be caused by the preponderance of one or another population type: *at the maxima the major contribution to the rotation is that of extreme population I and at the minima, that of population II.*

V. Discussion

In external galaxies measured radial velocities are generally those of the H II regions; velocities from absorption lines have so far been obtained essentially in nuclear regions. It may be argued that gas clouds are, conventionally, extreme population I components and would not show population II characteristics. However, high velocity clouds observed by the 21-cm neutral hydrogen line may change this conception; these clouds may well be the counterpart of the population II in the domain of gas clouds. Only a systematic study of the whole sky would reveal whether they are indeed falling into the galactic plane from all directions or not. Observations made from the northern hemisphere would show essentially negative velocities if the clouds are subject to the phenomenon of the asymmetrical drift and positive velocities would be observed in the opposite direction. This circumstance was pointed out earlier by this author, moreover it was suggested that whether falling in from all directions or subject to the asymmetrical drift, the clouds would show a slower average rotation and the net effect would be to lower the velocity of rotation at the interarm regions (Pişmiş, 1965).

Finally it is interesting to note, by turning the argument around, that if our interpretation is correct – at least for part of the observed amplitude – the waves observed in rotation curves may provide evidence for the existence of population types in galaxies.

Acknowledgements

My thanks are due to Dr. C. PAYNE-GAPOSCHKIN for a critical reading of the manuscript.

References

BRANDT, J.C.: Astrophys. J. **131**, 293 (1960).
BURBIDGE, E.M., BURBIDGE, G.R., PRENDERGAST, K.H.: Astrophys. J. **138**, 375 (1963).
BURBIDGE, E.M., BURBIDGE, G.R., PRENDERGAST, K.H.: Astrophys. J. **142**, 154 (1965a).
BURBIDGE, E.M., BURBIDGE, G.R., PRENDERGAST, K.H.: Astrophys. J. **142**, 641 (1965b).
BURBIDGE, E.M., BURBIDGE, G.R., PRENDERGAST, K.H.: Astrophys. J. **142**, 649 (1965c).
BURBIDGE, E.M., BURBIDGE, G.R.: Astrophys. J. **151**, 99 (1968).
CHINCARINI, G., WALKER, M.F.: Astrophys. J. **147**, 407 (1967).
DEMOULIN, M.-H., TUNG CHAN, Y.W.: Astrophys. J. **156**, 501 (1969a).
DEMOULIN, M.-H.: Astrophys. J. **157**, 75 (1969b).
DEMOULIN, M.-H.: Astrophys. J. **157**, 81 (1969c).
FEAST, M.W.: Monthly Notices Roy. Astron. Soc. **136**, 141 (1967).
KERR, F.J.: In F.J. KERR and A.W. RODGERS (eds.), The Galaxy and the Magellanic Clouds, IAU Symp. **20**, 81 (1964).
KRAFT, R.P., SCHMIDT, M.: Astrophys. J. **137**, 249 (1963).
MONNET, G.: In W. BECKER and G. CONTOPOULOS (eds.), The Spiral Structure of Our Galaxy, IAU Symp. **38**, 73 (1970).
PIŞMIŞ, P.: Bol. Obs. Tonantzintla y Tacubaya **4**, 8 (1965).
PIŞMIŞ, P.: In M.A. ARAKELJAN (ed.), Non-Stable Phenomena in Galaxies, IAU Symp. **29**, The Publ. House of the Academy of Sciences of Armenian S.S.R., p. 429 (1966).
ROGSTAD, D.H., SHOSTAK, G.S.: Abstract of paper presented at the third regular meeting of the Division of Dynamical Astronomy (1972).
RUBIN, V.C., BURBIDGE, E.M., BURBIDGE, G.R., CRAMPIN, D.J., PRENDERGAST, K.H.: Astrophys. J. **141**, 759 (1965a).
RUBIN, V.C., BURBIDGE, E.M., BURBIDGE, G.R., PRENDERGAST, K.H.: Astrophys. J. **141**, 885 (1965b).
SHANE, W.W., BIEGER-SMITH, G.P.: Bull. Astron. Inst. Neth. **18**, 263 (1966).
STRÖMBERG, G.: Astrophys. J. **59**, 228 (1924).
STRÖMBERG, G.: Astrophys. J. **61**, 353 (1925).
VARSAVSKY, C.M., QUIROGA, R.J.: In W. BECKER and G. CONTOPOULOS (eds.), The Spiral Structure of Our Galaxy, IAU Symp. **38**, 147 (1970).
WALKER, M.F., CHINCARINI, G.: Astrophys. J. **147**, 516 (1967).

On Galaxy Parameters as Derived from Primeval Turbulence

By N. DALLAPORTA and F. LUCCHIN
Istituto di Astronomia dell'Università di Padova, Italy

Abstract

The problem of galaxy formation through separation from the continuous matter background due to density fluctuations is reconsidered in the frame of the primordial turbulence theory, according to the general outline to the problem given by OZERNOY and collaborators (1968, 1969, 1970). The consideration of dissipation losses in the turbulence, which were neglected in the previous treatment of the problem, is now included. These dissipation effects are treated following HEISENBERG'S (1948a, 1948b) and CHANDRASEKHAR'S (1949) approach to the solution of the integro-differential equation of turbulence. With the modifications resulting from this inclusion, the main behaviour of the three border lines limiting the upper scale, the lower scale of the turbulence range, and expressing the equality of the turbulent velocity with sound velocity are followed along the three main phases of universal evolution: the radiation phase I, preceding equality between radiation density and matter density; the ionized matter phase II, following this time up to recombination time; and the neutral matter phase III, following recombination. The main physical assumption are KOLMOGOROFF'S law $v \sim l^{\frac{1}{3}}$ connecting velocity turbulence and scale range for subsonic regimes, and the extended KOLMOGOROFF'S law $v \sim l^n$ for supersonic regimes, where $n > \frac{1}{3}$ is a parameter connected to the compressibility law defined according to VON HOERNER'S (see WEIZSÄCKER, 1951) assumption by $\rho \sim l^{-3n+1}$. The dissipation laws are also directly connected with the value of n. The existence of density fluctuations enables one to derive a separation condition for the galaxies.

The results are somewhat modified in respect to those previously obtained without consideration of dissipation (DALLAPORTA and LUCCHIN, 1972). The whole scheme is still dependent on only three parameters, the maximum velocity of the turbulence V_L, the actual mean density of the universe ρ_{0m}, and the value of n. It is found that even by assuming the maximum turbulent velocity V_L as equal to sound velocity c_s at equality time $c_s = c/\sqrt{3}$, there is a bottleneck for the scale of turbulence at recombination which keeps open only for low density universes ($\rho \leqq 3 \times 10^{-30}$ g cm^3). The best fit for the power law connecting the mean angular momentum per unit mass with the mass for spiral galaxies yields a best fit value of $0.5 \sim 0.6$ for n, the compressibility parameter; and the best fit for both the maximum mass and maximum angular momentum per unit mass of spiral galaxies are in agreement with a mean universe density of around

$\rho \sim 10^{-30}$ g cm^3. The difficulty of fitting the present approach to larger values for the mean universe density appears as the most severe constraint imposed on this line of thought.

References

CHANDRASEKHAR, S.: Proc. Roy. Soc. A. **200**, 20 (1949).
DALLAPORTA, N. LUCCHIN, F.: Astron. Astrophys. **19**, 123 (1972).
HEISENBERG, W.: Z. Phys. **124**, 628 (1948 a).
HEISENBERG, W.: Proc. Roy. Soc. A. **195**, 402 (1948 b).
OZERNOY, L. M., CHERNIN, A. D.: Soviet Astron. **11**, 907 (1968).
OZERNOY, L. M., CHERNIN, A. D.: Soviet Astron. **12**, 901 (1969).
OZERNOY, L. M., CHIBISOV, G. V.: Soviet Astron. **14**, 615 (1970).
WEIZSÄCKER, C. F. VON: Astrophys. J. **114**, 165 (1951).

Evidence for the Existence of Second-Order Clusters of Galaxies

By M. Kalinkov
Department of Astronomy, Bulgarian Academy of Sciences, Sofia, Bulgaria

With 11 Figures

Abstract

The distribution of clusters of galaxies, contained in an area of 1933 square degrees and centered in the North Galactic Pole is examined. Zwicky's catalog and partly Abell's catalog are used. Some statistical tests and also surface smoothing and filtering are applied. Evidence for existence of second-order clusters of galaxies is given. The characteristic scale length of second-order clustering is about 45 Mpc.

I. Introduction

Many attempts for the investigation of high-order clustering of galaxies have been undertaken. A general review of the more important results was given recently by De Vaucouleurs (1971). It is known that various indications for the existence of second-order clusters have been found (e.g. Abell, 1958, 1961, 1965; Abell and Seligman, 1965, 1967; Karachentsev, 1966; Kiang, 1967; Kiang and Saslaw, 1969). But according to Zwicky and his collaborators, an important fact is the non-existence of genuine large clusters of clusters of galaxies (e.g. Zwicky, 1957; Zwicky and Rudnicki, 1963, 1966; Zwicky and Berger, 1965; Zwicky and Karpowicz, 1965, 1966; Karpowicz, 1967a, 1967b, 1970a, 1970b, 1971a, 1971b, 1971c). Yu and Peebles (1969) and Fullertone and Hoover (1972) suppose that if there exist second-order clusters only a small fraction of the rich clusters may be found in them. An acceptable assumption related to the apparent disagreement in the conclusions reached by various authors is expressed by De Vaucouleurs (1970). In the most recent extra-galactic monograph Vorontsov-Velyaminov (1972) writes that "The dispute for the existence or the non-existence of second-order clusters appears pointless at present".

The present investigation is based on an analysis of all clusters of galaxies having $b > +65°$ (new system) and is drawn out from Volumes I, II and III of Zwicky's catalog. Abell's catalog is partly used too. We submit evidence for the existence of second-order clusters.

Some statistical methods for testing the hypothesis for the existence of second-order clusters of galaxies have been applied. Several isopleth maps showing the surface distribution of clusters of galaxies have been constructed. Such maps for the apparent distribution of the galaxies have been constructed by many authors, e.g. the Lick Observatory counts of galaxies (SHANE and WIRTANEN, 1967), but the first maps for clusters of galaxies have been published recently (KALINKOV, 1972a). The cluster maps give the possibility to number those clusters, which belong to each large second-order cluster. Both the digital smoothing and the filtering of initial maps permit the rejection of any doubtful condensation. The next step is smoothing and filtering, not on a surface, but in space. All the results based on cluster maps may be verified through a correlation analysis.

II. Observational Material – Some Comments

ZWICKY'S catalog (ZWICKY *et al.*, 1961; ZWICKY and HERZOG, 1963, 1966) and also ABELL'S catalogue (1958) have been used. We use here the new galactic coordinates. All clusters having $b \geqq +65°.01$ were treated. Therefore the area selected lies around NGP (in order to avoid the galactic obscuration) and has $S = 1932.54\ \square°$, which is sufficiently large for certain conclusions. The clusters of galaxies are classified in ZWICKY'S catalog into three types: open (*o*), medium compact (*mc*) and compact (*c*); their rough angular diameters *D* and populations are also given. In this paper the area $\mathfrak{S}$ (in square degrees) covered by a cluster is used. The population (number of galaxies in each cluster as determined by ZWICKY) is denoted as $\mathcal{N}$. $\mathfrak{S}$ and $\mathcal{N}$ are treated as random variables. The estimated distances of the clusters are near (*N*) when $z < 0.05$, to extremely distant (*ED*) for $z < 0.2$.

The data for clusters in our sample are presented in Table 1 – the number of the clusters by types and distances, and in Table 2 – the total area (in square degrees) of the clusters in various types and distances. The Virgo cluster is excluded.

A comparison between Tables 1 and 2 on the one hand, and the results obtained by ZWICKY and RUDNICKI (1963), ZWICKY and BERGER (1965), ZWICKY and KARPOWICZ (1965, 1966), KARPOWICZ (1971a, 1971b) on the other hand shows that our sample is a representative one. It is interesting to note that the distribution of $\mathfrak{S}$ may be well represented by logarithmic normal distribution. In this sense ZWICKY'S considerations for the largest cluster area are not quite justified. The sample mode, median and mean for the lognormal distribution are given in Table 3.

Table 1. Number of clusters by types and distances in the sample (ZWICKY'S Catalog)

Distance	*o*	*mc*	*c*	All
N	40	39	6	85
MD	87	117	11	215
D	117	202	28	347
VD	100	310	148	558
ED	53	436	453	942
All	397	1104	646	2147

Table 2. Total area covered by clusters in the sample in square degrees (ZWICKY's catalog)

Distance	$\Sigma\,\mathfrak{S}\,\square^\circ$			
	o	*mc*	*c*	all
N	124.01	179.62	33.42	337.05
MD	72.35	71.74	6.52	150.61
D	32.48	64.45	7.79	104.72
VD	12.82	35.68	13.66	62.16
ED	3.55	23.03	17.33	43.91
All	245.21	374.52	78.72	698.45

Table 3. Sample mode, median and mean for lognormal distribution of the areas $\mathfrak{S}$ in square degrees

Distance	Mode	Median	Mean
N	0.54	2.2	4.5
MD	0.24	0.48	0.68
D	0.12	0.22	0.30
VD	0.042	0.080	0.112
ED	0.017	0.034	0.048

According to Table 3, the *D*, *VD* and *ED* clusters of the galaxies may be considered as points, if the sky area is divided into squares, larger than 1 square degree.

The sample drawn out of ABELL's catalog consists of 412 clusters of galaxies (Table 4). That sample is also a representive one.

Table 4. Number of clusters in the sample (ABELL's catalog)

Distance group	Clusters
1	3
2	2
3	7
4	18
5	168
6	213
7	1
All	412

III. Isopleths of Clusters of Galaxies

A view on an apparent distribution of cluster centres in ZWICKY's catalog may be obtained with isopleths. The area S was divided into zones of $\Delta l \times \Delta b = 10^\circ \times 2^\circ$ in which the cluster centres have been counted. The counts are reduced to density of clusters per square degrees and smoothed by averaging zone arrays of four (like SHANE and WIRTANEN, 1967). Isopleth maps are constructed for various distance categories. As unit in each case we use the quantity $\mathfrak{D} = S/\sum n$ clusters per square degrees, where $\sum n$ is the total number of clusters of a given category. The units are given in each figure. E.g. isopleth 2 in Fig. 3 stands for 2×0.49 cl/□°, and so on. A single map is only traced for *N*, *MD and D* clusters, since the corresponding numbers of the clusters are not large.

There are many condensations in Figs. 1–4. Which of them are second-order clusters? The first remark is that the Galaxy obscuration bears no relation to the traced condensations (in contradiction to the result obtained by ZWICKY, 1957). Secondly, we see that Fig. 4 is more uniform than for a given distance category. KIANG and SASLAW (1969) turned the attention to this fact.

Let us assume that all condensations in Figs. 1, 2 and 3, which are closed in the isopleths 2, are real second-order clusters. All well defined condensations in Figs. 1–4, having a large area are given in Table 5, containing coordinates of centres, limits along *l* and *b*, areas in square degrees, and coordinates of a maximum density.

Similar maps were constructed in the sample drawn out of ABELL's catalog for the distance groups 5 and 6, as well as for all clusters.

Table 5. Condensations in Figs. 1–4

Distance	No.	Center		Limits		Area □°	Max density	
		l	*b*	*l*	*b*		*l*	*b*
N+MD+D	1	181°	70°.1	172°–198°	<65°.0–71°.8	>30.6	175°	72
	2	194	77.4	182 –218	73.4–82.7	23.9	195	82
	3	221	70.9	213 –229	67.7–73.7	14.7	215	72
	4	252	68.3	242 –261	<65.0–70.2	>17.4	255	68
	5	264	74.1	253 –273	72.3–75.7	11.2	265	74
	6	338	70.6	322 –352	68.1–73.2	15.2	325	72
VD	1	26	69.2	19 – 31	66.7–71.2	10.8	25	70
	2	183	69.6	159 –202	<65.0–73.4	>56.1	185	72
	3	221	73.1	203 –231	68.8–76.4	18.4	215	74
	4	230	66	224 –238	<65.8–68.0	>12.9	235	66
ED	1	30	72	14 – 63	<65.0–77.8	>78.1	55	76
	2	45	66.8	38 – 52	<65.0–68.6	>18.2	55	66
	3	52	81.2	34 – 52	78.6–84.3	10.4	55	80
	4	162	70.4	142 –189	<65.0–73.4	>55.7	175	70
	5	298	82	278 –312	75.4–>78.8	21.6	285	82
All clusters	1	26	69.8	23 – 32	66.3–73.6	14.8	25	70
	2	184	69.4	168 –197	66.6–72.5	29.3	175	70

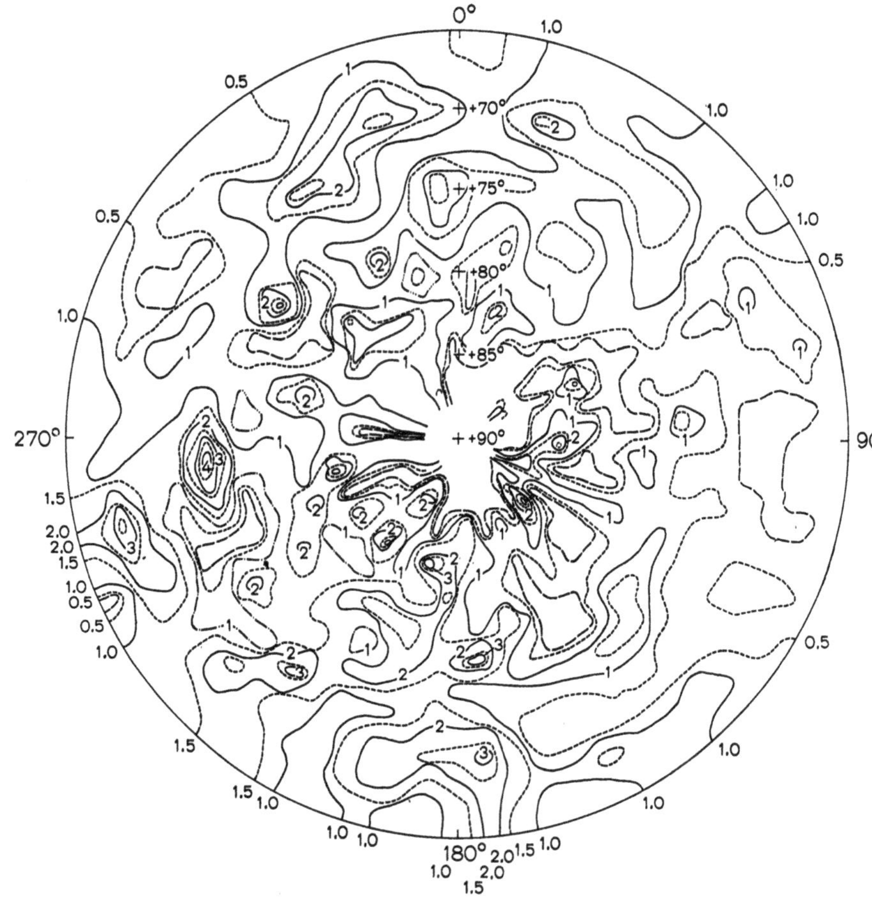

Fig. 1. Isopleth map for $\Sigma' = N + MD + D$ clusters in units of $\mathfrak{D} = 0.34$ clusters per square degree

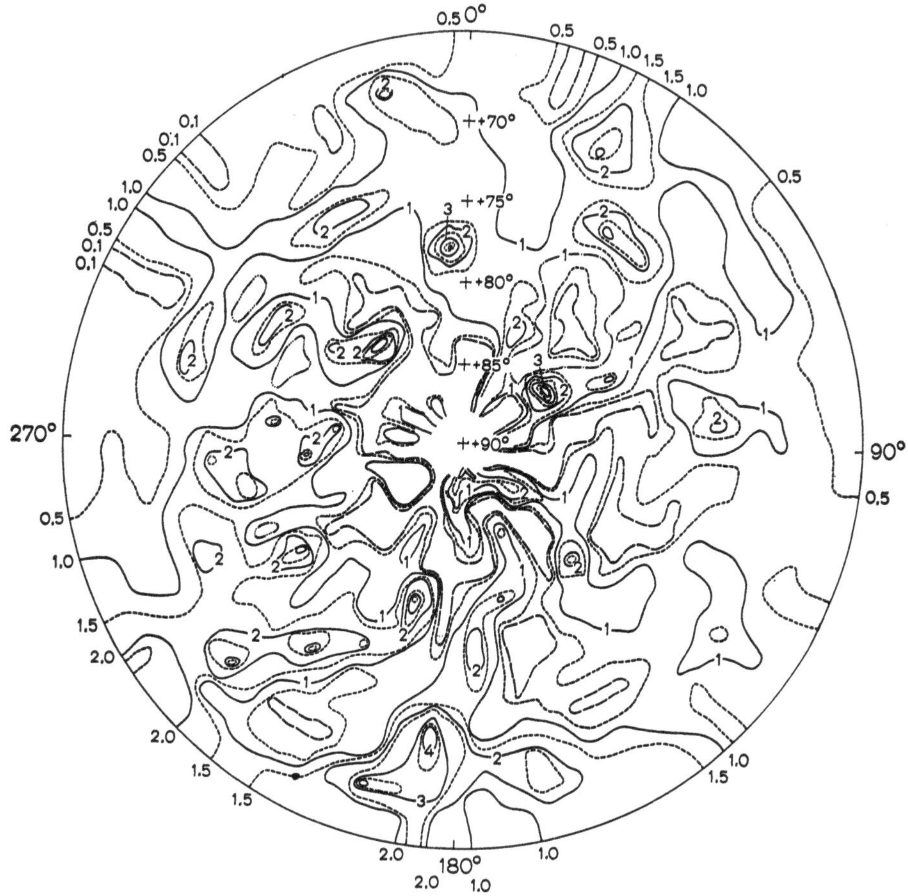

Fig. 2. Isopleth map for *VD* clusters in units of $\mathfrak{D}=0.29$ clusters per square degree

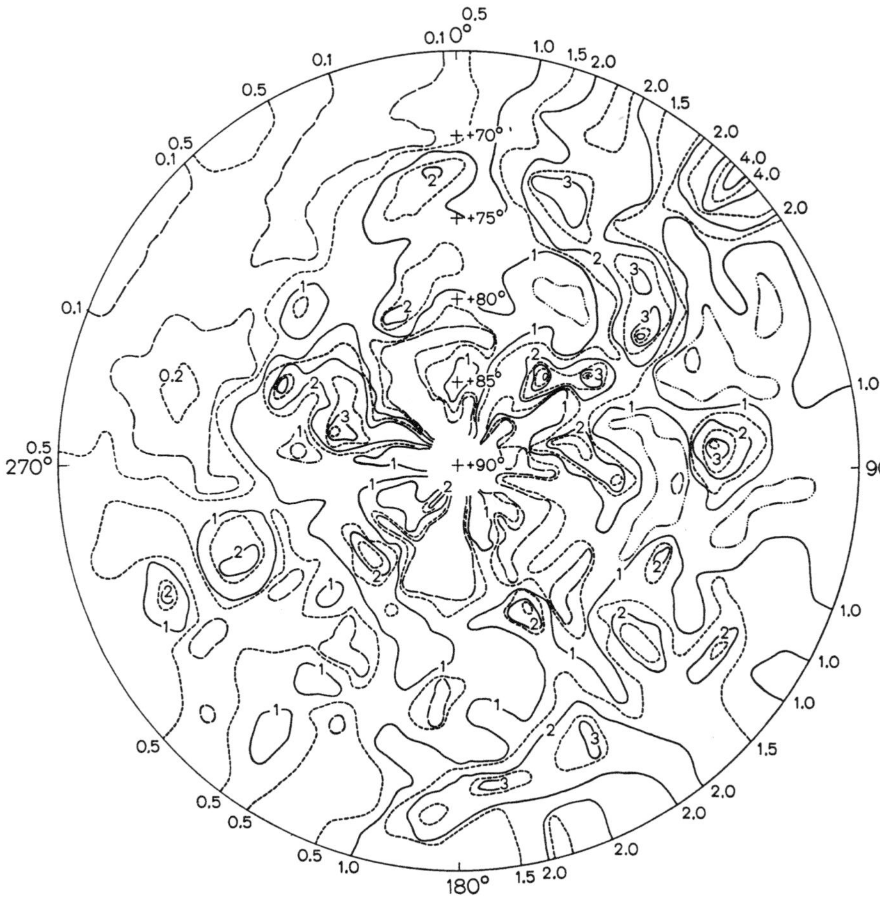

Fig. 3. Isopleth map for *ED* clusters in units of $\mathfrak{D}=0.49$ clusters per square degree

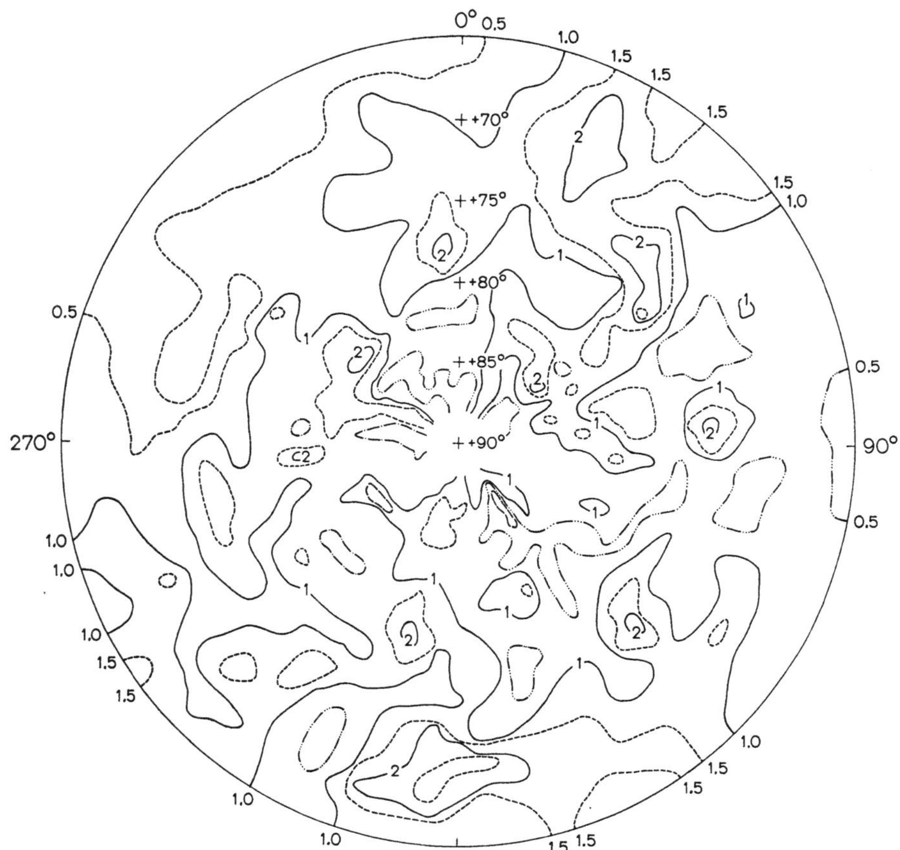

Fig. 4. Isopleth map for all clusters in units of $\mathfrak{D} = 1.11$ clusters per square degree

IV. Population of Clusters of Galaxies

We examine here the population of clusters $\mathcal{N}$ as a random variable. Let $\mathcal{N}_s$ be the value of $\mathcal{N}$ for the population of clusters located within isopleths 2, and $\mathcal{N}_n$ for other clusters. If the random variables $\mathcal{N}_s$ and $\mathcal{N}_n$ are not significantly different, then we may consider the two samples as being drawn out of the same population (here the term population has only a statistical meaning). Otherwise the two samples are drawn out from two essentially different populations.

Our zero hypothesis H_0 is: the two samples belong to the same population.

The testing of H_0 for N, MD and D is carried out for isopleths in Fig. 1. Observed distributions are given by KALINKOV (1972b). Numbers of clusters n_s and n_n along the distance categories are given in Table 6.

Table 6. Number of clusters inside (n_s) and outside (n_n) second-order clusters between $+65° < b < +85°$

Distance	n_s ($D \geqq 2\mathfrak{D}$)	n_n ($D < 2\mathfrak{D}$)
N	21	63
MD	49	158
D	87	253
$\Sigma' = N + MD + D$	157	474
VD	142	388
ED	319	592
$\Sigma'' = VD + ED$	461	980
Σ = all	618	1454

Testing of H_0

(i) *Median Test* (KENDALL and STUART, 1961; SARHAN and GREENBERG, 1962; OWEN, 1962; KALINKOV and SPASOVA, 1972). This non-parametric criterion is applicable for any continuous distribution function. The medians of $\mathcal{N}_s$ and $\mathcal{N}_n$ are given in Fig. 5 together with a 10% distribution – free confidence level. For MD, VD and ED the differences are significant, thus the hypothesis H_0 is rejected.

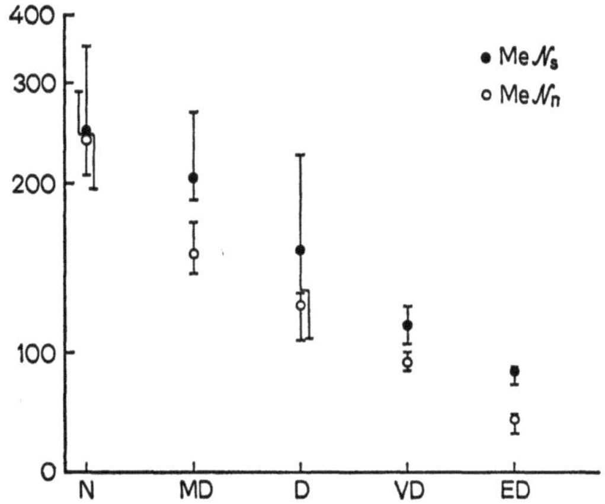

Fig. 5. Medians of $\mathcal{N}_s$ and $\mathcal{N}_n$ with a 90% confidence limits

Table 7. χ^2-test

Distance	ν	Observed distrib.	Log-normal	Loglog-normal
N	2	9.6(−1)[a]	8.7(−1)	9.9(−1)
	1	7.9(−1)	5.9(−1)	8.9(−1)
	1	8.0(−1)	7.9(−1)	1.0(−0)
MD	6	4.9(−3)	1.6(−2)	3.4(−3)
	4	2.2(−3)	4.9(−3)	2.9(−3)
	2	4.7(−4)	9.4(−4)	1.1(−3)
	1	9.1(−5)	7.6(−4)	1.4(−3)
	1	1.8(−3)	1.8(−3)	1.8(−3)
D	6	2.7(−3)	6.4(−3)	6.6(−4)
	4	8.4(−4)	1.3(−3)	1.3(−4)
	2	4.2(−4)	5.7(−4)	4.7(−5)
	1	9.6(−5)	1.2(−4)	1.8(−5)
	1	7.2(−3)	3.2(−2)	3.7(−2)
VD	8	1.7(−3)	2.0(−4)	9.2(−5)
	6	3.8(−4)	7.2(−5)	1.9(−5)
	4	1.7(−4)	2.7(−4)	6.7(−5)
	2	1.2(−4)	2.1(−4)	9.3(−6)
	1	5.5(−5)	3.0(−4)	2.1(−6)
	1	4.5(−3)	1.5(−3)	4.4(−3)
ED	8	2.4(−8)	2.4(−8)	2.6(−9)
	6	3.1(−8)	1.5(−7)	6.8(−9)
	4	4.8(−9)	1.8(−8)	2.0(−8)
	2	2.4(−10)	4.8(−9)	1.0(−8)
	1	2.9(−10)	2.0(−9)	8.6(−9)
	1	2.4(−6)	1.4(−6)	2.4(−7)
Σ'	8	8.8(−5)	2.1(−3)	1.3(−4)
	6	1.4(−3)	1.1(−3)	6.2(−5)
	4	2.6(−4)	2.1(−4)	1.5(−5)
	2	2.6(−5)	5.8(−4)	1.5(−5)
	1	4.5(−6)	1.1(−4)	2.8(−6)
	1	6.4(−3)	4.2(−2)	1.5(−3)
Σ''	8	1.3(−8)	7.1(−10)	6.3(−11)
	6	4.8(−9)	1.1(−8)	3.9(−9)
	4	4.4(−9)	5.6(−9)	4.0(−9)
	2	2.1(−7)	4.3(−6)	4.6(−8)
	1	3.6(−8)	1.8(−6)	1.1(−8)
	1	2.6(−6)	3.7(−5)	6.9(−7)
Σ	12	1.0(−5)	1.0(−4)	9.6(−6)
	10	2.2(−6)	4.6(−5)	2.7(−6)
	8	8.5(−7)	4.4(−5)	1.5(−5)
	6	1.4(−5)	3.7(−5)	2.0(−5)
	4	3.1(−6)	1.1(−5)	3.6(−6)
	2	1.9(−5)	2.8(−4)	3.0(−7)
	1	5.0(−5)	6.5(−5)	5.3(−6)
	1	9.2(−4)	1.7(−3)	7.2(−4)

[a] The numbers in parentheses are powers of 10.

(ii) χ^2 *Test* (LEHMANN, 1959; KENDALL and STUART, 1961; RAO, 1965). The observed distribution of $\mathcal{N}_s$ and $\mathcal{N}_n$ may be compared directly, using the χ^2 test. The results are presented in Table 7: the integral of χ^2, $P(\chi^2)$ is given where ν is the number of degrees of freedom. For $\nu=1$ we have the independence test for the 2×2 table (LEHMANN, 1959; KULLBACK, 1958). There are two values of $P(\chi^2)$ for $\nu=1$ since Med $\mathcal{N}_s$ and Med $\mathcal{N}_n$ are used (double dichotomy).

It is evident that the hypothesis H_0 is accepted only for clusters in the distance category N and is rejected in any other case.

(iii) SMIRNOV'S *Test* (SMIRNOV, 1936; KENDALL and STUART, 1961; BOLSHEV and SMIRNOV, 1968). This is another approach. The results are given in Table 8, where $K(y)$ is KOLMOGOROV'S distribution function.

H_0 is rejected in all cases except for the N clusters.

(iv) *Lognormal Approximation.* An approach to a more correct testing of the hypothesis H_0 is through the lognormal transformation of the random variables $\mathcal{N}_s$ and $\mathcal{N}_n$. The results of the χ^2 test are given in Table 7.

Table 8. SMIRNOV'S test

Distance	$1-K(y)$
N	9.1 (−1)[a]
MD	8.3 (−4)
D	1.3 (−4)
VD	1.8 (−3)
ED	9.2 (−9)
Σ'	3.0 (−6)
Σ''	1.4 (−8)
Σ	7.2 (−5)

[a] The numbers in parentheses are powers of 10

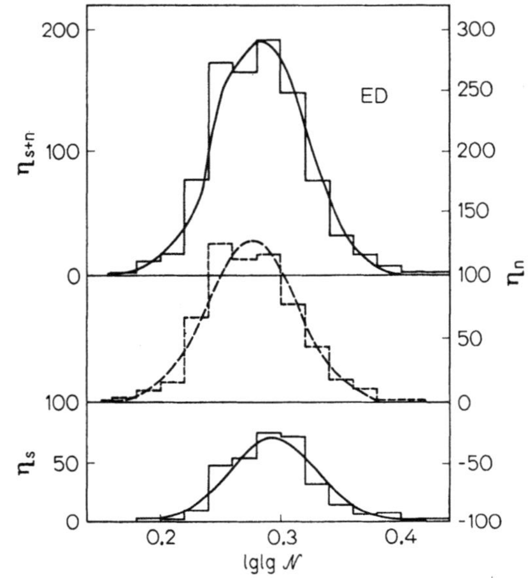

Fig. 6. Loglognormal distribution for *ED* clusters η_s – number of cases for $D \geqq 2\mathfrak{D}$; η_n for $D < 2\mathfrak{D}$ and η_{s+n} for the whole area

(v) *Loglognormal Approximation.* This is the next approximation. The fitting of *N*, *VD* and *ED* clusters with the loglognormal distribution has been accepted. In all other cases the new approximation appears as a not very good one, but in all cases $F=\sigma_1^2/\sigma_2^2\approx 1$. The final results are presented in Table 7. The agreement between the observed and theoretical distributions by *ED* clusters are given in Fig. 6.

From all the above tests it is evident that the hypothesis H_0 must be rejected. Therefore, the present results point out that the clusters within the isopleths 2 are different from the rest.

V. Smoothing and Filtering of the Surface Distribution of the Clusters

The whole sample area *S* was divided into equal squares (1° × 1°) and the cluster centres were counted. The new map allows digital smoothing and filtering. Here only the results for *ED* clusters are presented. The smoothing is carried out with normal filters, and the filtering – by differences between the smoothed fields. The characteristic scale response of the normal filters used in Fig. 7 is given. For squares 0°.5 × 0°.5 the characteristic scale cutoff is $L=1°$; for 1° × 1°, $L=2°$ and for 2° × 2°, $L=4°$. The filters have $\sigma=0°.5$ and $\sigma=1°$, denoted as [0.5] and [1]. The most simple case of filtering is the combination [0.5]–[1], which is valiable about $L\approx 3°$. Filter [1] for 1° × 1° squares gives the smoothed map in Fig. 8. The filtered map is given in Fig. 9.

Some of the established second-order clusters in Fig. 3 are retained after the smoothing and filtering. Hence, the second-order clusters do not appear as a result of the fluctuating density.

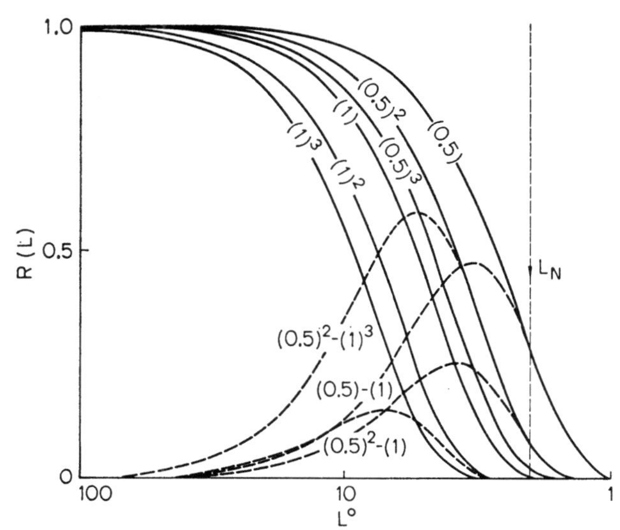

Fig. 7. Characteristic scale response of normal filters $\sigma=0°.5$ and $\sigma=1°$. The second application of filters are denoted with $[0.5]^2$ etc. L_N is the cutoff for squares 1° × 1°

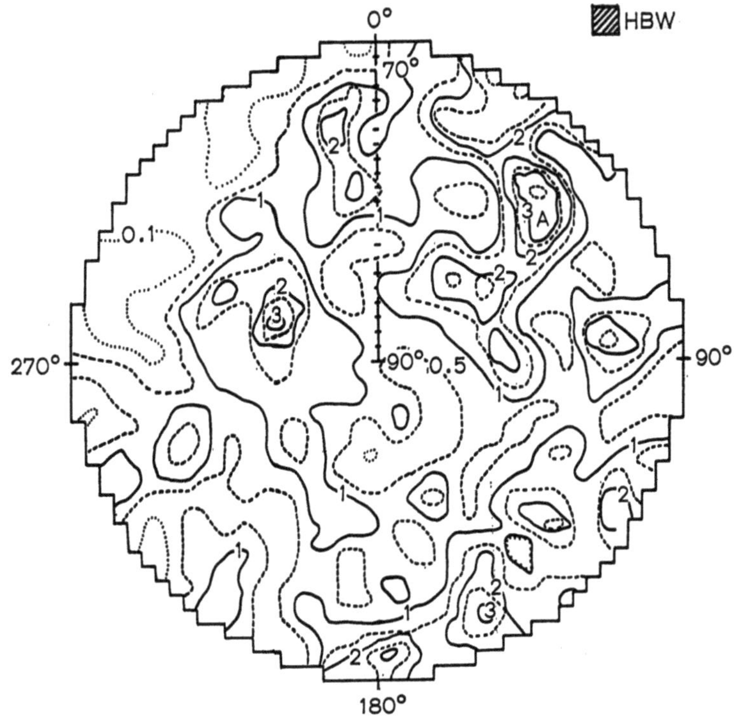

Fig. 8. Smoothed map for *ED* clusters 1° × 1° with [1]. *HBW* Half Beam Width of the smoothing function

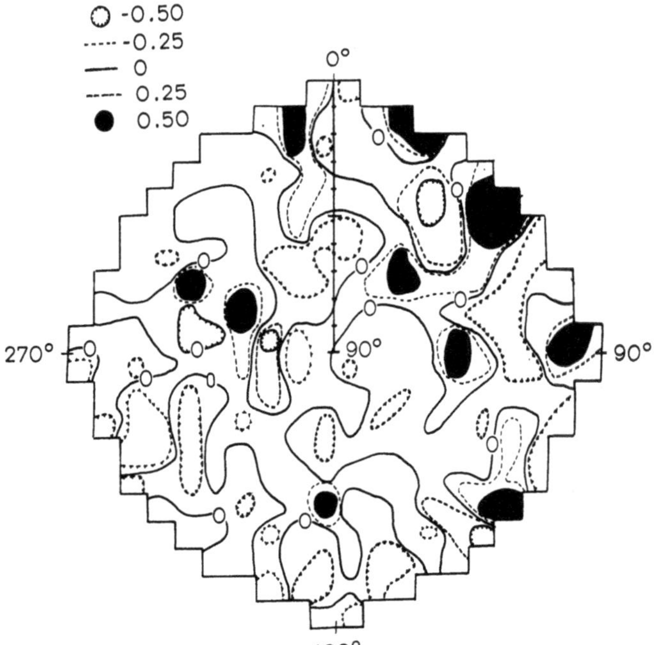

Fig. 9. Filtered map for *ED* clusters 2° × 2° with [0.5]–[1]

The smoothed map in Fig. 8 has been chosen for a more detailed treatment. All computations in Section IV have been repeated. Briefly, the obtained results are as follows:

(i) The difference between the two populations is confirmed, the boundary isopleth however is not 2, but 1.

Table 9. Clusters contained in a second-order cluster of galaxies (*A* on Fig. 8)

Clusters (Zwicky)	*l*	*b*	Type	Population (galaxies)	Area (square degrees)
1344.9+3143	55°9	+77°1	*mc*	111	0.070
1345.6+3101	53.0	77.1	*c*	102	0.039
1346.4+3124	54.2	76.9	*mc*	134	0.046
1347.4+3029	50.0	76.9	*c*	97	0.027
1348.7+3109	52.6	76.5	*c*	80	0.022
1350.6+3239	58.2	75.7	*mc*	97	0.027
1351.3+3156	55.3	75.7	*o*	69	0.070
1352.0+3212	56.2	75.5	*c*	103	0.033
1352.7+2935	45.6	75.8	*c*	100	0.054
1353.1+3010	48.0	75.7	*mc*	87	0.027
1353.2+3207	55.6	75.3	*o*	91	0.054
1353.6+3107	51.6	75.4	*mc*	149	0.132
1354.4+3234	57.0	74.9	*mc*	184	0.079
1355.6+3052	50.4	75.1	*mc*	274	0.145
1356.4+2912	43.9	75.1	*c*	47	0.004
1356.8+3006	47.3	74.9	*mc*	106	0.109
1356.8+3108	51.2	74.8	*c*	115	0.046
1357.3+2845	42.1	74.9	*c*	56	0.010
1357.5+2913	43.9	74.8	*mc*	57	0.013
1358.3+2847	42.2	74.7	*mc*	162	0.054
1358.9+2735	37.7	74.5	*mc*	75	0.027
1359.3+3147	53.2	74.1	*mc*	110	0.039
1359.4+2815	40°2	74°4	*c*	59	0.010
1401.2+2721	37.0	74.0	*mc*	125	0.054
1401.6+3237	55.8	73.5	*c*	91	0.033
1401.9+3218	54.6	73.5	*c*	70	0.022
1402.5+2743	38.4	73.7	*mc*	167	0.079
1402.6+3227	55.0	73.3	*c*	84	0.033
1403.1+3004	46.7	73.6	*c*	190	0.070
1403.8+2841	41.8	73.5	*mc*	55	0.017
1404.0+2650	35.4	73.3	*mc*	106	0.033
1404.0+2956	46.2	73.4	*c*	77	0.022
1404.5+2737	38.1	73.3	*mc*	279	0.280
1404.8+2802	39.6	73.2	*mc*	81	0.054
1404.8+2958	46.3	73.2	*c*	94	0.033
1405.8+2955	46.1	73.0	*c*	87	0.022
1405.8+3030	48.0	72.9	*mc*	74	0.061
1406.0+2828	41.1	73.0	*mc*	129	0.053
1406.5+3150	52.4	72.6	*c*	87	0.033
1406.8+2756	39.3	72.8	*c*	96	0.033
1407.0+3118	50.6	72.6	*c*	140	0.070
1407.4+2936	44.9	72.6	*mc*	137	0.121
1408.4+3006	46.5	72.4	*mc*	82	0.053
1410.7+2929	44.4	71.9	*c*	121	0.039

(ii) A fairly good approximation of the cluster population is given by means of a logloglognormal distribution.

(iii) The new determination of the cluster numbers among the second-order clusters is $n_s = 747$ clusters (from all 942 one).

(iv) After testing the hypothesis we find that its rejection is beyond any doubt.

(v) All *ED* clusters, contained in a second-order cluster for the area with a density $\geqq 2\mathfrak{D}$, noted in Fig. 8 by *A*, are listed in Table 9.

Smoothed and filtered maps for distance groups 5+6 from ABELL's catalog are constructed too (e.g. Fig. 10).

Some space smoothed maps have also been constructed.

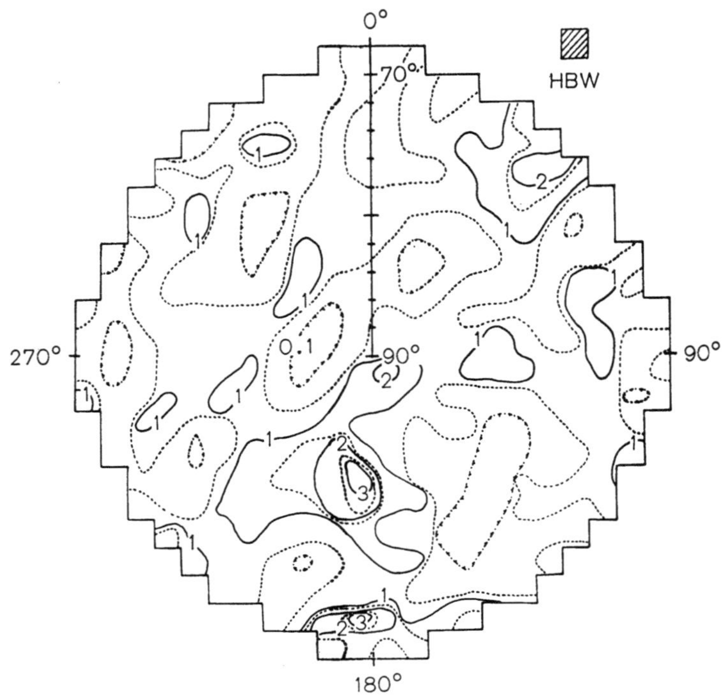

Fig. 10. Abell clusters 5+6, 2°×2°. Smoothed with [0.5]

VI. Correlation Analysis

Here we shall give the results of the correlation analysis of *ED* clusters, in an area 36°×36° centred in NGP. Two methods for the investigation of the correlation between the cluster centre number in spaced squares appear as possible.

Let us consider a square field with an area $w° \times w°$, divided into $n \times n$ equal squares, each of which contains d_{ik} numbers of cluster centres. Thus the observational material forms the following table

$$\begin{matrix} d_{11} & d_{12} & \dots & d_{1n} \\ d_{21} & d_{22} & \dots & d_{2n} \\ \cdot & \cdot & \cdot & \cdot \\ d_{n1} & d_{n2} & \dots & d_{nn} \end{matrix} \tag{1}$$

which is a two-dimensional discrete random field.

Method A. Let

$$\sum_{k=1}^{n} d_{ik} = d_{i.} \quad \text{and} \quad a_{ik} = d_{ik} - d_{i.}/n.$$

The values a_{ik} form a new table. If we examine all n consecutive lines in table (1) as random series, some information for the satistical structure of the field (1) may be obtained through the correlation functions

$$R(\tau, \Delta i) = \frac{\sum_{k=1}^{n-\tau} a_{i,k}\, a_{i+\Delta i,\, k+\tau}}{\left(\sum_{k=1}^{n-\tau} a_{i,k}^2 \sum_{k=1}^{n-\tau} a_{i+\Delta i,\, k+\tau}^2\right)^{\frac{1}{2}}} \tag{2}$$

where $\tau = 0, 1, 2, \dots$ is the lag in the index k, and $\Delta i = 0, \pm 1, \pm 2, \dots$ the lag in the index i. $R(\tau, \Delta i = 0)$ is the normalized autocorrelation function for a line, and $R(\tau, \Delta i \neq 0)$ is the normalized crosscorrelation function between two lines. The correlation matrix for a given τ has the form

$$\begin{matrix} R_{\tau 11} & R_{\tau 12} & \dots & R_{\tau 1n} \\ R_{\tau 21} & R_{\tau 22} & \dots & R_{\tau 2n} \\ \cdot & \cdot & \cdot & \cdot \\ R_{\tau n1} & R_{\tau n2} & \dots & R_{\tau nn}. \end{matrix} \tag{3}$$

The principal diagonal includes the values of the autocorrelation functions by lines and all other elements are values of crosscorrelation functions.

Let us define from (3) the means

$$\bar{R}_{\tau,0} = \sum_{l=m=1}^{n} R_{\tau lm}/n = +1,$$

$$\begin{aligned} \bar{R}_{\tau,+1} &= \sum_{l=1}^{n-1} R_{\tau,l,l+1}/(n-1), & \bar{R}_{\tau,-1} &= \sum_{m=1}^{n-1} R_{\tau,m+1,m}/(n-1), \\ \bar{R}_{\tau,+2} &= \sum_{l=1}^{n-2} R_{\tau,l,l+2}/(n-2), & \bar{R}_{\tau,-2} &= \sum_{m=1}^{n-2} R_{\tau,m+2,m}/(n-2), \\ &\dots & &\dots \\ \bar{R}_{\tau,+(n-1)} &= R_{\tau,1,n}, & \bar{R}_{\tau,-(n-1)} &= R_{\tau,n,1}. \end{aligned} \tag{4}$$

We form now the table of $\bar{R}_{\tau p}$, where p is the lag along lines and τ along columns.

Obviously for each random field (1) we have

$$\bar{R}_{0,0}=+1; \quad \bar{R}_{\tau,0}=\bar{R}_{-\tau,0}; \quad \bar{R}_{0,p}=\bar{R}_{0,-p}; \quad \bar{R}_{\tau,-p}=\bar{R}_{-\tau,p}; \quad \bar{R}_{\tau,p}=\bar{R}_{-\tau,-p} \tag{5}$$

and only for an isotropic random field

$$\begin{aligned} &\bar{R}_{0.0}=+1; \quad \bar{R}_{\tau,0}=\bar{R}_{-\tau,0}=\bar{R}_{0,p}=\bar{R}_{0,p} && \text{(for } \tau=p=1,2,\ldots); \\ &\bar{R}_{\tau,p}=\bar{R}_{-\tau,-p}=\bar{R}_{-\tau,p}=\bar{R}_{\tau,-p} && \text{(for all } \tau \text{ and } p). \end{aligned} \tag{6}$$

In the latter case we have

$$\bar{R}_{\tau,p}=\bar{R}\left(r=\sqrt{\tau^2+p^2}\right). \tag{7}$$

If we treat the quantities $R_{\tau lm}$ in (3), which are located on the s parallel above or below the diagonal, ($s=m-l$, or $s=l-m$) as random variables, then the variances $D[R]$ of $\bar{R}_{\tau p}$ for constant p can be determined. The limits of the 95% confidence intervals are

$$\bar{R}_{\tau p}-t_{0.05,n'-1}(D/n')^{\frac{1}{2}}, \qquad \bar{R}_{\tau p}+t_{0.05,n'-1}(D/n')^{\frac{1}{2}} \tag{8}$$

where n' is the correlation function number for the determination of $\bar{R}_{\tau p}$.

For an isotropic field, a quadrant

$$\bar{\bar{R}}_{\tau,p}=(\bar{R}_{\tau,p}+\bar{R}_{\tau,-p})/2 \quad \text{or even } \tfrac{1}{2} \text{ of a quadrant}$$

$$\bar{\bar{\bar{R}}}_{\tau,p}=(\bar{\bar{R}}_{\tau,p}+\bar{\bar{R}}_{p,\tau})/2 \quad \text{is sufficient}$$

$$\begin{array}{cccccccc} & & & \bar{\bar{R}}_{3,3}\cdots & & & \bar{\bar{R}}_{2,3}\cdots & \\ & & \bar{\bar{R}}_{2,2} & \bar{\bar{R}}_{2,3}\cdots & & \bar{\bar{R}}_{1,2} & \bar{\bar{R}}_{1,3}\cdots & \\ & \bar{\bar{R}}_{1,1} & \bar{\bar{R}}_{1,2} & \bar{\bar{R}}_{1,3}\cdots & \bar{\bar{R}}_{0,1} & \bar{\bar{R}}_{0,2} & \bar{\bar{R}}_{0,3}\cdots & \\ \bar{\bar{R}}_{0,0} & \bar{\bar{R}}_{0,1} & \bar{\bar{R}}_{0,2} & \bar{\bar{R}}_{0,3}\cdots & & & & \end{array} \tag{9}$$

Then $\bar{\bar{\bar{R}}}_{\tau,p}$ will be located at a distance $r=\sqrt{\tau^2+p^2}$ away from $\bar{R}_{0,0}=+1$ and the indices τ and p (they are equivalent). The values of (9) may be graphically plotted as a function of r only with the limits of 95% confidence intervals.

The above mentioned may be repeated for the columns (1) and hence the total results for both examinations will give a full information about the statistical structure of the field (1).

The final results for the area of 36° × 36°, centred in NGP, containing 651 *ED* clusters, give a correlation radius of about 3°. For a Hublle constant $H=$ 75 km sec^{-1} Mpc^{-1} this gives a characteristic length for second-order clusters of about 45 Mpc.

Method B. Let us have in (1) $\sum_{i,k=1}^{n} d_{ik}=d.$ and $c_{ik}=d_{ik}-d../n^2$. The correlation function for the random field c_{ik} may be computed for $\tau=r=1, \sqrt{2}, 2, \sqrt{5}, \sqrt{8}, 3, \sqrt{10}\ldots$. E.g. for $r=5$ we have

$$R(5)=\frac{\sum_{i=1}^{n}\sum_{k=1}^{n-5} c_{i,k}\,c_{i,k+5}+\sum_{i=1}^{n-5}\sum_{k=1}^{n} c_{i,k}\,c_{i+5,k}}{\left\{\left(\sum_{i=1}^{n}\sum_{k=1}^{n-5} c_{i,k}^2+\sum_{i=1}^{n-5}\sum_{k=1}^{n} c_{i,k}^2\right)\left(\sum_{i=1}^{n}\sum_{k=1}^{n-5} c_{i,k+5}^2+\sum_{i=1}^{n-5}\sum_{k=1}^{n} c_{i+5,k}^2\right)\right\}} \tag{10}$$

and the 95% confidence limits of $R(r)$ are $R(r)-1.96/n'$, $R(r)+1.96/n'$ where for $r=1$, $n'=2n(n-1)$; for $r=2$, $n'=2(n-1)^2$ etc.

For the same random field, $R(r)$ is plotted in Fig. 11, for squares $4°\times 4°$ – according to method A and method B. The dots are smoothed correlation functions with weights 0.25, 0.50, 0.25 (BLACKMAN and TUKEY, 1958).

Method B gives $r_{max}\approx 10°$, which corresponds to a characteristic scale length of about 150 Mpc and indicates the presence of third-order clusters of galaxies.

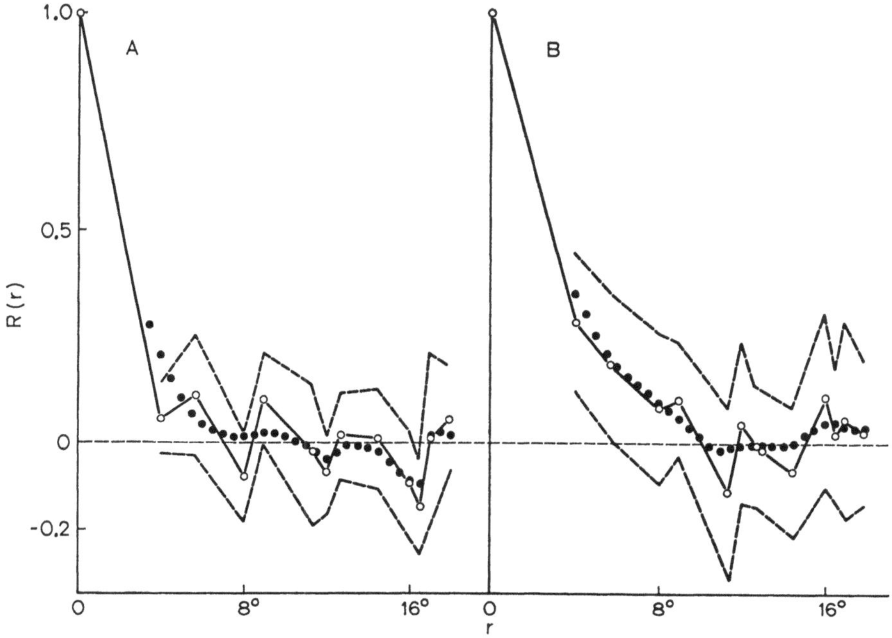

Fig. 11. Correlation functions *ED* clusters. Method A and B are applied

VII. Discussion

The random variable 𝔖 (area covered by clusters) defined in ZWICKY's catalog, may be approximated with a lognormal distribution. Therefore, the results for the apparent area, covered by the largest clusters in each category (ZWICKY and RUDNICKY, 1963; ZWICKY and BERGER, 1965; ZWICKY and KARPOWICZ, 1965, 1966; KARPOWICZ, 1971 a, 1971 b) must be reexamined.

The population of clusters of galaxies belonging to second-order clusters is larger than the population of the clusters belonging to the background (especially for *ED* clusters).

The digital smoothing and filtering of cluster centres offers a new approach to the investigation of clustering of galaxies and supports the above stated result.

The relative part of the clusters contained in second-order clusters is roughly 80% of all *ED* clusters. Consequently, second-order clustering is the rule.

It is known that correlation methods have been applied by many authors (NEYMANN, SCOTT and SHANE, 1953, 1954; SCOTT, SHANE and SWANSON, 1954; LIMBER, 1954; RUBIN, 1954; KALINKOV and RAKOVA, 1968; TOTSUJI and KIHARA,

1969) for first-order clustering, and KARACHENTSEV (1966), KIANG (1967), KIANG and SASLAV (1969) and also KALINKOV (1972a), KALINKOV and JANEVA (1972), for second-order clustering. It is important to note the difference between the two methods: the first method gives information about the microstructure of the random field (1), but the second-method gives information about the macrostructure of (1). The first method establishes the existence of second-order clustering and the second method indicates the existence of a third-order clustering. The results obtained by DE VAUCOULEURS (1971) for high-order clustering are the same as those of the present.

References

ABELL, G.O.: Astrophys. J. Suppl. **3**, 211 (1958).
ABELL, G.O.: Astron. J. **66**, 607 (1961).
ABELL, G.O.: Ann. Rev. Astron. Astrophys. **3**, 1 (1965).
ABELL, G.O., SELIGMAN, G.E.: Astron. J. **70**, 317 (1965).
ABELL, G.O., SELIGMAN, G.E.: Astron. J. **72**, 288 (1967).
BLACKMAN, R.B., TUKEY, J.W.: The Measurement of Power Specta, Dover, New York (1958).
BOLSHEV, L.N., SMIRNOV, N.V.: Statistical Tables, Moscow (1968).
DE VAUCOULEURS, G.: Astrophys. Letters 5, 219 (1970).
DE VAUCOULEURS, G.: Publ. Astron. Soc. Pacific 83, 113 (1971).
FULLERTONE, W., HOOVER, P.: Astrophys. J. **172**, 9 (1972).
KALINKOV, M.: Bull. Section Astron. **5**, 165 (1972a).
KALINKOV, M.: Bull. Section Astron. **6**, 71 (1972b).
KALINKOV, M.: Compt. Rend. Acad. Bulg. Sci. (in press) (1972c).
KALINKOV, M., JANEVA, N.: Compt. Rend. Acad. Bulg. Sci. (in press) (1972).
KALINKOV, M., RAKOVA, S.: Compt. Rend. Acad. Bulg. Sci. **21**, 99 (1968).
KALINKOV, M., SPASOVA, N.: Bull. Section Astron. **5**, 149 (1972).
KARACHENTSEV, I.D.: Astrofizika **2**, 307 (English translation in Astrophysics **2**, 159) (1966).
KARPOWICZ, M.: Z. Astrophys. **66**, 301 (1967a).
KARPOWICZ, M.: Z. Astrophys. **67**, 139 (1967b).
KARPOWICZ, M.: Acta Astron. **20**, 391 (1970a).
KARPOWICZ, M.: Acta Astron. **20**, 395 (1970b).
KARPOWICZ, M.: Acta Astron. **21**, 103 (1971a).
KARPOWICZ, M.: Acta Astron. **21**, 115 (1971b).
KARPOWICZ, M.: Acta Astron. **21**, 391 (1971c).
KENDALL, M.G., STUART, A.: The Advanced Theory of Statistics, Griffin, London, Vol. **2** (1961).
KIANG, T.: Monthly Notices Roy. Astron. Soc. **135**, 1 (1967).
KIANG, T., SASLAW, W.C.: Monthly Notices Roy. Astron. Soc. **143**, 129 (1969).
KULLBACK, S.: Information Theory and Statistics, Willey, New York (1958).
LEHMANN, E.L.: Testing Statistical Hypotheses, Wiley, New York (1959).
LIMBER, D.N.: Astrophys. J. **119**, 655 (1954).
NEYMANN, J., SCOTT, E.L., SHANE, C.D.: Astrophys. J. **117**, 92 (1953).
NEYMANN, J., SCOTT, E.L., SHANE, C.D.: Astrophys. J. Suppl. Ser. **1**, 269 (1954).
OWEN, D.B.: Handbook of Statistical Tables, Addison-Wesley, Palo Alto-London (1962).
RAO, C.R.: Linear Statistical Inference and its Applications, Wiley, New York (1965).
RUBIN, V.C.: Proc. Nat. Acad. Sci. **40**, 541 (1954).
SARHAN, A.E., GREENBERG, B.G.: Contributions to Order Statistics, Wiley, New York (1962).
SCOTT, E.L., SHANE, C.D., SWANSON, M.D.: Astrophys. J. **119**, 91 (1954).
SHANE, C.D., WIRTANEN, C.A.: Publ. Lick Obs. **22**, part I (1967).
SMIRNOV, N.V.: Compt. Rend. Acad. Sci. Paris **202**, 449 (1936).
TOTSUJI, H., KIHARA, H.: Publ. Astron. Soc. Japan **21**, 221 (1969).
VORONTSOV-VELYAMINOV, B.A.: Extra-Galactic Astronomy, Moscow, p. 339 (1972).
YU, J.T., PEEBLES, P.G.E.: Astrophys. J. **158**, 103 (1969).

ZWICKY, F.: Morphological Astronomy, Springer Verlag, Berlin (1957).
ZWICKY, F., BERGER, J.: Astrophys. J. **141**, 34 (1965).
ZWICKY, F., HERZOG, E.: Catalogue of Galaxies and of Clusters of Galaxies, California Institute of Technology, Pasadena, Vol. 2 (1963).
ZWICKY, F., HERZOG, E.: Catalogue of Galaxies and of Clusters of Galaxies, California Institute of Technology, Pasadena, Vol. 3 (1966).
ZWICKY, F., HERZOG, E., WILD, P.: Catalogue of Galaxies and of Clusters of Galaxies, California Institute of Technology, Pasadena, Vol. 1 (1961).
ZWICKY, F., KARPOWICZ, M.: Astrophys. J. **142**, 625 (1965).
ZWICKY, F., KARPOWICZ, M.: Astrophys. J. **146**, 43 (1966).
ZWICKY, F., RUDNICKI, K.: Astrophys. J. **137**, 707 (1963).
ZWICKY, F., RUDNICKI, K.: Z. Astrophys. **64**, 245 (1966).

Discussion

LYNDEN-BELL, D.:
Is there any apparent correlation length caused by the fact that Palomar plates are taken to slightly different limiting magnitudes? One might expect 6° to come out in such an analysis.

KALINKOV, M.:
For squares $1° \times 1°$, the autocorrelation function is slightly larger for the values of r between 4° and 5°. But for $2° \times 2°$ and $4° \times 4°$ (Fig. 11A) this effect is non-significant, since $R(r > 4°) \approx 0$. As for Method B (Fig. 11B) this effect vanishes. Consequently this effect does not change the results (naturally, only for the selected area – around NGP).

The Stability of Stellar Masses in General Relativity

(Invited Lecture)

By S. CHANDRASEKHAR
University of Chicago, Chicago, Illinois, USA

The results of the stability analysis of stellar masses under various circumstances, both in the frameworks of Newtonian theory and of general relativity, are contrasted.

The following problems are considered:

1) The stability with respect to radial modes of oscillation of spherical stars.

2) The effect of a small uniform rotation on the modes of radial pulsation of a spherical star and on its stability.

3) The criterion for the onset of stability of rapidly rotating stars for axisymmetric perturbations.

4) Secular instability in rotating systems induced by the presence of dissipative mechanisms.

On the Newtonian theory, the results of the analysis of the foregoing problems are as follows:

1) Stability with respect to radial pulsations depends on an average value of the adiabatic exponent γ which relates the fractional (Lagrangian) change, $\Delta p/p$, in the pressure to the corresponding fractional change, $\Delta\rho/\rho$, in the density as we follow a fluid element during its motion:

$$\frac{\Delta p}{p}=\gamma\frac{\Delta\rho}{\rho}. \tag{1}$$

Thus, instability will occur if

$$\bar{\gamma}=\frac{\int_0^M \gamma\, p\, dM(r)}{\int_0^M p\, dM(r)}<\frac{4}{3} \qquad (dM(r)=4\pi\,\rho\, r^2\, dr); \tag{2}$$

and the e-folding time of the instability is of the same order as the period of pulsation ($\sim[I/|\omega|]^{\frac{1}{2}}$ where I denotes the moment of inertia and ω the gravitational potential energy of the star); and for this reason the instability is said to be "dynamical."

2) The effect of a slow uniform rotation (Ω) is to replace the condition (2) by

$$\bar{\gamma}-\frac{4}{3}+\text{constant}\,\frac{\Omega^2}{\pi G\bar{\rho}}<0, \tag{3}$$

where $\bar{\rho}$ denotes the mean density of the star. More generally, an exact expression can be given for the fundamental frequency (σ) of axisymmetric pulsation of a slowly rotating star in terms of the frequency (σ_0) and the associated proper amplitude (ξ) of radial pulsation of the non-rotating star and the ($l=0$)-distortion of the star caused by rotation. Thus

$$\sigma^2 = \sigma_0^2 + \Omega^2 \sigma_1^2 + O(\Omega^4), \tag{4}$$

where, as stated, σ_1^2 depends only on the proper solution ξ belonging to σ_0 and the spherically symmetric part of the rotational distortion of the star.

3) When the rotation can no longer be considered as slow, the global instability that occurs for radial perturbations can be extended, via the criterion (3), into the domain of finite Ω by considering modes of oscillation that maintain the axisymmetry of the configuration. The instability for these modes sets in via $\sigma^2 = 0$, i.e. via a mode of neutral deformation and a marginal state of neutral stability.

4) With respect to the possible onset of secular instability of rapidly rotating stars, the only fully understood case concerns that which occurs along the Maclaurin sequence of rotating homogeneous masses. Along this sequence secular instability by viscous dissipation occurs at the point where the Jacobian sequence bifurcates. In the absence of viscous dissipation, no instability occurs at this point; but if viscous dissipation is present, then instability occurs (with an e-folding time depending on the magnitude of the prevailing viscosity) by the neutral mode which transforms the Maclaurin spheroid into a Jacobian ellipsoid at the point of bifurcation.

In the framework of general relativity, the solutions of the same problems are as follows:

1) The theory of radial oscillations of spherical non-rotating stars in general relativity parallels the Newtonian theory very closely and is in all essentials as simple. The criterion for instability is again an inequality for an average value of γ, though it cannot in general be expressed as simply as in Eq. (2). By actual numerical solutions of the pulsation equation, it has been shown that instability can occur under circumstances in which the Newtonian theory predicts stability. This *relativistic instability* is of particular importance when $\bar{\gamma}$ is close to, but greater than, $\frac{4}{3}$ as will be the case when radiation pressure is dominant (as in massive stars) or when the constituent particles move with velocities comparable to the velocity of light, c (as in degenerate configurations near their limiting mass). Thus, if γ should be a constant through the star and exceed $\frac{4}{3}$ by a small amount, then it follows from the theory that instability will occur when the radius of the star

$$R < \frac{2GM}{c^2} \frac{K}{\gamma - \frac{4}{3}}, \tag{5}$$

where K is a constant which, while it depends on the structure of the star, is generally of order unity. This last formula shows very clearly that effects arising from general relativity can cause instability even under circumstances when its effects on the structure of the equilibrium configuration are entirely negligible.

The relativistic instability plays an important role in the evolution of supermassive objects in the range 10^4–10^8 solar masses. It is also responsible for setting a lower limit to the periods of radial oscillation of degenerate configuration; this is an important consideration in ruling out white dwarfs as an explanation for pulsars.

2) The stability of uniformly rotating stars in general relativity also parallels the Newtonian theory.

For the case of slow rotation, a formula, exactly analogous to Eq. (4) that obtains in the Newtonian theory, has been derived; and σ_1^2 depends, as in the Newtonian theory, only on the proper solution belonging to the fundamental mode of radial pulsation of the non-rotating star and the functions describing the $(l=0)$-distortion caused by the rotation. The reason why this problem of the axisymmetric pulsation of a slowly rotating star allows an exact solution is that under the circumstances of the problem (slow rotation!) we do not need to include any terms arising from the emission of gravitational radiation: they are all of order higher than any that need to be retained for a determination of σ^2.

3) On the assumption that instability sets in (as in the Newtonian theory) via a mode of neutral deformation, the criterion for its onset can be written down in terms of a general variational expression for σ^2 (since no complication from the existence of gravitational radiation arises). This last criterion for the existence of a neutral mode of deformation can be readily specialized to apply to a stationary axisymmetric vacuum-solution external to a black hole by simply suppressing all terms that occur with ε (the energy density) and/or p (the pressure) as factors. In this manner, a necessary and sufficient condition can be established that the initial stationary solution will allow a non-trivial neutral deformation. An examination of this condition shows that *no such deformation is possible;* and this is essentially the content of CARTER'S theorem which provides the principal basis for the current belief that the Kerr metric describes the black hole that will result from the collapse of a rotating massive star. (The Kerr metric is an exact axisymmetric solution of EINSTEIN'S vacuum field-equations that is asymptotically flat. It has two parameters, the mass M and the angular momentum J $(=aM)$ and it reduces to the Schwarzschild metric when $a=0$; and it has, like the Schwarzschild metric, an event horizon.)

(The results summarized in the two foregoing paragraphs have been obtained in some recent work carried out by the author jointly with his student, JOHN FRIEDMAN.)

4) In general relativity dissipation of both energy and angular momentum during the oscillations of rapidly rotating stars is built into the theory. Accordingly, the possibility of secular instability arises, caused by this dissipation and resulting from the existence of radiation-reaction terms in the hydrodynamical equations. In the framework of the $2\frac{1}{2}$-post-Newtonian approximation, which provides these radiation-reaction terms, the secular instability of the Maclaurin spheroid has been analyzed. And it has been shown that while radiation-reaction causes instability for the same value of the eccentricity (for which viscosity causes instability in the Newtonian framework) the mode by which the instability sets in is not the one which leads to the Jacobian sequence: instead, it is the one which leads to the Dedekind sequence. (The Dedekind ellipsoid, in contrast to

the Jacobi ellipsoid, is stationary in the inertial frame and derives its ellipsoidal shape from internal motions of uniform vorticity.)

In view of the results stated under 4, the question whether along the Kerr sequence secular instability might set in becomes a matter of great interest. In some unpublished work (together with JOHN FRIEDMAN) a criterion for the occurrence of a Dedekind-like point of bifurcation has been established. If by the application of this criterion it can be shown that there is a point along the Kerr sequence where secular instability sets in, then, the astrophysical consequences will be immense in view of the existence of mechanisms (such as those which PENROSE first described) by which energy of far larger amounts than by nuclear processes can be extracted from the rotational energy of the black hole.

Remarks on Schwarzschild Black Holes

By G.C. McVittie
Canterbury, Kent, U.K.

With 1 Figure

Abstract

The Kruskal and Künzle coordinates for the Schwarzschild space-time are obtainable by a direct solution of Einstein's equations. In addition to the length-equivalent, m, of the mass of the central body, the coefficients of the metric involve three other constants of integration. Various physical identifications of these constants are possible. Circular orbits for material particles and radial motions of photons are worked out in terms of Kruskal coordinates when the central body has a physical radius less than $2m$. An external observer will recognize the region interior to radius $2m$ as being singular, even though this property will not be apparent by inspection of the coefficients of the metric.

I. Black Holes and Coordinate Systems

The simplest type of black hole is the Schwarzschild black hole which may be described thus. A spherically symmetric body of mass M is surrounded by a vacuous region in which the gravitational effect of any energy that may be present is regarded as neglibly small compared with the gravitational field of the body itself. The energy-tensor is therefore zero in the vacuous region. The field in the region is described by the well-known solution of Einstein's equations of general relativity

$$\begin{aligned} ds^2 &= (1-2m/r)\,dt^2 - \frac{1}{c^2}\frac{dr^2}{1-2m/r} - \frac{r^2}{c^2}\,d\Omega^2, \\ d\Omega^2 &= d\theta^2 + \sin^2\theta\, d\phi^2, \end{aligned} \tag{1}$$

where $r=0$ is the centre of the body. The coordinates (t, r, θ, ϕ) are the statical curvature coordinates. The process of solution of Einstein's equations throws up, in this coordinate system, two constants of integration only. One is a scale constant for the time t and is relatively trivial, whereas the other is important and is $2m=2GM/c^2$. This second constant expresses the mass of the central body in terms of the length-units used for r and for $c\,t$.

In a normal situation the physical radius r_b of the central body is far larger than $2m$ and the field (1) is valid for $r>r_b$ only. This follows, of course, from the fact that (1) refers to a region in which the energy-tensor is zero, whereas it is clearly not zero in the material of the body. Thus $r=2m$ is a surface deep inside the body and indeed is merely a mathematical way of stating that, in the external vacuous region, the field is produced by the body. However it is conceivable, if the physics of the material involved is ignored, that the body could be in such a high state of compression that $r_b \leqq 2m$. This situation is the Schwarzschild black hole. It is illustrated in Fig. 1 for the case when $r_b < 2m$. In region I the energy-tensor does not vanish and therefore (1) does not apply. In region II $(2m \geqq r > r_b)$ there is also no material and therefore (1) might be supposed to apply. But the coefficients of dt^2 and of dr^2 have changed sign as if r/c were now a kind of time and $c\,t$ a space coordinate. NOVIKOV (1961) has used this notion to devise a form of the metric which might apply in region II. In region III, where $r>2m$ the metric (1) applies without difficulty. The boundary between regions II and III is characterized by a zero value of the coefficient of dt^2, and an infinite value of that of dr^2, which constitutes the singularity. However, it is often said that this singularity may be spurious and can be eliminated by a change of coordinate system. Among many such alternative coordinate systems, those of KRUSKAL (1960) and of KÜNZLE (1967) have received much attention from this point of view. The methods used for finding these systems by the two authors will be discussed later. For the moment we point out that they can be obtained by a direct solution of EINSTEIN's equations[1] for the vacuum by starting from a metric of the form

$$ds^2 = e^{2\lambda(z)}\left(dT^2 - \frac{1}{c^2}\,du^2\right) - \frac{r^2(z)}{c^2}\,d\Omega^2, \tag{2}$$

where u is the radial coordinate, T the associated time coordinate and λ and r are functions of the physically dimensionless variable

$$z = \frac{u^2 - c^2\,T^2}{l^2}. \tag{3}$$

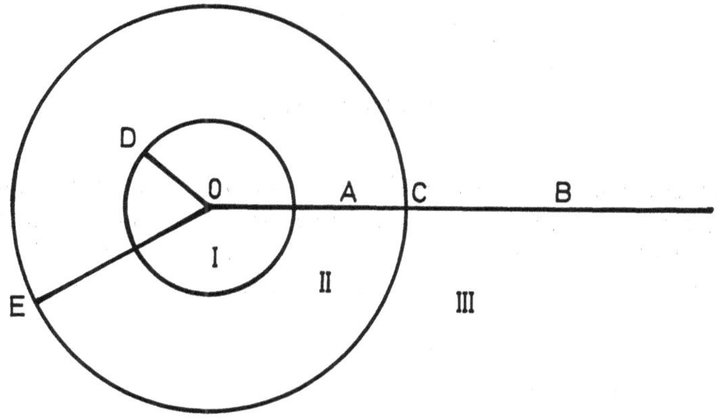

Fig. 1. Schematic diagram of a black hole. The central body occupies region I, of radius $OD = r_b < 2m$, and region II has an outer radius of $OE = OC = 2m$

[1] The equations to be solved are Eqs. (2.10) of CAHILL and MCVITTIE (1970) with $\gamma = \alpha = \lambda$, $x^4 = T$, $x^1 = u$ and $G^\mu_\nu = 0$. The Eq. (3.1), with $R^3_{232} = 2m/r(u, T)$ must also be used in order to find one of the constants of integration.

Here l is an unspecified constant length. The solution of Einstein's equations gives

$$\left\{\left(\frac{r}{2m}-1\right)\frac{2m}{L_2}\right\}^{2m/L_1} e^{r/L_1}=\frac{u^2-c^2\,T^2}{l^2},$$
$$e^{2\lambda}=4\,\frac{L_1^2}{l^2}\left(\frac{L_2}{r}\right)\left\{\left(\frac{r}{2m}-1\right)\frac{2m}{L_2}\right\}^{1-2m/L_1} e^{-r/L_1}, \qquad (4)$$

where $2m$, L_1, L_2 are the essential constants of integration produced by the solution operations. The coefficients of the metric (2) are thus obtained as functions of u and T. If however it is required to convert the metric to the statical curvature coordinate form (1), then the function t given by

$$\tan\mathrm{h}(c\,t/2L_1)=c\,T/u$$

must be used in conjunction with the first of Eqs. (4). But this function is in no way involved in the process of deriving (4) from Einstein's equations. In any case it is clear that t is defined only for $c\,T \leqq u$, which means that $r \geqq 2m$, whereas no such restriction on the values of T and u is implied by the formulae (4).

Künzle's (1967) method of deriving (4) consisted of a topological analysis apparently equivalent to the statement that the metric (2) should be invariant in form under Lorentz transformations between u and T. This is seen by inspection to be true of (4). The physical meaning of Lorentz transformations between a time, and a radial, coordinate is however somewhat obscure.

II. Identification of Constants

Clearly if $L_1 \neq 2\mathrm{m}$, the coefficient $e^{2\lambda}$ of the metric is either zero or infinite at $r=2m$ and the singularity persists. But it apparently disappears if, and only if, the constant L_1 is arbitrarily equated to $2m$. The coefficients of the metric for such (Kruskal) coordinates are therefore given by

$$\left(\frac{r}{2m}-1\right)e^{r/2m}=\frac{L_2}{2m}\left(\frac{u^2-c^2\,T^2}{l^2}\right),$$
$$e^{2\lambda}=\frac{8m\,L_2}{l^2}\left(\frac{2m}{r}\right)e^{-r/2m}. \qquad (5)$$

This result was originally derived by Kruskal (1960) by finding the transformation equations from (t, r) to (T, u). He also assumed that $L_2=2m$, $c=1$ and that u, $c\,T$ and m were multiples of the (unknown) length l. In fact Kruskal's coordinates may be written as $\bar{v}$, $\bar{u}$, with the associated constant $\bar{m}$, given by

$$\bar{v}=c\,T/l, \qquad \bar{u}=u/l, \qquad \bar{m}=m/l.$$

The metric (2) then becomes

$$ds^2=\frac{32\bar{m}^3}{r}\,e^{-r/2\bar{m}}\,(d\bar{v}^2-d\bar{u}^2)-r^2\,d\Omega^2,$$

with $r(\bar{v},\bar{u})$ given by the solution of the equation

$$e^{r/2\bar{m}}\,(r/2\bar{m}-1)=\bar{u}^2-\bar{v}^2.$$

Thus the scale factor l is concealed in the coordinate system. The choice made by KRUSKAL for the identification of L_2 and the non-identification of l is not mandatory. A method analogous to the use of the A.U. in celestial mechanics will now be developed. To do this the metric (2) with coefficients (5) is transformed to the physically dimensionless variables

$$\sigma=\frac{c}{l}s, \quad \eta=\frac{cT}{l}, \quad \xi=\frac{u}{l}, \quad x=\frac{r}{2m}. \tag{6}$$

Then (2) becomes

$$d\sigma^2=\frac{k_2}{x\,e^x}(d\eta^2-d\xi^2)-k_3^2\,x^2\,d\Omega^2, \tag{7}$$

where x is the function of $(\xi^2-\eta^2)$ given by

$$k_1\,e^x(x-1)=\xi^2-\eta^2, \tag{8}$$

and the (dimensionless) constants are

$$k_1=\frac{2m}{L_2}, \quad k_2=\frac{8m\,L_2}{l^2}, \quad 4k_3^2=k_1\,k_2=4\left(\frac{2m}{l}\right)^2. \tag{9}$$

Consider an observer in the space-time (7) whose orbit is a time-like geodesic in the plane $\phi=constant$. Such a geodesic is defined by well-known equations (MCVITTIE, 1965a) which reduce to

$$\frac{d\theta}{d\sigma}=\frac{h}{x^2}, \quad (h=constant), \tag{10}$$

$$\frac{k_2}{x\,e^x}\left\{\left(\frac{d\eta}{d\sigma}\right)^2-\left(\frac{d\xi}{d\sigma}\right)^2\right\}=1+\frac{k_3^2\,h^2}{x^2}, \tag{11}$$

and, if y stands for *either* η or ξ,

$$\frac{d}{d\sigma}\left(\frac{1}{x\,e^x}\frac{dy}{d\sigma}\right)-\frac{1}{4}\frac{1}{x\,e^x}\left(\frac{1}{k_3^2}\frac{x+1}{x}+\frac{h^2}{x^2}\frac{x+3}{x}\right)y=0. \tag{12}$$

Suppose that the orbit is "circular", which means that along it

$$x(\eta,\xi)=X=R/2m=constant. \tag{13}$$

Then an appropriate solution of (12) is

$$\eta=A\sinh K\sigma, \quad \xi=A\cosh K\sigma, \tag{14}$$

where A is a constant of integration and

$$K^2=\frac{1}{4}\left\{\frac{1}{k_3^2}\frac{X+1}{X}+\frac{h^2}{X^2}\frac{X+3}{X}\right\}, \tag{15}$$

so that K^2 is a positive constant. But the last two equations must yield quantities that also satisfy (8) and (11) with $x=X$. These give, with (9) also,

$$\begin{aligned} A^2&=k_1\,e^X(X-1)=\frac{1}{L_2}\,e^{R/2m}(R-2m),\\ K^2&=\frac{1}{2k_3^2}\frac{X}{2X-3}=\frac{l^2}{8m^2}\frac{R}{R-3m},\\ h^2&=\frac{l^2}{16m^3}\frac{R^2}{R-3m}. \end{aligned} \tag{16}$$

Hence K and h are real and finite only if $R>3m$, as can also be proved otherwise (McVittie, 1965b). It is also known (Whittaker, 1927) that circular orbits with $R>6m$ are stable and that there exist unstable circular orbits if R lies in the range $6m>R>4m$. The observer is now free to make the two independent assumptions that

$$l=R, \quad \xi_{\eta=0}=1, \quad \text{or} \quad u_{T=0}=R,$$

which produce $A=1$ in (14). Hence by (16)

$$L_2=(R-2m)\, e^{R/2m},$$

and this identifies L_2 in terms of R and $2m$. The first assumption means that the observer is following the practice of celestial mechanics in using the "radius" of his orbit as the unit of length. The second assumption fixes the scale of u so that, at $T=0$, the orbit in the (T, u) system shall have $u=R$ also.

III. Radial Motion of a Photon

The next point to be considered is the extent to which the use of the Kruskal coordinates employed in (5) "abolishes" the singularity at $r=2m$. When the coefficients of the metric are alone considered it may be said with Graves (1971) that "nothing very unusual happens at this point [$r=2m$]... but at $r=0$... we encounter the unavoidable singularity". Thus the singularity appears to have been pushed back to $r=0$, where it would also occur in Newtonian gravitational theory. Quite a different conclusion emerges as soon as the physical predictions deduced from the metric with coefficients (5) are investigated. The Eqs. (16) for the circular orbit of a free material particle have already indicated that the approach to $r=2m$ produces peculiar results, which certainly do not follow from Newtonian theory. Further examples of this kind arise from a study of the radial motions of photons, whose governing equations are simpler to deal with than are those for material particles. Consider a photon that starts moving radially from the point A of Fig. 1, this point being in region II. The photon's motion is governed by a null geodesic with $\theta=constant$ and $\phi=constant$, its path being represented by the line $OACB$ of Fig. 1. The remaining equations of the null geodesic are easily derived from the textbook results (McVittie, 1965c). They are, for the metric in the form (7),

$$\frac{d}{d\mu}\left(\frac{1}{x\, e^{x}}\frac{dy}{d\mu}\right)=0, \quad \left(\frac{d\eta}{d\mu}\right)^2-\left(\frac{d\xi}{d\mu}\right)^2=0, \tag{17}$$

where μ is a non-zero parameter varying along the null geodesic and y stands for either η or ξ, as before.

(i) Consider first an outward motion of the photon which starts at A, where $\xi=\xi_A$, at time η_A. By an outward motion will be meant one in which both x and ξ increase as η increases. The orbit of the photon is therefore, from Eqs. (17),

$$\xi-\xi_A=\eta-\eta_A. \tag{18}$$

Suppose that at time η_P ($>\eta_A$) the photon has reached some point $P(\xi_P)$ lying between A and C ($x=1$). Then the overall restriction (8) applies at both A and P points for which both x_A and x_P are less than unity and $x_P > x_A$. Thus

$$\begin{aligned} \eta_A^2 - \xi_A^2 &= k_1 f(x_A) > 0, \\ \eta_P^2 - \xi_P^2 &= k_1 f(x_P) > 0, \\ f(x) &= (1-x)\, e^x. \end{aligned} \tag{19}$$

Hence $\eta_A > \xi_A$, $\eta_P > \xi_P$. At this point appeal may be made to EINSTEIN's notion that all coordinate systems should be qualitatively as good as one another. A property of radial coordinates, whether of the r of (1) or the u of (2), is that they are necessarily positive[2]. Thus η_A cannot be zero, which immediately alerts the investigator to a peculiar feature of region II. Again (18) yields $\eta_P = \xi_P + (\eta_A - \xi_A)$ which may be used to eliminate η_P from (19). The result is

$$2(\xi_P - \xi_A)(\eta_A - \xi_A) = k_1 \{f(x_P) - f(x_A)\} < 0,$$

the inequality following from the fact that $f(x)$ is a decreasing function of x. Hence $\xi_P < \xi_A$ which contradicts the hypothesis on which (18) was based, namely, that ξ increases as the photon moves from A to C. The photon therefore cannot travel outwards from A.

(ii) Consider next an inward motion of the photon from A, the point P now lying between A and the boundary of region I. Eq. (18) is replaced by

$$\xi - \xi_A = \eta_A - \eta,$$

so that $\eta_P = (\eta_A + \xi_A) - \xi_P$. The Eqs. (19) are still valid except that now $x_P < x_A$. If η_P is eliminated as before there comes

$$2(\eta_A + \xi_A)(\xi_A - \xi_P) = k_1 \{f(x_P) - f(x_A)\} > 0.$$

Hence $\xi_A > \xi_P$ as it should be. Thus the photon can move inwards from A, a conclusion still valid if A is identical with C. At C, $x=1$, $f(1)=0$ and $\xi_C = \eta_C$. Hence a photon starting from C can move inwards in accordance with the relations

$$\begin{aligned} \xi_P - \xi_C &= \xi_C - \eta_P, \\ 4\xi_C(\xi_C - \xi_P) &= k_1 f(x_P). \end{aligned} \tag{20}$$

(iii) Suppose next that the photon starts from a point B beyond $r=2m$, so that $x_B > 1$, and moves outwards. Its orbit is

$$\xi - \xi_B = \eta - \eta_B.$$

When the photon has reached a point P, to the right of B in Fig. 1, for which therefore $x_P > x_B > 1$, the condition (8) gives

$$\begin{aligned} \xi_B^2 - \eta_B^2 &= k_1 g(x_B) > 0, \\ \xi_P^2 - \eta_P^2 &= k_1 g(x_P) > 0, \\ g(x) &= (x-1)\, e^x. \end{aligned} \tag{21}$$

Hence $\xi_B > \eta_B$, $\xi_P > \eta_P$ and it is now possible to assign the value zero to η_B. When this is done and ξ_P is eliminated there comes

$$\eta_P = \frac{k_1}{2\xi_B} \{g(x_P) - g(x_B)\} > 0,$$

[2] This conclusion is at variance with that normally made for Kruskal coordinates (ANDERSON, 1967).

since $x_P > x_B$ and $g(x)$ is an increasing function of x. As is to be expected, the photon can move outwards from B. In particular, suppose that B lies close to C so that $x_B = 1+\varepsilon$, $(\varepsilon > 0)$, where ε is small compared with unity. Then

$$\xi_B = \{k_1\, g(1+\varepsilon)\}^{\frac{1}{2}} = (k_1\, e\, \varepsilon)^{\frac{1}{2}}$$

and so

$$\eta_P = \frac{1}{2}\left(\frac{k_1}{e}\right)^{\frac{1}{2}} \frac{1}{\varepsilon^{\frac{1}{2}}} \{g(x_P) - (k_1\, e\, \varepsilon)^{\frac{1}{2}}\}.$$

Thus the η-time to reach an external point P increases indefinitely as ε tends to zero, a result similar to that found for the lapse of t-time when the (t, r) coordinates of (1) are employed (Hilton, 1964).

(iv) Finally consider the inward motion of a photon which starts at time $\eta_B = 0$ from the point B of Fig. 1. Its orbit is

$$\xi_B - \xi = \eta, \tag{22}$$

and the Eqs. (21) still hold but with $\eta_B = 0$ and $x_B > x_P$. The motion to the point C is of particular interest. The condition (8) gives

$$\xi_B^2 = k_1\, g(x_B), \qquad \xi_C^2 - \eta_C^2 = 0.$$

Since, by (22), $\xi_B - \xi_C = \eta_C$, it follows from the last two equations that

$$\eta_C = \tfrac{1}{2}\{k_1\, g(x_B)\}^{\frac{1}{2}}.$$

The photon starting at B at time $\eta_B = 0$ reaches C in a finite time $\eta_C > 0$. It could then be regarded as a photon "starting from C" and moving inwards into region II, as described in (ii) above. This result appears to be at variance with Rosen's (1970) conclusion that the photon would be reflected at the surface $r = 2m$. It is true that Rosen worked in terms of the t-time of (1) and with a radial coordinate u defined in terms of r by $u^2 = 8m(r - 2m)$.

IV. Conclusions

These calculations in terms of Kruskal coordinates may therefore be said to show that the region for which $r \leqq 2m$ is a very peculiar one indeed from the point of view of an observer outside the Schwarzschild black hole. As has already been indicated, circular orbits are impossible if their radii are less than $3m$ and that they are unstable if the radii lie between $6m$ and $4m$. Moreover photons cannot emerge from the surface $r = 2m$, or from its interior, though when they fall radially from outside onto the surface they can apparently penetrate it and proceed towards region I. Indeed Kruskal coordinates may be said to conceal the peculiarity of region II, a peculiarity at once laid bare by the study of the motions of material particles or of photons. These results do not however touch the basic problem of the black hole, which a continuum theory like general relativity cannot solve. The solution to the problem would entail a physical theory of the state of matter in region I of Fig. 1 where the density could be far in excess of the nuclear density of some $10^{14}\ \mathrm{g\,cm^{-3}}$.

References

ANDERSON, J.L.: Principles of Relativity Physics, Academic Press, New York, pp. 389–392 (1967).
CAHILL, M.E., MCVITTIE, G.C.: J. Math. Phys. **11**, 1382 (1970).
GRAVES, J.C.: The Conceptual Foundations of Relativity Theory, Mass. Inst. of Tech. Press, Cambridge, Mass., p. 229 (1971).
HILTON, E.: Proc. Roy. Soc. London **A 283**, 491 (1964).
KRUSKAL, M.D.: Phys. Rev. **119**, 1743 (1960).
KÜNZLE, H.P.: Proc. Roy. Soc. London **A 297**, 244 (1967).
MCVITTIE, G.C.: General Relativity and Cosmology, Chapman and Hall, London. Equations (2.807) and (2.808) (1965a).
MCVITTIE, G.C.: Ibid, p. 89 (1965b).
MCVITTIE, G.C.: Ibid, Equations (2.309) and (2.809) (1965c)
NOVIKOV, I.D.: Soviet Astron. A. J. **5**, 423 (1961).
ROSEN, N.: In M. CARMELI, S.I. FICKLER and L. WITTEN (eds.), Relativity, Plenum Press, New York, p. 229 (1970).
WHITTAKER, E.T.: Analytical Dynamics, University Press, Cambridge, 3rd. Ed., p. 393 (1927).

Classical Fields in the Vicinity of a Schwarzschild Black Hole

By S. Persides
University of Thessaloniki, Thessaloniki, Greece

Abstract

Classical test fields satisfying the scalar wave equation and Maxwell equations are analysed in Schwarzschild's space-time. The time independent and time dependent cases are treated separately and the radial equations are studied and classified. Special emphasis is given to the behavior of the field on the event horizon. It is found that only the scalar static field can possibly blow up on the horizon while the time dependent fields behave normaly there, thus eliminating the possibility of destruction of the black hole. Finally, some comments are made on the problem of the appropriate boundary conditions.

I. Introduction

In flat space-time geometrical optics has preceded studies of wave phenomena. The same is true in the curved spaces of general relativity. The geometry of null geodesics was among the first subjects to be investigated with the well-known deflection of a light ray in the gravitational field of a star. However, studies of fields coexisting with the gravitational field met insurmountable difficulties because of the contribution of the additional field to the curvature of space-time and the mathematical complications arising from it.

During the last decade general relativity has been influenced by other branches of physics and astronomy and the emphasis has been shifted from the unified-field problem to the test-field problem. Since, by its definition, a test field does not affect the geometry of space-time, the assumption of having a test field is by itself an approximation. On the other hand, the problem and its mathematics become simpler and we have a chance in solving the problem.

Many problems concerning classical test fields have been considered resently in the simplest space-time of general relativity, that of Schwarzschild. Matzner (1968) has studied the steady-state scattering of scalar waves by a Schwarzschild black hole, Lapiedra (1970) some singular electromagnetic fields and plane electromagnetic waves in Schwarzschild's space-time and Mo and Papas (1971) the first order effects of the curvature on the propagation of electromagnetic waves. More recently, Price (1972a, 1972b) has studied the radiation of scalar, electromagnetic and gravitational multipoles during gravitational collapse and

MISNER *et al.* (1972), DAVIS *et al.* (1972) and CHITRE and PRICE (1972) have examined the possibility of emission of gravitational synchrotron radiation by a particle moving in an almost circular orbit near a Schwarzschild black hole.

However, comparing our knowledge on the behavior of classical fields in Schwarzschild's space-time to that in flat space-time, we recognize that a thorough and systematic investigation of the subject is still needed. At the University of Thessaloniki, research is being done towards that direction. This contribution reports the first results. In Sec. 2 we formulate the problem and present the differential equations and the questions to be answered. In Sec. 3 we examine the fields near the horizon and study some of the direct consequencies such as the radiation of multipoles and the infinite stresses on the horizon. Finally, Sec. 4 is devoted to the examination of the boundary conditions.

II. The Problem and the Radial Equations

We are interested in studying classical fields in the space-time of a Schwarzschild black hole. The whole study is based on two assumptions. First, the geometry is described *everywhere* by the metric tensor

$$g_{\mu\nu}=\mathrm{diag}\left[\left(1-\frac{r_s}{r}\right)c^2,\left(1-\frac{r_s}{r}\right)^{-1},-r^2,-r^2\sin^2\theta\right] \tag{1}$$

in the Schwarzschild coordinates t, r, θ, φ, where $r_s=2GM\,c^{-2}$ is the Schwarzschild radius. This means that the black hole *has been formed* and no collapse is taking place. This point of view is similar to that of COHEN and WALD (1971) and different from that of PRICE (1972a, 1972b), who considers sources anchored on the collapsing star *during* collapse.

Second, we assume that the field is weak enough to neglect its contribution to the curvature of space-time. We have a *test-field*. Consequently, the field equations are those of flat space-time written in covariant form with the covariant operations taken with respect to the metric tensor (1) (MISNER *et al.*, 1973). If the flat-space equations are linear, then the Schwarzschild-space equations will be linear.

In increasing order of complexity we will consider the following fields:

I. Scalar static field.
II. Electrostatic field.
III. Scalar time dependent field.
IV. Electromagnetic time dependent field.

In cases I and III the field is represented by a single scalar quantity Ψ which is a function of r, θ, φ and, in case III, of t. To study the electromagnetic field (cases II and IV) the Newman-Penrose formalism is employed (NEWMAN and PENROSE, 1962; JANIS and NEWMAN, 1965). The field is represented by three complex scalar quantities Φ_0, Φ_1, Φ_2, which are essentially the only independent projections of the electromagnetic tensor $F_{\mu\nu}$ on four null vectors. Φ_0, Φ_1, Φ_2 are functions of r, θ, φ and, in case IV, of t. The use of scalar quantities for the description of the fields makes the results and conclusions independent of the coordinate system. The electrostatic field (case II) has been studied in detail by

COHEN and WALD (1971) using vector representation, but is included here for completeness and comparison.

Our attitude in studying the problem differs from most of the previous investigators in three respects. First, we consider in each case the full non-homogeneous problem. The source is regarded as an essential part of the physical system and the solutions of the non-homogeneous field equations are sought. Second, we are interested in analytical solutions and not in results established partially or totaly by numerical analysis through a computer. Third, we consider the interior of the black hole and the phenomena in it as important as the exterior Schwarzschild space and its phenomena.

In studying the test fields I–IV we have set a number of objectives which can be described as follows:

1) To separate the variables in the field equations and obtain the radial equation.

2) To study the behavior of the solutions near the origin $r=0$, the event horizon $r=r_s$ and at infinity $r=+\infty$.

3) To study the global behavior of the solutions through a "matching" technique, namely, to express the solution near a singular point as a linear combination of the solutions near another singular point.

4) To find the appropriate boundary conditions for a physical problem.

5) To find the field of a point source.

6) To examine whether or not the fall of a point source into the black hole is accompanied by radiation of the multipole moments of the test field and whether the energy is radiated towards infinity or into the black hole.

7) To find if there is any possibility of destruction of the black hole, because of infinite fields on the horizon during the fall.

The method of solving the field equations is that of separating the variables. For cases I and II we assume that the field quantities $\Psi, \Phi_0, \Phi_1, \Phi_2$ are linear combinations of $R_{lm}(r)\, Y_{lm}(\theta,\varphi)$, $R_{0lm}(r)\, Y^1_{lm}(\theta,\varphi)$, $R_{1lm}(r)\, Y^0_{lm}(\theta,\varphi)$ and $R_{2lm}(r)\cdot Y^{-1}_{lm}(\theta,\varphi)$, respectively, where $Y^1_{lm}(\theta,\varphi)$, $Y^0_{lm}(\theta,\varphi)=Y_{lm}(\theta,\varphi)$ and $Y^{-1}_{lm}(\theta,\varphi)$ are the spherical harmonics of spin weight $+1, 0, -1$. For cases III and IV the field is first Fourier-analyzed in time. The analysis adds a factor $e^{-i\omega t}$ in the expressions for $\Psi, \Phi_0, \Phi_1, \Phi_2$. Since the field equations do not contain the time t explicitly, but only derivatives with respect to it, the factor $e^{-i\omega t}$ cancels from the differential equation leaving behind the wave number $k=\omega/c$. Thus, an ordinary linear differential equation of second order is obtained in each case for R, R_0, R_1, R_2 (the indices lm have been suppressed). This is the *radial equation.* If for static fields (I and II) we set $\rho=r/r_s$ and for time-dependent fields $x=k\,r$ and $x_s=k\,r_s$, then the homogeneous differential equations are as follows:

Case I (PERSIDES, 1973a)

$$\rho(\rho-1)\frac{d^2R}{d\rho^2}+(2\rho-1)\frac{dR}{d\rho}-l(l+1)\,R=0. \tag{2}$$

Case II (COHEN and WALD, 1971)

$$\rho(\rho-1)\frac{d^2R_n}{d\rho^2}+(4\rho-n-2)\frac{dR_n}{d\rho}+\left[2\left(1-\frac{\delta_{2n}}{\rho}\right)-l(l+1)\right]R_n=0, \tag{3}$$

with $n=0, 1, 2$ and δ_{mn} Kronecker's symbol.

Case III (PERSIDES, 1973b)

$$x(x-x_s)^2 \frac{d^2 R}{dx^2} + (x-x_s)(2x-x_s)\frac{dR}{dx} + [x^3 - l(l+1)(x-x_s)]\,R = 0. \tag{4}$$

Case IV (PERSIDES and XANTHOPOULOS, 1973)

$$\begin{gathered} x(x-x_s)^2 \frac{d^2 R_0}{dx^2} + (x-x_s)(4x-2x_s)\frac{dR_0}{dx} \\ + \{x^3 + i\,x(2x-3x_s) + [2-l(l+1)](x-x_s)\}\,R_0 = 0, \end{gathered} \tag{5}$$

$$x(x-x_s)^2 \frac{d^2 R_1}{dx^2} + (x-x_s)(4x-3x_s)\frac{dR_1}{dx} + \{x^3 + [2-l(l+1)](x-x_s)\}\,R_1 = 0, \tag{6}$$

$$\begin{gathered} x(x-x_s)^2 \frac{d^2 R_2}{dx^2} + (x-x_s)(4x-4x_s)\frac{dR_2}{dx} \\ + \left\{x^3 - i\,x(2x-3x_s) + \left[2\left(1-\frac{x_s}{x}\right) - l(l+1)\right](x-x_s)\right\} R_2 = 0. \end{gathered} \tag{7}$$

Note that the scalar static field and the electrostatic field satisfy different equations contrary to the flat-space case. Also in time-dependent situations the scalar and the electromagnetic equations are different.

The curved character of the space-time is represented by the fact that the constant r_s is different from zero. This constant enters only in the radial equations and in most cases in the expression of the source. It does not affect the angular part of $\Psi, \Phi_0, \Phi_1, \Phi_2$. Consequently, the solution of the radial equation is the essential step in understanding wave phenomena in Schwarzschild's space-time.

III. Behavior on the Horizon and its Consequencies

Eqs. (2) and (3) for the static fields have three regular singular points at $r=0$, $r=r_s$ and $r=+\infty$. They are, or can be reduced, to special cases of the hypergeometric differential equation and their solutions can be written explicitly in terms of the Legendre functions of first and second kind (PERSIDES, 1973a; COHEN and WALD, 1971; PERSIDES and XANTHOPOULOS, 1973). Eqs. (4)–(7) for the time-dependent fields have two regular singular points at $r=0$ and $r=r_s$ and one irregular singular point at $r=+\infty$. It appears that they cannot be related to any known differential equation of mathematical physics. Their solutions can be obtained in the form of converging power series in the neighborhood of the regular singular points and in asymptotic expansions near the irregular singular point (PERSIDES, 1973b; PERSIDES and XANTHOPOULOS, 1973).

The existence of the singular point $r=r_s$ is a direct consequence of the Schwarzschild metric. We have not anything similar in flat space-time for comparison. Consequently, the behavior of the solutions near $r=r_s$ can be considered as an essential property of waves in Schwarzschild's space-time. If

$$A = e^{i x_s \ln|x-x_s|}, \tag{8}$$

then the most important terms in the expansion of the field are given in Table 1. For each case, the variable representing the field is given in the left column. For simplicity, the electrostatic field (II) is represented by the only non-vanishing component A_0 of the vector potential A_μ. In the next two columns, the behavior of the two linearly independent solutions is given. We observe that in case I we have a logarithmic blow up, which does not appear in the corresponding flat-space case. Since A oscillates with $|A|=1$ as $r \to r_s$, the scalar time-dependent field remains finite everywhere. The electromagnetic time-dependent field has at $r=r_s$ a diverging term in Φ_0 because of the factor $(x-x_s)^{-1}$. However, we expect that the source will eliminate this singularity, although this point remains to be proved rigorously.

Of great importance is the fact that one solution for A_0 goes to zero at $r=r_s$. From this property COHEN and WALD (1971) have concluded that all the multipole moments of the field of a point source (except the monopole) are radiated during the slow fall of the source into the black hole. In case I, however, the solutions do not exhibit this property and the effect of the source has to be taken into account. In fact, with an appropriate choice of source, we can have static fields in which the multipoles are or are not radiated, when the source is lowered slowly into the black hole. This ambiguity is resolved, if we consider the fall as a time dependent phenomenon. Then we are in case III, where the logarithmic singularity does not appear and the multipoles are radiated during the fall. For the electromagnetic field (case IV) there are strong indications that the conclusions drawn from electrostatics will be verified. However, this work remains to be completed.

Concerning the creation of infinite stresses on the horizon, it is obvious from Table 1 that only Ψ in case I and Φ_0 in case IV can become infinite. Whether or not the field will become infinite at $r=r_s$ depends on the source, but in any case the danger is eliminated when we take into account the time dependence. Similarly, in case IV, the source is expected to eliminate the factor $(x-x_s)^{-1}$ and the possibility of destruction for the black hole. The above results are summarized in Table 2.

Table 1. Behavior* near $r=r_s$

Field		1st solution	2nd solution
I.	Ψ	$\sim$ const.	$\sim \ln(r/r_s-1)$
II.	A_0	$\sim(r-r_s)$	$\sim$ const.
III.	Ψ	$\sim A$	$\sim A^{-1}$
IV.	Φ_0	$\sim A$	$\sim(x-x_s)^{-1}A^{-1}$
	Φ_1	$\sim A$	$\sim A^{-1}$
	Φ_2	$\sim A$	$\sim(x-x_s)A^{-1}$

(* $f(x)\sim g(x)$ as $x\to x_0$ if and only if $\lim_{x\to x_0}(f/g)$ exists and differs from zero.)

Table 2. Fall of the source into the black hole

Field	Are multipoles radiated away?	Do the fields become infinite on the horizon?
I	yes and no	possibly
II	yes	no
III	yes	no
IV	yes?	no?

IV. Boundary Conditions

In the solution of a physical problem in Schwarzschild's space-time, the field equations will be supplemented by the *boundary conditions*, namely a set of equations containing the values of the field and its derivatives on some surfaces. Consequently, the question of finding the appropriate boundary conditions on $r=0$, $r=r_s$ and $r=+\infty$ is raised.

At a regular point we can ask the field to be finite, continuous with continuous first derivatives. However, at the singular points such restrictions do not express reasonable physical requirements and sometimes no solution of the field equations exist, which can satisfy these conditions. Moreover, in Schwarzschild's space-time we can have observers which cannot communicate among themselves and, possibly, different boundary conditions will apply for each observer.

We start from the surface $r=+\infty$, where it is easy to specify the boundary conditions. We ask that the static field be finite and the time-dependent field be finite and outgoing at $r=+\infty$ (provided we deal with a source of finite energy and dimensions).

On the horizon we want the static or time-dependent field to be finite (except perhaps at the position of a point source). The static fields can be continuous there, but we cannot impose this condition to time dependent fields (at least with respect to an observer at infinity). The fact that a disturbance from any point $r>r_s$ will reach $r=r_s$ after infinite time (for the distant observer) shows that for any finite t the field on the horizon and inside the black hole is zero. If we consider that infinite time has passed and we assume incoming fields (MATZNER, 1968) at $r=r_s$, then we face the problem of propagation in the interior of the black hole. Consequently, we have to find some scheme to describe phenomena that happen after the freely falling observer crosses the horizon without resorting to the time of that observer (for whom everything happens in a finite time). Moreover, this scheme must be such that transient solutions of the time-dependent equations (representing changes in the source of finite duration and energy) reduce to solutions of the static equations when $t\to+\infty$.

Coming to the last singular point $r=0$, it appears that we have to abandon even the requirement that the field is finite there. We would like to have a *physically preferable* field, which remains finite at $r=0$, but there is no reason to reject a priori a field, which becomes infinite there, but has finite energy in a finite volume. Such fields may well be *physically acceptable*.

References

CHITRE, D. M., PRICE, R. H.: Phys. Rev. Letters **29**, 185 (1972).
COHEN, J., WALD, R.: J. Math. Phys. **12**, 1845 (1971).
DAVIS, M., RUFFINI, R., TIOMNO, J., ZERILLI, E.: Phys. Rev. Letters **28**, 1352 (1972).
JANIS, A. I., NEWMAN, E. T.: J. Math. Phys. **6**, 902 (1965).
LAPIEDRA, R.: Ann. Inst. Poincaré, **12**, 183 (1970).
MATZNER, R. A.: J. Math. Phys. **9**, 163 (1968).

MISNER, C.W., BREUER, R.A., BRILL, D.R., CHRZANOWSKI, P.L., HUGHES, H.G. III, PEREIRA, C.M.: Phys. Rev. Letters **28**, 998 (1972).
MISNER, C.W., THORNE, K., WHEELER, J.: Gravitation, Univ. of Maryland (preprint) (1973).
MO, T.C., PAPAS, C.H.: Phys. Rev. D **3**, 1708 (1971).
NEWMAN, E.T., PENROSE, R.: J. Math. Phys. **3**, 566; **4**, 998 (1962).
PERSIDES, S.: J. Math. Anal. and Applic. (to appear) (1973a).
PERSIDES, S.: On the radial wave equation in Schwarzschild's space-time (to be published) (1973b).
PERSIDES, S., XANTHOPOULOS, B.: The electromagnetic field of a point source in Schwarzschild's space-time (to be published) (1973).
PRICE, R.H.: Phys. Rev. D **5**, 2419 (1972a).
PRICE, R.H.: Phys. Rev. D **5**, 2439 (1972b).

Discussion

MCVITTIE, G.C.:
If the infinities in the fields are accompanied by infinite energies, how can you continue to use the Schwarzschild space-time which, by definition, describes a region where the energy-tensor is zero?

PERSIDES, S.:
If the field at a point different from the source and the energy in a finite volume become infinite, then most certainly we cannot speak of test-fields in SCHWARZSCHILD's space-time. Even the case where we have infinite field but not infinite energy has to be reexamined in the framework of a more general theory.

Stability of Non-Radial Vibrational Modes of Relativistic Neutron Stars

By P. CAZZOLA
Istituto di Fisica dell'Università, Padova, Istituto di Fisica dell'Università, Ferrara
and L. LUCARONI
Istituto di Fisica dell'Università, Ferrara

Abstract

One of the theoretical problems in connection with neutron stars is the vibrational stability with respect to different oscillation modes in a fully relativistic context.

The problem has got a complete and satisfactory answer for purely radial modes, namely in the monopole case $l=0$. Difficulties arise for $l>1$, essentially due to the occurrence of gravitational radiation by the star. Our aim is to consider the zero temperature cases $l>1$ in a fully relativistic context, for nonrotating neutron stars. A paper by THORNE (1969) furnished to us the starting point. In few words, one can "explore", so to speak, the proper vibrational modes of the star by a monocromatic ficticious external gravitational wave, giving rise to forced oscillations in the star.

To be more precise, in the framework of a given multipole l, we consider an impinging wave of circular frequency ω giving rise to a reflected wave, so that a system of stationary waves takes place. In the Gauss ω-plane the proper complex frequencies ω_n of the star are represented by points lying in the upper half-plane or on the real axis, if the corresponding modes are stable. Conversely, if the representative point of a mode lies below the real axis, the mode is unstable[1]. Now a resonance occurs in our ideal scattering of gravitational waves when the external (real) frequency ω approaches the real part of one of the ω_n, if the corresponding point is close enough to the real axis.

In order to develop the above ideas we borrowed a formalism extensively worked out by several authors regarding the modern quantum mechanical treatment of one-particle resonant scattering in a spherically symmetric static potential field. In the framework of one-particle potential scattering with angular momentum l (to be compared in the neutron star case with the multipole of order l) the scattering is described by the so called S-function defined by

$$S_l(k)=\exp[2i\,\delta_l(k)],$$

where δ_l is the scattering phases-shift and k is the particle wave number. $S_l(k)$ was shown by JOST (1947) and DE ALFARO and REGGE (1965) to be analytical in the complex k-plane, except for singular points k_n. Roughly speaking, these points k_n describe quasi-bound or resonant states of the particle if they are located below the real k-axis, while they describe true bound states if the k_n are located

1 Here we refer to a time dependence of the perturbation of the type $\exp(i\,\omega_n\,t)$.

in the upper half k-plane, in the imaginary axis. In other words, there is a strong analytical connection between scattering on one hand and bound or quasi-bound states on the other. The analyticity has turned out to be a very powerful tool in order to understand one particle scattering. It is also convenient to introduce the so called Jost function $F_l(k)$ which, apart from a multiplicative constant, is essentially the Wronskian function formed with the regular stationary solution, ψ_l, and a purely outgoing solution, $\psi_l(+)$.

The Wronskian function is defined as

$$W=\psi_l\,\psi_l^{(+)\prime}-\psi_l'\,\psi_l^{(+)},$$

where the prime denotes derivative with respect to the distance r from the field center. Such distance is to be compared, in the neutron star case, with the distance from the star center, while the potential field is to be compared with the equilibrium structure of the star (depending on the chosen equation of state). The Jost function has the following relation with the S_l-function:

$$S_l(k)=F_l(k)/F_l(-k);$$

so, if we exclude the existence of poles of $F_l(k)$ (to make here things quite simple) the poles of $S_l(k)$ arise from zeroes of $F_l(-k)$.

Having the Gauss k-plane in mind, it may happen that for increasing potential strength (if this is of course as a whole attractive) one or more poles in the lower k-plane, describing quasi-bound states, move to the upper half-plane necessarily along the imaginary axis, thus becoming true bound states; the transition point is necessarily $k=0$.

A similar treatment by CAZZOLA and LUCARONI (1972) was carried out in the case of gravitational waves impinging on a fully relativistic object at a zero temperature. In this case however, the wave number k is replaced by the circular frequency ω and the concepts of bound state and quasi-bound state are to be replaced respectively by the concept of unstable mode and stable mode of oscillation. Except for this, the parallelism is fairly close (we conveniently introduce functions similar to the S-function and Jost function). Also in this case "stable" poles may move to the instability region $\operatorname{Im}\omega<0$. In the present work, to be published soon, we show that the transition point is again $\omega=0$, similarly to what happens in wave mechanics where, as already stated, the transition point is $k=0$.

So, in order to analyze the possible instability, we should look for the changes of sign of the Wronskian function at $\omega=0$, for increasing central densities of the star.

The stability problem is thus anchored to the zero-frequency scattering problem. Numerical computations are going on to test the nonradial stability of various equilibrium configurations.

References

CAZZOLA, P., LUCARONI, L.: Phys. Rev. D**6**, 950 (1972).
DE ALFARO, V., REGGE, T.: Potential Scattering, North-Holland Publ. Co. (1965).
JOST, R.: Helv. Phys. Acta **20**, 256 (1947).
THORNE, K.S.: Astrophys. J. **158**, 1 (1969).

On Some Properties of Relativistic MHD Flows

By I. LUKAČEVIĆ
Faculty of Sciences and Institute for Geomagnetism *, Belgrade, Yugoslavia

Abstract

In the first part of this communication are exposed briefly general relations of the relativistic magnetohydrodynamics of a perfect thermodynamic fluid, well-known from works and monographs of LICHNEROWICZ (1967) and PHAM MAU QUAN (1956, 1971). We make, in the second part, the assumption that the magnetic field is steady, and obtain therefrom several consequences; with an additional, purely thermodynamic, condition we obtain results which allow us to compare differences, for such a fluid, between the general relativistic and the special relativistic case. In the third part of the paper we analyze several properties of the vorticity (or the rotation) tensor, using some of the results obtained by EHLERS (1961) for the kinematics of perfect fluids (see also TAUB and SHEPLEY (1967)) without any assumption for the magnetic field. In the last section we discuss the properties of an irrotational flow in the absence of shear and dilatation; we consider then the possibility of "isorotation", in the sense of classical MHD, which has been also considered by YODZIS (1971), and comment on his result.

I. General Relations

We consider a relativistic perfect fluid with an energy tensor of the form

$$T_{\alpha\beta}=(\rho+p+\mu h^2)\, u_\alpha u_\beta+(p+\tfrac{1}{2}\mu h^2)\, g_{\alpha\beta}-\mu h_\alpha h_\beta. \tag{1.1}$$

Space-time metric has the signature +2. ρ is the proper density of energy for the mechanical part of the energy tensor, p the pressure, h_α the magnetic field vector, μ the magnetic permeability (assumed to be constant), u_α the velocity unit vector with $u^\alpha u_\alpha=-1$. The magnetic field vector is space-like and orthogonal to u_α. The electric field vector, also orthogonal to u_α, vanishes, the fluid being magnetodynamic, but as the electric conductivity σ is infinite, the consequence is that, in the electric current, the part which represents conduction is undetermined (see LICHNEROWICZ (1967) and CREMONESE (1971)).

The equations of motion are

$$\nabla_\alpha T^\alpha_\beta=0, \tag{1.2}$$

* This communication is, in part, the result of my cooperation with the group working, in the Institute for Geomagnetism in Grocka (Belgrade), on questions of theoretical MHD.

and the equation of conservation of matter

$$\nabla_\alpha (r\, u^\alpha) = 0 \tag{1.3}$$

(∇ denotes covarian derivative). r is the proper density of matter. We assume, as it is made in TAUB and SHEPLEY (1967), that the relation between thermodynamic quantities ρ and r is

$$\rho = r(1+\varepsilon), \tag{1.4}$$

where ε denotes the internal energy (for unit proper-volume). We shall not discuss in detail thermodynamic relations.

In the MHD case the Maxwell equations of the electromagnetic field take the form

$$\nabla_\alpha (h^\alpha u^\beta - h^\beta u^\alpha) = 0 \tag{1.5}$$

and

$$\eta^{\alpha\beta\gamma\delta} (\nabla_\beta u_\gamma \cdot h_\delta + u_\gamma \nabla_\beta h_\delta) = Y^\alpha = \gamma\, u^\alpha + \lambda^\alpha \tag{1.6}$$

$\eta^{\alpha\beta\gamma\delta}$ is the Ricci permutation tensor, Y^α the electric current vector, in which λ^α denotes the undetermined part due to conduction, and γ the specific density of charge per unit proper-volume. Viscosity and thermal conductivity are neglected.

We denote by

$$w^\alpha \equiv u^\beta \nabla_\beta u^\alpha \tag{1.7}$$

the relativistic acceleration vector. Characteristic kinematical quantities that we shall use are

$$\omega_{\alpha\beta} \equiv \tfrac{1}{2} (\nabla_\beta u_\alpha - \nabla_\alpha u_\beta + u_\beta w_\alpha - u_\alpha w_\beta), \tag{1.8}$$

the rotation tensor, with

$$v^\alpha \equiv \eta^{\alpha\beta\gamma\delta} u_\beta\, \omega_{\gamma\delta} \tag{1.9}$$

the vorticity vector, and

$$\sigma_{\alpha\beta} \equiv \tfrac{1}{2} (\nabla_\alpha u_\beta + \nabla_\beta u_\alpha + u_\alpha w_\beta + u_\beta w_\alpha) - \tfrac{1}{3} \nabla_\gamma u^\gamma (g_{\alpha\beta} + u^\alpha u^\beta) \tag{1.10}$$

is the shear tensor which expresses the deformation of an element of the medium considered. These quantities were introduced by ALFVEN-FÄLTHAMMAR (1963) (see also TAUB and SHEPLEY (1967) and LINET (1971)). Both rotation and shear tensors, contracted with u^β give identically zero.

II. The Case of a Steady Magnetic Field

We shall suppose, in this part of the paper, that the magnetic field is steady, i.e. that it does not depend on proper time,

$$u^\alpha \nabla_\alpha h_\beta = 0. \tag{2.1}$$

From the first group of the Maxwell equations (1.5), we obtain immediately, contracting with u_β,

$$\nabla_\alpha h^\alpha = 0 \tag{2.2}$$

and

$$h^\alpha w_\alpha = 0. \tag{2.3}$$

Then the equations of motion become

$$\mu h^\alpha \nabla_\alpha h_\beta = (\rho + p + \mu h^2) w_\beta + (\rho + p + \mu h^2) \nabla_\alpha u^\alpha u_\beta \\ + u^\alpha \partial_\alpha (\rho + p) u_\beta + \partial_\beta (p + \tfrac{1}{2} \mu h^2). \tag{2.4}$$

By contraction of (2.4) with h^β we obtain

$$h^\alpha \partial_\alpha p = 0. \tag{2.5}$$

The pressure remains constant along the magnetic field lines. Eqs. (1.5) become

$$h^\alpha \nabla_\alpha u_\beta = \nabla_\alpha u^\alpha \cdot h_\beta, \tag{2.6}$$

or the absolute derivative of the four-velocity, taken along the magnetic field lines, is proportional to the magnetic field vector. By contraction of the equations of motion with w^β, we have

$$\mu h^\alpha w^\beta \nabla_\alpha h_\beta = (\rho + p + \mu h^2) w^\beta w_\beta + w^\beta \partial_\beta (p + \tfrac{1}{2} \mu h^2). \tag{2.4'}$$

Having in mind (2.1) and its consequences (2.2) and (2.6), we have, differentiating along a world line

$$u^\alpha \nabla_\alpha (h^\alpha u^\beta \nabla_\gamma h_\beta) = w^\beta h^\gamma \nabla_\gamma h_\beta + u^\alpha h^\beta u^\gamma \nabla_\alpha \nabla_\beta h_\gamma$$

and also

$$u^\alpha \nabla_\alpha (u^\beta h^\gamma \nabla_\gamma h_\beta) = -u^\alpha \nabla_\alpha (h^\gamma \nabla_\gamma u^\beta h_\beta) = -h^2 u^\alpha \nabla_\alpha \nabla_\beta u^\beta.$$

If we set

$$\theta \equiv \nabla_\alpha u^\alpha \tag{2.7}$$

(we shall call θ the velocity dilatation) the preceding relation takes the form

$$-h^2 u^\alpha \partial_\alpha \theta = h^\gamma w^\beta \nabla_\gamma h_\beta + u^\alpha h^\beta u^\gamma \nabla_\alpha \nabla_\beta h_\gamma.$$

Similarly we obtain

$$u^\alpha h^\beta u^\gamma \nabla_\beta \nabla_\alpha h_\gamma = -h^\gamma \nabla_\gamma u^\alpha \cdot u^\beta \nabla_\alpha h_\beta = h^2 \theta^2.$$

Making use of the Ricci identity

$$\nabla_\beta \nabla_\alpha h_\gamma - \nabla_\alpha \nabla_\beta h_\gamma = R_{\delta\gamma\beta\alpha} h^\delta$$

we have from the above relations

$$h^\alpha w^\beta \nabla_\alpha h_\beta = -h^2 (u^\alpha \partial_\alpha \theta + \theta^2) + R_{\alpha\beta\gamma\delta} h^\alpha u^\beta h^\gamma u^\delta.$$

Putting this expression in (2.4′) we obtain

$$(\rho + p + \mu h^2) w^\alpha w_\alpha + w^\alpha \partial_\alpha (p + \tfrac{1}{2} \mu h^2) + \mu h^2 (u^\alpha \partial_\alpha \theta + \theta^2) \\ - \mu R_{\alpha\beta\gamma\delta} h^\alpha u^\beta h^\gamma u^\delta = 0. \tag{2.8}$$

In the absence of the velocity dilatation

$$\theta = 0 \tag{2.9}$$

we have immediately from (1.3)

$$u^\alpha \partial_\alpha r = 0 \tag{2.10}$$

and from (2.6)

$$h^\alpha \nabla_\alpha u^\beta = 0. \tag{2.11}$$

It is well-known (see LICHNEROWICZ (1967) and TAUB and SHEPLEY (1967)) that, in a perfect thermodynamic fluid, entropy S is always conserved along world-lines, under the classic thermodynamic condition

$$T\,dS = d\varepsilon + p\,d\left(\frac{1}{r}\right). \tag{2.12}$$

If we add the assumption (2.10), all the thermodynamic quantities are conserved along stream-lines. The consequence is (2.9) and, for a steady magnetic field, also (2.11). So if in a perfect MHD fluid, proper density of matter is conserved along stream-lines, the magnetic field being steady, the velocity field u^α consists of families of vectors, parallel to the magnetic field lines. Relation (2.8) reduces then to

$$(\rho + p + \mu h^2)\, w^\alpha w_\alpha + w^\alpha \partial_\alpha (p + \tfrac{1}{2}\mu h^2) = \mu R_{\alpha\beta\gamma\delta} h^\alpha u^\beta h^\gamma u^\delta. \tag{2.8'}$$

In General Relativity, when the Riemannian curvature is null on the two-dimensional surface S, locally determined by u^α and h^α, and, obviously, also in Special Relativity, we see that the scalar $p + \frac{1}{2}\mu h^2$, representing the MHD total pressure, must be variable along the acceleration vector. But if the total pressure remains constant in the direction of the (non-vanishing) acceleration, a "compensation" is given by the Riemannian curvature of S, which is proportional to $R_{\alpha\beta\gamma\delta} h^\alpha u^\beta h^\gamma u^\delta$, and the flow is not locally geodesic. Conversely, for a geodesic flow of a MHD perfect fluid, if both magnetic field and thermodynamic variables do not depend on proper time (or if all thermodynamic variables can be expressed as functions of one parameter), the Riemannian curvature of the surface S, determined by world lines and magnetic field lines, vanishes.

In the case of a geodesic flow in a steady magnetic field, without the previous thermodynamic limitation, the Riemannian curvature for S is, from (2.8), equal to

$$K \equiv \frac{R_{\alpha\beta\gamma\delta} h^\alpha u^\beta h^\gamma u^\delta}{(g_{\alpha\gamma} g_{\beta\delta} - g_{\alpha\beta} g_{\gamma\delta}) h^\alpha u^\beta h^\gamma u^\delta} = -(u^\alpha \partial_\alpha \theta + \theta^2). \tag{2.13}$$

We see that K is the function of the velocity dilatation and of its time-like derivative only. If θ is not zero, curvature K will vanish also for

$$\theta = \frac{1}{\tau + c(x_0^1, x_0^2, x_0^3)} \tag{2.14}$$

where τ is the proper time (x_0^1, x_0^2, x_0^3 are space-like coordinates).

All the preceding conclusions are, of course, only limiting cases appearing as a consequence of our first assumption (2.1). Questions concerning the possibility of a simultaneous statement of all these limitations, including geodesic flow, especially from the standpoint of the algebraic classification of space-times containing different kinds of energy tensors, can be the object of further research.

III. The Rotational Flow

Let us consider the rotational flow of a MHD perfect fluid, without any assumption on the magnetic field.

The divergence of the vorticity vector has the value

$$\nabla_\alpha v^\alpha = w_\alpha v^\alpha \tag{3.1}$$

(see EHLERS, 1961). If we assume that the vorticity vector is steady,

$$u^\alpha \nabla_\alpha v^\beta = 0. \tag{3.2}$$

We have, from the fact that v^β is always orthogonal to u_β,

$$u^\alpha \nabla_\alpha (u^\beta v_\beta) = 0.$$

Wherefrom, because of (3.2),

$$v^\beta w_\beta = 0. \tag{3.3}$$

This means that, if the vorticity vector is steady, its divergence vanishes. Consider also the case when the rotation tensor is steady,

$$u^\alpha \nabla_\alpha \omega_{\beta\gamma} = 0. \tag{3.4}$$

Working in the same way, we have

$$u^\alpha \nabla_\alpha (u^\beta \omega_{\beta\gamma}) = 0$$

and

$$w^\beta \omega_{\beta\gamma} = 0. \tag{3.5}$$

w^β being orthogonal to u_β, must be parallel to v^β which represents an algebraic solution of the system (3.5), and then simultaneous steadiness of v^β and of $\omega_{\alpha\beta}$ lead us, for a rotational flow, to the conclusion that the acceleration must vanish. The same result appears, because of (3.1), if the vorticity tensor is steady and the vorticity vector has a null divergence. Let us retain the second possibility for a rotational geodesic flow,

$$\nabla_\alpha v^\alpha = 0. \tag{3.6}$$

The equations of motion take the form

$$\mu h^\alpha \nabla_\alpha h_\beta + \mu \nabla_\alpha h^\alpha \cdot h_\beta = (\rho + p + \mu h^2) \nabla_\alpha u^\alpha \cdot u_\beta + u^\alpha \partial_\alpha (\rho + p + \mu h^2) u_\beta + \partial_\beta (p + \tfrac{1}{2} \mu h^2),$$

which, multiplied by h^β become

$$\mu h^2 \nabla_\alpha h^\alpha = h^\alpha \partial_\alpha p.$$

From the Maxwell equations we have

$$u^\alpha u^\beta \nabla_\alpha h_\beta + \nabla_\alpha h^\alpha = 0.$$

This reduces, in our case, to

$$\nabla_\alpha h^\alpha = 0. \tag{3.7}$$

We get immediately

$$h^\alpha \partial_\alpha p = 0. \tag{3.8}$$

In the case of a rotational geodesic flow, which is the consequence of (3.4) and (3.6), the dilatation of the magnetic field vector is null and the pressure remains constant along magnetic field lines.

If we make the additional assumption that thermodynamic quantities remain constant along the stream-lines, we shall obtain also relations (2.9) and (2.10), and Eqs. (1.5) reduce to the form

$$u^\alpha \nabla_\alpha h_\beta - h^\alpha \nabla_\alpha u_\beta = 0. \tag{3.9}$$

IV. Irrotational Flow and Isorotation

The possibility of obtaining a relativistic extension of FERRARO's theorem of isorotation (see ALFVEN-FÄLTHAMMAR (1963)) was considered by YODZIS (1971), under the assumption that the magnetic field vector and the rotation tensor were steady, and that the dilatation scalar and the shear tensor, defined by (1.9), vanished. The analysis of these conditions led us first to the conclusion that the magnetic field and the vorticity vectors are parallel, then to the absence of acceleration, and in the case of a Minkowski metric, to irrotational flow. These conclusions can be verified using the results of the preceding sections.

If we put $\sigma^2 \equiv \frac{1}{2}\sigma^{\alpha\beta}\sigma_{\alpha\beta}$ and $\omega^2 \equiv \frac{1}{2}\omega^{\alpha\beta}\omega_{\alpha\beta}$, we shall have (see EHLERS, 1961)

$$R_{\alpha\beta}\,u^\alpha u^\beta = u^\alpha \partial_\alpha \theta + \tfrac{1}{3}\theta^2 - \nabla_\alpha w^\alpha + 2(\sigma^2 - \omega^2), \tag{4.1}$$

where $R_{\alpha\beta}$ is the Ricci curvature tensor. This relation shows that assumptions made by YODZIS (1971) lead to irrotational flow, in the case when the metric is pseudo-Euclidean. But in the case of General Relativity we have, from the equations of gravitation,

$$R_{\alpha\beta} - \tfrac{1}{2}R\,g_{\alpha\beta} = -\kappa\,T_{\alpha\beta} \tag{4.2}$$

for a MHD energy tensor of the form given by (1.1),

$$\omega^2 = \tfrac{1}{2}\kappa(\rho + 3p + \mu h^2). \tag{4.3}$$

Then we must add to the previous assumptions two more conditions:

1) We have a thermodynamic equation of state $\rho = \rho(p)$ (which is, because of (1.3), in accordance with the assumption that $\theta = 0$).

2) The magnetic field vector has a constant intensity along its field lines (but with possibly different values for the different lines).

Then, by differentiation of (4.3) along the magnetic field lines, we have from (2.5),

$$h^\alpha \partial_\alpha \omega^2 = 0, \tag{4.4}$$

which represents formally a relativistic extension of FERRARO's theorem of isorotation. Let us remark that we obtain, for the Riemannian curvature, properties we have seen in section 2. But v^α and h^α being parallel, the case is even more special.

FERRARO's theorem was obtained in classical MHD under the assumptions that:

a) The magnetic field is steady.

b) The angular velocity has the same value for points equidistant with respect to the rotation axis, and having the same z-coordinate (z is calculated along the rotation axis).

c) The magnetic field is independent of the cyclic coordinate.

In the special relativistic case assumption (a) is not necessary, and for (b) we can take

$$\theta = 0, \qquad \sigma = 0, \tag{4.5}$$

which corresponds to flow without deformation. For (c) we have to take

$$h^\alpha \nabla_\alpha \nabla_\beta w^\beta = 0, \tag{4.6}$$

which is not easy for physical interpretation, but after some examining it has similarity with (c) of classical MHD. Then we shall have (4.4). In the general relativistic case, we must take in consideration also the steadiness of the magnetic field (2.1), because we have (4.3), and also assumptions (1) and (2), in order to obtain (4.4).

References

ALFVÉN, H. and FÄLTHAMMER, G.-G.: Cosmical Electrodynamics, Oxford, New York (1963).

EHLERS, J.: Akad. Wiss. Mainz. Abh. Math. Naturwiss. No 11 (1961).

LICHNEROWICZ, A.: Relativistic Hydrodynamics and Magnetohydrodynamics, Benjamin, New York (1967).

LINET, B.: Ann. Inst. H. Poincaré **15**, 141 (1971).

LUKAČEVIĆ, I.: Publ. Math. Inst. Belgrade, t 7 (21), p. 51 (1967).

PHAM MAU QUAN: Arch. Rat. Mech. Anal. **5**, 475 (1956).

Relativistic Fluid Dynamics (C.I.M.E. session 1970), Edizioni Cremonese, Roma (1971).

TAUB, A. and SHEPLEY, L.: Commun. Math. Phys. **5**, 237 (1967).

YODZIS, P.: Phys. Rev. D. **3**, 2941 (1971).

Astrophysical Implications of Extragalactic Radio Sources

(Invited Lecture)

By MARTIN J. REES
Astronomy Centre, University of Sussex, Falmer, Brighton, U.K.

With 4 Figures

Abstract

The observations of powerful extragalactic radio sources are briefly reviewed, and some of their implications are discussed.

I. Introduction

This article deals with the possible relevance of the ideas discussed by Dr. PACINI – and applied by him to pulsars and the Crab Nebula – to some larger-scale extragalactic phenomena. Most current attempts to understand quasars and explosive galactic nuclei are, not surprisingly, carried out within the framework of conventional physics (which in this context includes general relativity). Because nuclear energy is capable of releasing only ~ 0.008 of the rest mass energy of the material involved, it seems more likely that their power is gravitational in origin, and derives from *either* (a) the contraction of a single supermassive star or (b) the gravitational attraction of millions of stellar-mass objects in an environment where they are much more closely packed than stars in the solar vicinity. There are so far no data which permit a definite decision between these two classes of model.

a) Models involving a "superstar" of mass $\gtrsim 10^6\ M_\odot$ were investigated quite thoroughly as soon as quasars were discovered. The difficulty with this idea is that, according to general relativity, a spherically symmetric superstar would become unstable, and collapse catastrophically to a black hole, before releasing much gravitational energy in useful form (and perhaps before even becoming hot enough for nuclear reactions to occur in the interior). This instability results from the dominance of radiation pressure over gas pressure in a massive star. Such an object (with γ only slightly exceeding $\frac{4}{3}$) is so highly compressible that it is only marginally stable against radial oscillations in Newtonian theory; and the post-Newtonian terms of general relativity, even when they are comparatively small, can destroy this delicate equilibrium. It is now realised that these instabilities can be suppressed to some extent by rotation. Indeed, a differentially rotating disc may remain stable even when its gravitational binding energy amounts to $\sim 30\%$ of Mc^2. (Though if it is too highly flattened, it could be unstable – as discussed by SALPETER (1971) – against fragmentation into "sub-discs".) This type of explanation for quasars and related phenomena is thus still very much "in the field". One way of testing it observationally is by searching for periodicities (such as might arise from rotation or pulsation) in the optical or radio brightness fluctuations. The detection of any well-defined period would indicate that one coherent massive object was involved, and that the emission did not come from a multitude of independent smaller objects.[1]

b) Among the various proposals involving millions of stellar-mass objects are the following:

(i) *That Stellar Collisions in a Compact Cluster may Trigger Violent Explosions of the Individual Stars.* In such a collision a compression wave would propagate through the interior of each colliding star. When the shock waves complete their passage through the stars, they would accelerate a small fraction of the material to very high velocities. This suggestion has been explored by GOLD and his associates. Physical processes in dense star systems are reviewed by SPITZER (1971).

(ii) *That Stellar Collisions may Lead to Coalescence, and thence to the Formation of Stars of* $\sim 100\,M_\odot$. Coalescence, rather than disruption, occurs when two stars collide at speeds $\lesssim$ the escape velocity from their surfaces. Stars of $\sim 100\,M_\odot$ would evolve in only $\sim 10^6$ years to the supernova stage. The total energy emitted by a supernova-type event might be $\sim 10^{53}$ erg. This is some three orders of magnitude more than the estimated bolometric luminosity integrated over the duration of observed supernovae, but the *kinetic* energy of the expanding remnant may be larger. However the optical emission line spectra of quasars imply the presence of gas with particle density $\sim 10^6\,\text{cm}^{-3}$ (compared to ~ 1 for the interstellar medium) and it is certainly possible that the ejecta are rapidly braked by their dense surroundings, their kinetic energy being efficiently transformed into radiation (COLGATE, 1967). It also cannot be ruled out that the stellar-mass objects formed in the hot dense conditions prevailing in a galactic nucleus differ significantly from ordinary stars in their structure and evolution. One specific possibility that will be considered below in connection with the radio emission from strong extragalactic sources is:

(iii) *That Galactic Nuclei may Contain Large Numbers of Compact, Spinning, Pulsar-Like Bodies whose Rotational Energy is being Extracted Via Electromagnetic Torques (see Section II).*

All the above variants of alternative (b) appeal to gravitation – either the binding energy of the cluster as a whole, or the binding energy of the individual members – as the primary energy source. Moreover, (ii) and (iii) seem to require that some unspecified process should cause $\sim 10^8\,M_\odot$ of material to condense into stars during an interval $\lesssim 10^6$ years. One general test of this type of model would be to see whether the position of the centre of activity "jitters" as the intensity fluctuates.

The final state of either a single superstar or a dense star cluster would almost inevitably seem to be a massive "black hole". This consideration has prompted

---

[1] This would not, however, be a completely watertight conclusion: it is conceivable – though arguably unlikely – that many stellar-mass objects could be embedded in an extensive gas cloud which pulsates (and so varies periodically in density and opacity), or that the objects could themselves constitute a cluster which exhibits some periodic dynamical behaviour.

some recent suggestions that much of the non-stellar activity in active galactic nuclei may be energized by accretion of surrounding gas onto a preexisting "black hole". If the accreted material possesses sufficient angular momentum, it will form a differentially rotating disc composed of fluid elements which gradually spiral in towards the centre. The efficiency depends on the nature of the central object, but for a Kerr black hole with the maximum possible angular momentum up to 40% of the rest mass energy of infalling matter can be transformed into radiation (BARDEEN, 1970) – though any estimate of the predominant waveband in which this radiation emerges would involve thorough investigations of the gas-dynamical problems involved (see LYNDEN-BELL, 1969; LYNDEN-BELL and REES, 1971; and NORMAN and TER HAAR, 1972; for some discussions of this question).

II. Extended Extragalactic Radio Sources

I should now like to concentrate on one particular manifestation of "violent activity" in external galaxies – the strong extragalactic radio sources which emit 10^3–10^6 times more strongly than normal galaxies like our own, and which comprise the bulk of the sources revealed by radio surveys away from the direction of the Galactic Plane. Of these sources, several hundred have been identified. In most cases, the associated extragalactic object is either a *bright elliptical galaxy* or a *quasar*. The properties of these radio sources span a wide range in all respects, especially in power output and linear dimensions. But there seems *no* very clear-cut correlation between any *radio* property and the nature of the optical counterpart.

The luminosity function of extragalactic radio sources stretches all the way from normal galaxies (with power $\sim 10^{38}$ erg s^{-1}) up to objects like Cygnus A whose radio output amounts to several times 10^{44} erg s^{-1}. The space density drops off steeply with increasing radio power, but the more powerful sources are of course detectable out to greater distances and so are well represented in surveys down to a limiting flux density. In fact most of the sources in radio catalogues are powerful objects with significant redshifts.

The linear sizes of radio sources range from ~ 0.5 Mpc down to $\lesssim 1$ pc, and I shall consider the more extended sources first. The most remarkable feature is the prevalence of linear structure, the radio emission being concentrated in two blobs, one lying on each side of the visible object. This characteristic "dumb-bell" structure has been known about for a long time, but recently much better structural information has come from the aperture synthesis technique, which gives us images with resolution of a few arc seconds. (See BRANSON *et al.*, 1972; for impressive examples of such maps made with the Cambridge "1 mile" telescope).

Most features are well examplified by the radio galaxy Cygnus A (shown in Fig. 1), which has the highest flux density of any extragalactic object in the radio sky, and is also one of the most *intrinsically* powerful radio sources (it has a redshift $z = 0.05$). The two blobs are fairly symmetrically disposed on either side of the associated galaxy (at distances 50–100 kpc, depending on the assumed Hubble constant). There is evidence for fine structures within the radio-

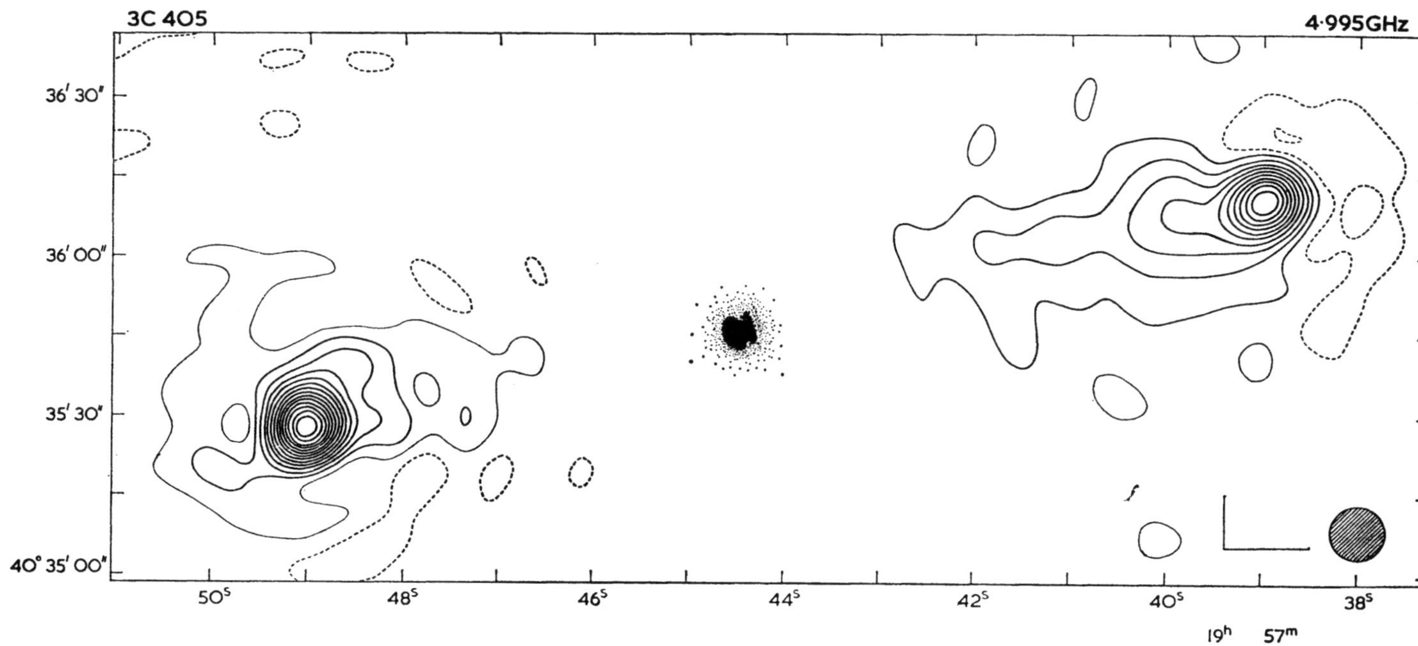

Fig. 1. The structure of Cygnus A at 5000 GHz. The vertical scale has been compressed so as to make the effective beam (shown at bottom-right) circular. (Reproduced from MITTON and RYLE, 1969)

emitting regions on scales $\sim 1''$. In many cases, there appears to be a "bridge" of emission extending from the blobs towards the central object. The predominance of double structure is restricted to the stronger sources (radio power $\gtrsim 10^{42}$ erg s^{-1}): weaker sources display a more varied and complex morphology.

MACKAY (1971 a) has analysed a statistically complete sample of 200 3C sources. Of the ~ 50 well-resolved double sources in his sample, the intensity ratio (at 1407 MHz) was closer to equality than 1.6:1 in half the cases. This cannot be attributed to selection effects, and indicates a genuine degree of symmetry. It was also found that the *fainter* component (at 1407 MHz) tended: to be *further* from the optical object; to have a *steeper spectrum* than the brighter component; and to be *less compact* than the brighter component, both along and perpendicular to the source axis.

What, then, is understood about extended radio sources of this kind? One point on which a general consensus *does* exist concerns the nature of the radiation mechanism. The spectra are power-laws $S(\nu) \propto \nu^{-\alpha}$, α typically being about 0.7, and the emission is attributed to synchrotron-type radiation by relativistic electrons. The two essential constitutents of the radio blobs are therefore generally believed to be relativistic (Bev) electrons with a differential power law spectrum $N(E) \propto E^{-\gamma}$ ($\gamma \simeq 2.4$), and magnetic fields. The weaker the field is, the more numerous must be the electrons (and the higher their individual energies) to yield the observed flux—and conversely. The total energy content in these two forms would be minimised if there were rough equipartition between field energy and particle energy. After making an estimate of the effective emitting volume, one can therefore calculate the minimum energy content. For Cygnus A this turns out to be $\sim 10^{58}$ erg (the corresponding magnetic field being $\sim 3 \times 10^{-4}$ gauss), and it is still more for many other sources with larger emitting volumes. Furthermore, these are likely to be serious *under*-estimates of the actual energies involved. There is no good reason for expecting equipartition between field and particle energies. Furthermore, these estimates make no allowance for relativistic protons, "inefficiency factors" in the particle acceleration, adiabatic losses, and other forms of energy. The realisation that extended radio sources entailed these colossal energy requirements was specially important when these estimates were made about 15 years ago (see BURBIDGE, 1958) because it was really the *first* evidence for extragalactic "violent events" on a large scale—an idea to which we have become more accustomed since the discovery of quasars and related phenomena.

Quite apart from the origin of the energy, and other such general considerations, the double radio sources raise some specific *dynamical* problems. From the appearance of the map of Cygnus A (Fig. 1), the natural inference is that the components have somehow "propagated out" from the parent galaxy. But a blob containing *nothing but* relativistic particles and magnetic fields would expand at a speed $\sim c$, because the sound speed—and perhaps the Alfven speed as well—would be relativistic. So even if the blobs moved out from the galaxy at a speed $\sim c$ it is hard to see how they can have such small sizes in relation to their separation. This ratio is about 1:15 in Cygnus A, less than 1:30 in 3C33, and may be even smaller in other objects, since many radio components are still unresolved, at least in the transverse direction. Something must therefore be impeding the free expansion of the blobs, so that

$$\frac{\text{transverse expansion velocity of blobs}}{\text{velocity away from central object}} \lesssim \frac{d}{D} = \frac{\text{component size}}{\text{component separation}}.$$

Two main possibilities have been considered. The first is that the blobs contain a large mass of *non*-relativistic plasma, which provides extra inertia without much extra pressure. The magnetic field would "glue" this plasma to the relativistic particles (yielding a non-relativistic sound speed for the composite fluid) so the expansion velocity could be lower. (This hypothesis requires $\gtrsim 10^9\ M_\odot$ of material in the case of Cygnus A.) The *kinetic energy* of this non-relativistic matter, however, would exceed the relativistic energy content E in the blobs by a factor $\sim (D/d)^2$, implying total energies above 10^{60} erg in Cygnus A. A second possibility, somewhat more economical in terms of energy, involves appealing to an *external medium:* the intergalactic or intracluster gas often invoked by theorists, whose existence could be compatible with all our present knowledge, but which has always frustratingly eluded positive detection. The thermal pressure $\frac{3}{2} n k T$ is probably negligible compared to the internal pressure, rendering a static equilibrium unlikely. However if the blobs are moving *hypersonically* with velocity V through gas of density ρ_{ext}, then they will feel a much larger "ram pressure" $\rho_{\text{ext}} V^2$ which is independent of the external temperature. (There would of course be a stand-off shock ahead of the blob, behind which the intergalactic gas would be heated to a temperature such that the sound speed $\sim V$.) Consistent models can be devised with $V \simeq 0.1\,c$ and $\rho_{\text{ext}} \simeq 10^{-28}\ \text{g cm}^{-3}$ (not altogether an implausible density within clusters of galaxies). The main energy requirement is, in this model, associated with the work done in pushing against the interstellar medium, and exceeds E by a factor $\sim (D/d)$ – but not $(D/d)^2$ as in the "inertial confinement" model.

Granted that the radio components have been shot out from the central optical object, and consist of clouds of plasma rather than gravitationally bound objects, it is hard to think of any alternative to one or other of the two confinement mechanisms mentioned above at least for the stronger sources. Other possibilities – for example, *thermal* pressure of the external medium – may, however, be important in lower-luminosity sources.

Detailed models based on ram pressure confinement have been worked out by DE YOUNG and AXFORD (1967), CHRISTIANSON (1969), MILLS and STURROCK (1970) and DE YOUNG (1971). All these authors assumed that the components contain non-relativistic plasma, whose inertia allows the blobs to plough outward through the external medium without being decelerated too rapidly. MILLS and STURROCK devised a detailed model to explain the shape of the radio components. In this particular theory, turbulence, rather than relativistic particles and magnetic fields, provided the main internal pressure, with a consequent further increase in the energy requirements. BLAKE (1972) has discussed Rayleigh-Taylor and

[2] The hypothesis that the blobs are moving outward *ultra*-relativistically and that their dimensions remain small because of time dilation effects is untenable because, for random orientations of the source axis relative to the line of sight, it would lead to greaters asymmetries between the two components than are observed (RYLE and LONGAIR, 1967).

Kelvin-Helmholtz instabilities which might occur at the interface between the components and the confining medium. He shows that the growth time for Rayleigh-Taylor instabilities in the kind of radio blob considered by DE YOUNG and AXFORD is small compared with the age. Fragmentation would occur into sub-blobs too small to be resolved with existing telescopes. However, the sub-blobs may subsequently be decelerated at different rates, and this may partially account for the apparent elongation of the components along the source axis.

These ram pressure models, if correct, tell us about the current dynamical situation within the radio-emitting blobs. But they leave the nature of the explosive phenomenon which energised the radio source more mysterious than ever. In most published versions of the theory, this hypothetical explosion is required to have flung out $\sim 10^{60}$ erg (or more) in two oppositely-directed "bullets", the energy being *partly* in the form of magnetic fields and relativistic particles, but *mainly* in bulk kinetic energy of the non-relativistic matter whose inertia enables the blobs to plough on out through the intergalactic gas until they have attained distances $\gtrsim 100$ kpc from the parent object. Moreover, in some versions of the theory the timescale of the explosion is envisaged as $\ll$ the age of the source. In RYLE and LONGAIR'S (1967) model, for instance, the explosive energy production occupies $\lesssim 10^3$ years, a corollary being that these authors are forced to invoke a *power* output of $\gtrsim 10^{50}$ erg s^{-1}, which far exceeds anything normally contemplated by the high energy astrophysicist. One may ask not merely: Why does one not see evidence for explosions of this violence (since even one such object on our light cone would surely outshine "ordinary" quasars and radio sources)? but also: By what mechanism is the required power output, and directivity, achieved? STURROCK has suggested that enormous amounts of magnetic energy (deriving originally from gravitational contraction) can accumulate and then be suddenly released by "tearing mode" instabilities, and has recently outlined (STURROCK and BARNES, 1972) a theory according to which repetitive flares would occur in a differentially spinning disc. PIDDINGTON (1970, and earlier references cited therein) has speculated that the magnetic field of a differentially rotating supermassive object is wound up and amplified, and is then expelled (rather like a coil spring) along the rotation axis. A somewhat analogous situation was computed by LE BLANC and WILSON (1970). These authors calculated what happens when the inner part of a spinning magnetised star collapses. Amplified by the resulting differential rotation, the magnetic field bursts out in two well-collimated jets along the rotation axis (it is not clear, however, to what extent the qualitative features of these striking results depend on the axisymmetry which was built into the computations). It is also somewhat disappointing that the astronomical systems to which these computations were supposed to apply – supernova remnants – often tend to be remarkably spherical! Perhaps some relativistic effects, not yet understood, associated with strong gravitational fields could be responsible for the directivity and double structure of the explosions. And there are many other possibilities connected with supermassive objects, some of which are discussed by Dr. OZERNOY in his contribution to this meeting. (See also MORRISON and CAVALIERE, 1971.)

Another constraint on tenable theories is that even if pressure confinement *does* work for a source with the present parameters of Cygnus A, it could *not* have been equally effective when the components were smaller (and nearer the optical galaxy) unless the energy content of the blobs was much *less* than it is now.

The ideas summarised above constitute the general consensus – insofar as there is one – on the interpretation of double radio sources. But I would like to outline briefly some rather different ideas which I have discussed more fully elsewhere (REES, 1971, 1972; BLANDFORD and REES, 1972). This proposal differs from others in that the energy content of the radio components must (in the simplest version of the theory, anyway) be *supplied continuously* by the nucleus of the galaxy in a specific form, instead of having been mainly liberated earlier in the history of the source. It also differs in that *no thermal plasma* is required within the components to provide inertia. Nor need any magnetic field, in the usual sense, be present. This model also suggests how the directivity of ejection implied by the observed double structure might arise.

This model for extragalactic sources is motivated by general considerations similar to those already discussed by Dr. PACINI in the context of pulsars. An object which contracts, conserving its angular momentum, feeds an increasing fraction of its gravitational binding energy into kinetic energy of rotation. Rotational braking of compact spinning objects can thus extract a large fraction of their gravitational binding energy: up to 10% of Mc^2 for a neutron star; and as much as 40% of the rest mass energy for a spinning object undergoing complete gravitational collapse. If the collapsing object is magnetised, the rotational torques can be predominantly electromagnetic. Current understanding of pulsars shows that this kind of process can indeed be very efficient, and that the energy may then emerge in electromagnetic or Alfven waves at the rotation frequency. If, therefore, the energy output from galactic nuclei is due to a succession of stellar-mass objects – which either end their lives by forming spinning neutron stars, or else undergo complete collapse to black holes, losing their angular momentum by electromagnetic (and not gravitational) radiation – then the power will emerge as *electromagnetic waves in the general frequency range 1–1 000 Hz.* These waves cannot propagate to us, being below the plasma frequency in galactic gas (and perhaps in intergalactic gas as well). Their indirect effects, however, can be very striking: indeed it seems that the single hypothesis that such waves constitute the *primary energy output from galactic nuclei* leads quite naturally to an explanation of all the main features of extragalactic radio sources.

For any electromagnetic wave of angular frequency ω, we can define an equivalent magnetic field H_{eq} with the property that $H_{eq}^2/8\pi$ equals the energy density ε of the wave. If the electron gyrofrequency in a field H_{eq} is Ω, then we can define a Lorentz-invariant quantity $f=\Omega/\omega$ as the "strength parameter" of the wave. For ordinary electromagnetic radiation, in the laboratory as well as in cosmic situations, this parameter is $\ll 1$. For the very low values of ω that concern us here, however, f can be $\gtrsim 1$ even for energy densities such that $H_{eq} \simeq 10^{-4}$ gauss (the condition >1 implying merely that the gyroradius is less than one wavelength). *Any electron exposed to a wave with $f \gtrsim 1$ becomes relativistic, with $\gamma \gtrsim f$,* and this is important in the application to radio sources. Another property of these waves is that, because of their low frequency, they may not be able to propagate at all. The propagation condition is less stringent than the usual requirements $\omega > \omega_p$, where $\omega_p = (4\pi n_e e^2/m_e)^{\frac{1}{2}}$ because the electrons, being relativistic, have larger effective masses. The actual condition for propagation is $\omega > \omega'_p$ where $\omega'_p = \omega_p \langle 1/\gamma \rangle^{\frac{1}{2}}$ (γ being the electron Lorentz factor).

As a preliminary step towards a radio source model, consider the effect of introducing a localized source of low frequency electromagnetic waves into a homogeneous plasma of density n_e (and corresponding plasma frequency ω_p). Suppose that the wave source radiates at an angular frequency ω *less* than ω_p, and that, after switching on, it maintains a constant luminosity $L(\omega)$. Neglecting attentuation, the strength parameter $f \propto 1/R$. Where R is small enough that $f > (\omega_p/\omega)^2$ the waves will accelerate all the electrons to a value of γ ($\gtrsim f$) such that $\omega'_p < \omega$. They can then propagate outwards at $\sim c$, making all the electrons they encounter relativistic. But when they have attained a value of R such that f is below this critical value, the waves can no longer propagate through the matter. They will instead expel the surrounding matter at a rate governed by the balance between internal (wave and particle) pressure and external ram pressure, the wave energy going mainly into accelerating relativistic electrons at the boundary of the cavity.

The gas within a galactic nucleus probably has a density of up to 10^6 cm^{-3} ($\omega_p/2\pi \simeq 10^7$ Hz) and extends out to a distance of ~ 10 pc (these are the rough parameters inferred for the emission-line region in active galaxies. However the situation is probably more complicated, the gas being irregularly distributed, with a rather small "filling factor" $\langle n_e \rangle^2 / \langle n_e^2 \rangle$). Even if this gas has rather little angular momentum per unit mass, it is likely to be rotationally flattened into a disc. Thus a more realistic picture might resemble that schematically illustrated in Fig. 2. Waves from each collapsing star create a separate expanding cavity,

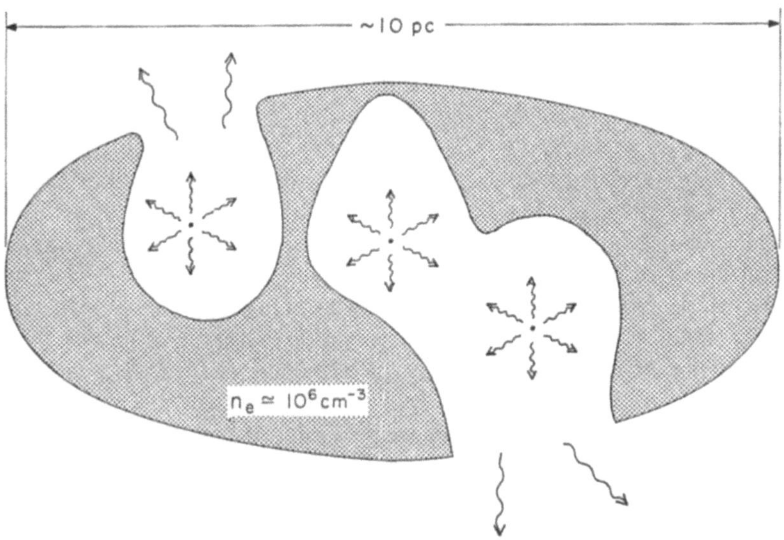

Fig. 2. Schematic illustration of the situation inside an active galactic nucleus according to the synchro-Compton theory of radio sources (REES, 1971). Stellar mass objects collapse at a rate of ~ 1 per year to form either spinning neutron stars or "mini-spinars". Low frequency electromagnetic waves generated by these objects produce expanding cavities in a disc of ionized gas (with density $n_e \simeq 10^6$ cm^{-3}) in which they are embedded. The waves "break out" preferentially in directions perpendicular to the plane of the disc

which is not spherical but tends to expand fastest in directions of decreasing density, and to "break out" of the nuclear disc preferentially in directions perpendicular to the axis. The wave energy emerging from the nucleus will thus tend to bifurcate into two beams. These beams would not, however, be well collimated. Even after escaping from the nucleus, the wave propagation will still be impeded by the lower-density plasma in the ordinary interstellar medium of the galaxy and in the halo, and even by the very tenuous intergalactic gas. The waves will therefore still have to inflate a cavity, but this cavity will be *conical* and not spherical. This situation is illustrated in Fig. 3. Another crucial property of the waves is that they tend to be *self-focussing*. This is because the electrons are accelerated to *higher* γ in places where the waves are stronger, and f higher. This makes the electrons "stiffer", lowers the effective plasma frequency, and *raises* the refractive index $\mu=(1-\omega_p'^2/\omega^2)^{\frac{1}{2}}$. The waves are thus focussed towards the directions where they are already strongest (analogously to the self-trapping phenomena observed in lasers). This effect therefore narrows down the cone angle, and improves the degree of collimation of the waves.

So long as the production of low frequency waves from the nucleus continues (and collapse of stellar-mass objects at a rate ~ 1 per year could yield a power $\sim 10^{46}$ erg s^{-1}) the cavity continues to grow. If it eventually penetrates beyond the galaxy, then one might establish a situation resembling that in Fig. 4. The cavity takes the form of a long tube, along which the waves are channelled. Wave energy is deposited at the end of the tube (accelerating relativistic particles) and the tube lengthens at a rate governed by balance between the wave momentum flux and the external ram pressure. The width of the tube grows relatively slowly,

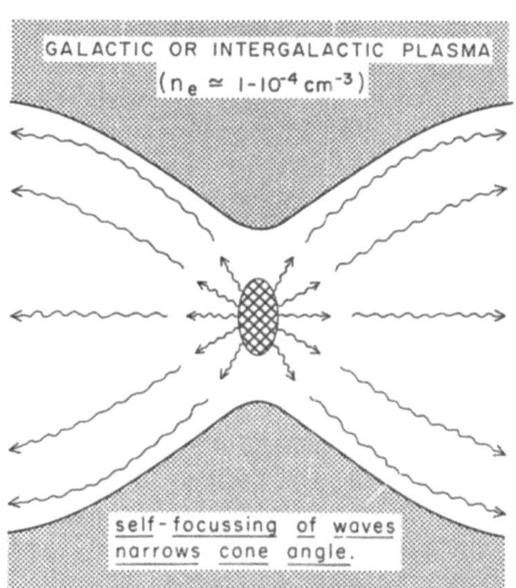

Fig. 3. The two beams of low frequency waves emerging from the nuclear disc will be "self-focussed" as they propagate outwards, because the refractive index is larger (i.e. closer to unity) where the waves are more intense

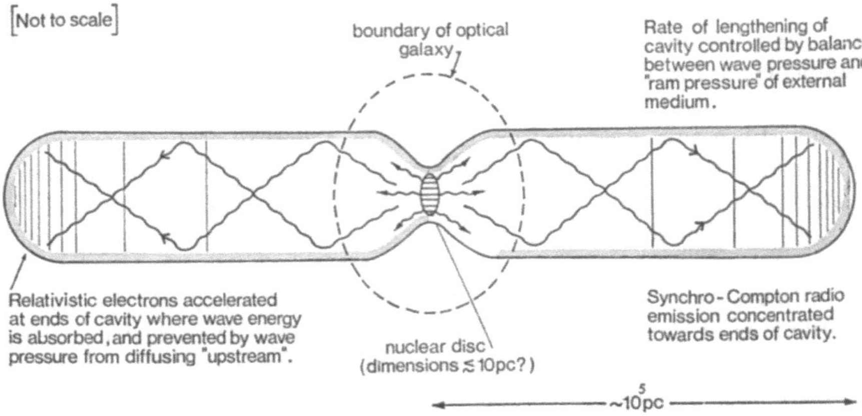

Fig. 4. Possible structure of double radio source. The radio emission can all be due to the "synchro-Compton" process. But it is conceivable that a relativistic wind from the nucleus may carry a magnetic field, which accumulates in the components. In this case, the electrons may predominantly emit ordinary synchrotron radiation, even if low frequency waves constitute the main energy output from the nucleus

because once the collimation has been achieved the transverse pressure is very small. Before reaching the ends of the cavity, the waves may suffer several glancing-incidence reflections off the sides of the tube. Since these reflections can occur where μ is still close to unity, little absorption need be associated with them.

What about the radio emission? As well as being responsible for accelerating relativistic electrons, the waves can also play the role usually ascribed to a magnetic field. Relativistic electrons exposed to strong waves radiate by a process intermediate between the synchrotron process and ordinary inverse Compton radiation —called "synchro-Compton radiation". In its gross features, this radiation is the same as would be emitted by an electron in a static magnetic field with the same strength as the wave H-vector. We therefore identify the ends of the tubular cavity in Fig. 4 with the components of a typical double source like Cygnus A (Fig. 1). The structure of the cavity boundary has not yet been calculated except in some very idealised cases. Ideally, one would like to know the spectrum of the relativistic electrons, but at the moment it is possible to estimate only the characteristic particle energy. A test electron released in a *monochromatic* strong wave of strength parameter f attains a γ in the range f to f^2 (depending on wave polarization, etc.) and thereafter executes periodic motion in its guiding centre frame. However, if it is instead exposed to isotropic *broad band* radiation, *stochastic acceleration* raises its energy until $\gamma \simeq f^{\frac{2}{3}}(\lambda/r_e)^{\frac{1}{3}}$, r_e being the classical electron radius, when the statistical acceleration is counter-balanced by synchro-Compton radiation losses (BLANDFORD, 1973). This acceleration is very much more rapid for electrons than for protons. The radiation reaching the ends of the "tubes" in Fig. 4 would in effect be broad-band radiation, since it would have travelled, along many paths with different travel-times, from many different collapsing objects. (It would not be isotropic, but this is easily taken into account in the calculation by transforming to a frame in which it *is* isotropic.) If we take numbers appropriate to Cygnus A, the strength parameter at the components would be ~ 1000/(characteristic wave frequency in Hz). This is an uncertain parameter, but even if $f \simeq 1$ there is no difficulty in stochastically accelerating electrons to the Bev energies required for the centimetre wavelength radio emission.

If this type of model has any virtues as compared with more conventional ideas, they would seem to be the following:

(i) The large-scale magnetic field, whose origin poses one of the major problems in conventional theories, becomes quite superfluous (but see remarks below)[3].

(ii) The waves provide continuous, and very efficient, acceleration of electrons "in situ", thus relieving problems connected with adiabatic expansion.

(iii) The energy content of extended sources is built up steadily over their lifetime. There is thus no need to invoke powers outside the range $\lesssim 10^{46}$ erg s^{-1} for which we have other evidence.

High resolution maps of the intensity and polarization distribution in double sources will soon be available. If the radio emission is ordinary synchrotron radiation, the polarization tells us the degree of ordering and preferred direction of the magnetic field. The synchro-Compton mechanism in most respects preserves the polarization characteristics of the low frequency waves. Even if these waves are *un*polarized, however, (as would be expected if they come from many spinning objects whose rotation axis are randomly oriented), the synchro-Compton radiation can be linearly polarized if the particle density in the emitting region is sufficient to make the refractive index (for the low frequency waves themselves) significantly less than unity. In these circumstances, the $E:B$ ratio is increased, and the synchro-Compton radiation is polarized perpendicular to the Poynting vector of the low frequency waves – in other words, in general perpendicular to the axis of a double source. There is no obvious way of understanding the polarization *along* the source axis (as is often found at the outer edges of the components) on the basis of pure synchro-Compton theory. However the transverse static magnetic field indicated by these observations can be incorporated in a more complex – and perhaps more realistic – theory where the main energy output from the nucleus still emerges as low frequency waves from spinning objects, but there is also an associated wind of relativistic particles containing a "frozen in" magnetic field. This field would "pile up", and be compressed and amplified, at the cavity boundary. Even if the magnetic field comprised only a small fraction of the emerging energy, it might nevertheless build up until its energy density in the components achieved a value comparable with that of the waves. This non-oscillatory magnetic field would then determine the polarization characteristics of the radio emission.

It is of course quite possible that more detailed future observations of extended sources will suggest an interpretation along quite different lines. One thing which is already hard to square with *any* of the conventional views is MACKAY's (1971 b) statistical evidence that the axes of double sources tend to lie in the *equatorial*

[3] Even if this field is as strong as the magnetic field associated with the waves, there is a wide range of $n_e \langle 1/\gamma \rangle^{-1}$ which is low enough to allow wave propagation, but nevertheless *high* enough for the electrons to be able to carry the currents associated with a "frozen in" field. This is because the length-scale over which the field changes direction, which determines the current density, is only $\sim c/\Omega$ for a wave field, but is much longer for a typical "frozen in" magnetic field (GUNN and REES, 1973).

plane of the associated elliptical galaxy—rather than being aligned with the rotational (minor) axis as most theories would lead one to expect. Unless one can make a plausible case that the spin of the nuclear disc or central supermassive object should *not* be aligned with the angular momentum of the whole galaxy, this evidence (if confirmed by more extensive studies) would suggest some radically different model-involving fission of a supermassive object, for example. SALPETER (1971) has considered the hierarchical fragmentation of a supermassive disc into smaller discs, and proposes that this process may continue until the "sub-discs" are stellar-mass objects which, if magnetised, would exhibit the behaviour envisaged by REES (1971). The time-scale for each stage in the fragmentation process is, however, estimated by SALPETER as up to $\sim 10^8$ times longer than the characteristic rotation periods. Perhaps, therefore, this process does not develop beyond the first stage of fragmentation, because dynamical interactions between members of the resulting flattened "few-body" system quickly eject one member, the remaining part of the system (perhaps just a binary) recoiling in the opposite direction. This process could yield double sources with axes aligned in the manner found by MACKAY (though it might be hard to reconcile this idea with the symmetry of the two components). The discovery of very compact components in extended sources, or of evidence for a double sub-structure within each component, would lend credence to this view.

Before leaving extended sources, it may be helpful to enumerate a few key observations which would be especially crucial to the elucidation of their true nature:

(i) *Direct Evidence for Intergalactic Gas.* So far the only evidence for emission or absorption by the intergalactic gas which plays an essential role in some theories of the confinement of radio source components is either negative or else ambiguous. Some indirect evidence that ram pressure by intergalactic gas is dynamically important in extended sources comes from a study by DE YOUNG (1972), who finds that the components of sources *in* clusters of galaxies have a mean separation which is only half that for sources *not* in clusters—just as might be expected if the gas were concentrated into clusters.

(ii) *Evidence for—or Limits to—Thermal Gas within Radio Blobs.* The large mass of material that must be invoked (in some theories) to provide inertia may be a detectable source of bremsstrahlung X-rays, or of optical emission. (Emission lines would not be readily detectable because of the large unknown mean velocity of the blob, and the likelihood of large internal turbulent velocities as well.) It would also cause internal Faraday rotation (and hence depolarization) of the synchrotron radiation. Studies of the radio polarization could also tell us about the structure and degree of ordering of the magnetic field.

(iii) *Detection of Non-Thermal Optical Emission from the Blobs.* If the power-law radio spectrum extended, without a cut-off, up to the visible band, the surface brightness of many radio components would be high enough for them to be detectable optically. Crude estimates of the colour of any such optical emission should be sufficient to indicate whether it is more likely to be bremsstrahlung or synchrotron radiation. Optical synchrotron radiation would indicate the presence of relativistic electrons of $\sim 10^{12}$ eV within the components. The radiative lifetimes of these very high energy particles would be too short for them to have survived for the whole age of the source. Their presence would therefore imply continuous "in situ" acceleration, either (as in the synchro-Compton model) by low frequency waves, or else by compact objects in the components.

(iv) *Detection of Fine Structure or "Hot Spots" ($\ll 1''$ in Size) in Blobs.* This would imply regions whose energy density greatly exceeded the mean, and would therefore aggravate the confinement problem. (Unless the "hot spots" were merely transient features which did not maintain their identity).

III. Cosmological Interferences

The theories of extended radio sources so far proposed all have the consequence that a particular source starts off small, expands outward from the parent object, and eventually fades. This suggests that perhaps the wide scatter in observed source properties can be interpreted as different stages in the evolution of a single more standardized phenomenon. To this end, many authors have constructed diagrams in which size, power, surface brightness, etc. are plotted against each other. The results are discouraging, in that no very clear-cut correlations emerge. However even if such diagrams revealed some obvious correlations, instead of looking more like pure "scatter diagrams", one should be wary of over-interpreting the results, for the following reasons:

(i) Some areas of such diagrams are bound to be blank because of selection effects. For example, the higher-power sources in any survey are bound to be all at substantial distances; and therefore they are less likely to be resolved.

(ii) The space density of the *very* powerful sources is so low that those in the sample will have significant cosmological redshifts. Therefore the correlation diagram should perhaps be convolved with some *cosmological* effect whereby the mean properties of sources depend on cosmic epoch.

(iii) To interpret any band in such a diagram as an evolutionary track could be as misleading as interpreting a globular cluster *H-R* diagram as representing the evolutionary track of an individual star.

One common feature of most theories is that the active lifetime of a strong source is $\lesssim 10^7$ years (though this obviously does not preclude repeated outbursts in the same galaxy, nor the possibility that the sources may continue for a longer period at a lower radio luminosity). The fact that this is much less than the Hubble time means that the sources observed at large redshifts are not individually younger – in the sense of having experienced shorter active lifetimes – than nearby sources. Thus the apparent epoch-dependence of radio source properties, inferred from the radio source counts and from the redshift-distribution of *QSRS*, must be attributed to an epoch-dependence in the source birth rate *and/or* in the form of the evolutionary track traced out by a typical source.

Another direct inference from the short lifetime of individual sources is that dead sources must greatly outnumber the living. This means that the integrated effect of all generations of radio sources is important for the energy budget of intergalactic space. This conclusion is even stronger when the higher inferred comoving density of strong sources at early epochs is taken into account. In particular, radio sources may produce a universal flux of cosmic rays; and a flux of relativistic electrons which, via inverse Compton scattering of microwave background photons, contribute to the X-ray background. In all ram pressure confinement models the energy expended in pushing against (and heating) the external medium exceeds, by a factor at least $\sim(D/d)$, the integrated radio luminosity. This energy—which is provided either by the kinetic energy of thermal material in the blobs or by the continuous thrust of low frequency waves channelled from the galactic nucleus—may therefore be important for heating the intergalactic medium, and could even make it so hot that it is no longer confined within clusters. This also means that the trapping of individual particles need not be very efficient in the "wave field" models, since there is enough energy available in the form of low frequency waves to replenish the relativistic electrons $\sim(D/d)(c/V)$ times, when V is the outward velocity of the components.

IV. Compact and Variable Sources

So far I have concentrated on radio sources whose characteristic dimensions exceed the size of an optical galaxy, and which can be mapped (crudely, at least) using the aperature synthesis technique. But there also exist radio sources only a few parsecs in size, which—when they are identified—always appear to *coincide* with their optical counterpart. The existence of these compact components was first inferred from the presence of low frequency cut-offs in some radio spectra.

Interpreting these cut-offs as an effect of synchrotron self-absorption, which would occur when the radio surface brightness temperature became comparable with the kinetic temperature $\gamma m_e c^2/3k$ of the relevant relativistic electrons, one derives angular sizes of only $\sim 10^{-3\prime\prime}$. This corresponds to linear dimensions of no more than a few parsecs, even when the source has a cosmological redshift $z \simeq 1$. Some sources possess both extended and compact components: indeed, it is possible that all deviations from a straight power-law radio spectrum are attributable to the presence of small self-absorbed components.

The small size deduced from the spectral shape was confirmed when *radio variability* was discovered in 3C273 (DENT, 1965) and subsequently in many other objects (KELLERMANN (1971) and references contained therein). In many cases the radio variability can be analysed into a succession of fairly similar outbursts. Each contributes a spectrum $S(\nu)$ peaking at some frequency ν_{max}. Generally ν_{max} drifts towards lower frequencies, and $S(\nu_{max})$ falls, on a timescale ~ 1 year. The most obvious way whereby an optically thick source of limited surface brightness can increase its luminosity is by increasing its area, and the fairly extensive data on variable sources have often been interpreted in terms of a very simple model first suggested by SHKLOVSKY (1965). This model involves an expanding cloud of relativistic particles, optically thick below the frequency ν_{max} which itself decreases during the expansion.

Detailed studies, however, revealed an apparent snag with this interpretation. The minimum angular size in seconds of arc is

$$\theta''_{\text{self-absorption}} \simeq 1.5 \times 10^{-3} \left(\frac{H}{10^{-4}\,\text{gauss}}\right)^{\frac{1}{4}} S_{\text{f.u.}}^{\frac{1}{2}}\, \nu_{\text{GHz}}^{-\frac{5}{4}}.$$

Taking the redshifts of the sources to be cosmological, the value of $\dot{\theta}$ deduced from the variability can be translated into a rate of apparent linear expansion. If this rate were assumed to be $<c$, this led in practice to an upper limit on H which was far below the equipartition strength, and in fact so low that the overall particle energy requirements became very high, and the synchrotron emission would be swamped by inverse Compton radiation. In fact the assumption that inverse Compton losses are at most comparable with the synchrotron losses leads (in conjunction with the above relation) to the result that we must have $\theta \gtrsim \theta_{\text{inverse Compton}}$, where

$$\theta''_{\text{inverse Compton}} \simeq 1.3 \times 10^{-3} S^{\frac{1}{2}}_{\text{f.u.}} \nu^{-1}_{\text{GHz}}.$$

These results could, however, be reconciled by appealing to the fact that apparent rate of increase of linear size of a relativistically expanding object can exceed c by a factor of the order of the bulk Lorentz factor γ. (The possibility of apparent "superluminous" effects in astronomy was pointed out by CLAUDERC (1939) in connection with "light echoes" from the Nova Persei observed by KAPTEYN in 1901. The possibility of similar effects in fast-moving material objects was discussed by REES (1966) with specific application to these variable radio sources.) Taking these straightforward kinematic effects into account, we obtain

$$\theta''_{\text{variability}} \lesssim \frac{10^{-4}\, t_{\text{years}}}{z} \times \begin{Bmatrix} \text{relativistic} \\ \text{factor } \sim\gamma \end{Bmatrix}$$

where t is the timescale of the variability and z the source redshift. This formula holds only for $z \lesssim 1$: if z is very large one must specify a particular world model.

The situation in the years 1966–68 was that *if* the radio emission from compact components *was* synchrotron-type incoherent emission from relativistic electrons, then θ had to be changing at a calculable rate $\gtrsim \dot{\theta}_{\text{inverse Compton}}$, and this rate implied "superluminous" expansion. At that time, these angular sizes were not directly measurable, but in the last three or four years the technique of Very Long Baseline Interferometry (VLBI) has achieved angular resolutions down to $0.4 \times 10^{-3\prime\prime}$ at centimetre wavelengths. As is now well known, these studies did indeed find the predicted changes in angular structure, suggesting that material is moving with Lorentz factor of up to ~ 5. This supports the hypothesis that an incoherent process is responsible for the emission. The inferred values of γ are not really surprising—after all, the only material which is inferred to be travelling at this speed is relativistic plasma, whose sound speed is itself relativistic. It is of course conceivable that no actual expansion is involved in any of these sources, and that we are instead witnessing the flaring and fading of independent high-surface-brightness objects. However this would be as likely to mimic a contraction as an expansion, and will therefore be testable when more examples have been studied. Even if the successive outbursts originate in different locations, the debris from each must still display "superluminous" expansion if the radio emission is incoherent.

Models can be contrived in an infinite number of ways to fit the presently available VLBI and variability data on these sources. There is no reason to expect conditions to be specially simple deep within the nucleus of an active galaxy, and the consistency of the observations with very simple models is probably symptomatic of the paucity of data rather than the excellence of the models. It seems less worthwhile to construct elaborate models to explain specific cases than to concentrate on basic geometry-independent quantities like surface brightness and timescales.

What is the physical nature of these radio outbursts in variable sources? The radio power, integrated over the duration of a single outburst, in no case amounts to more than $\sim 10^{53}$ erg, and could therefore come from a stellar-mass object provided the conversion into radio emission was reasonably efficient. The expanding cloud of relativistic electrons could be associated with a shock wave from a supernova-type event. Alternatively, a burst of low frequency strong waves from a newly formed pulsar or collapsing "mini-spinar", propagating through a plasma of density $\sim 100\ \mathrm{cm}^{-3}$, could accelerate the ambient electrons to relativistic energies. These electrons, emitting synchro-Compton radiation in the field of the waves, would produce a radio outburst whose time-dependence agrees with that observed (BLANDFORD and REES, 1972)[4].

The character of radio variability perhaps lends some support to interpretations of galactic nuclei involving many independent energy sources.

It is obviously of great interest if all extragalactic radio emission is due to incoherent processes. The fact that all sources seem at least marginally resolved by VLBI indeed lends some positive support to this view. However there is a straightforward limitation on VLBI which makes it difficult to settle this question by ground-based telescopes. The smallest angle that can be resolved by an interferometer whose baseline is of the order of the Earth's diameter is

$$\theta''_{\mathrm{VLBI}} \simeq 1.5 \times 10^{-3}\, \nu_{\mathrm{GHz}}^{-1}.$$

This corresponds to an upper limit on the brightness temperature, *independent of* ν, of $T_{\mathrm{VLBI}} \simeq 10^{12}\, S_{\mathrm{f.u.}}$ °K. This happens to be about the maximum temperature which one would expect from an incoherent process in the radio band. Thus one cannot settle this question interferometrically without using (say) an Earth-Moon baseline. There is, however, an indirect technique: one may search for variations at *low* frequencies, which would be too rapid to be explained by synchrotron theory without invoking bulk motions with an implausibly large γ-factor. Such variations could be due either to a *scaled-up* mechanism of the type operative in pulsars or other celestial masers, *or* (as discussed by LOVELACE and BACKER (1972)) to interstellar scintillation. Interstellar scintillations would not be observed unless the source angular size was $\ll \theta_{\text{self-absorption}}$, so this latter interpretation of variability would also imply coherent emission.

As was done above for extended sources, it may be useful now to list some specially important observations pertaining to compact radio sources.

[4] If the synchro-Compton interpretation of extended sources is correct, one might wonder why extended sources do not *always* have variable central components. It is possible that radio emission from many nuclei may suffer free-free absorption, and that variable sources are detected only when there is a low-density "lane" along our line of sight. In any case, the example of the Crab Nebula—where the pulsar provides the essential energy input for the whole system, despite being a very weak emitter compared with the nebula itself—warns us that the hypothesis that events in the nucleus energise extended sources is fully compatible with absence of conspicuous activity in the nucleus itself.

(i) *Confirm or Refute "Superlight" Expansions.* It should soon be feasible to accumulate enough observations, with several different baselines, of the strongest variable sources, and to correlate these with intensity measurements at various frequencies to build up a more detailed picture of the source structure. This information should reveal whether this is genuine expansion and also whether the compact sources show directionality or double structure. It is possible to distinguish between different source shapes (e.g. rings, doubles, ellipses, etc.) with existing VLBI data, and it is not out of the question that it will soon be possible to cover enough of the $u-v$ plane to produce a reliable model of the central 0.1″ of a compact source with millisecond-of-arc resolution.

(ii) *Search for Rapid Variability at $\lesssim 100$ MHz, either Caused by Interstellar Scintillations or Intrinsic to the Source.* This would indicate coherent emission. It would seem especially worthwhile to study 3C273, which has a compact component whose intensity (according to the evidence from lunar occultation) rises steeply towards low frequencies in a fashion strikingly reminiscent of pulsar spectra. It would indeed be exciting if the story of the compact source in the Crab Nebula were to repeat itself on an extragalactic scale.

(iii) *Detect "Jitter" in Radio Position (Indicating Many Independent Stellar-Mass Explosions) or Periodicity in Variations (Suggesting One Supermassive Object).* Studies of the polarization may also indicate whether the field directions were the same in all outbursts. If this *were* so, however, it would not necessarily imply repeated outbursts of a single object: particles could be being ejected from independent objects into a pre-existing field; or else the explosions could be directional (as would be the case for synchro-Compton emission in the situation illustrated in Fig. 2).

High resolution radio studies of these compact sources are perhaps the most direct clues we are likely to have on the nature of the mysterious regions where the energy for all non-stellar activity in extragalactic objects is generated. But when one tries to extend the chain of inference from the radiation mechanism (the most direct deduction that one can make from observations) to the particle acceleration processes, the dynamics, and the primary source of energy, one soon encounters the realm of conjecture and speculation. It is probably over-optimistic to expect any very rapid breakthrough in this fundamental topic.

The *extended* sources, on the other hand, provide scope for useful work on theoretical problems which – though perhaps less fundamental – are at least well-posed. This is because, if we are right in thinking that interaction with an external medium plays a key role in these larger sources, then their structure and evolution may in some respects be insensitive to what is happening in the nucleus itself; so the gas-dynamical aspects of the phenomenon can perhaps be tackled profitably even in complete ignorance of the form of the energy supply. The relevance of the powerful double sources to other branches of high energy astrophysics would lend added interest to any new insight. But the characteristic *binary structure* of these sources, and the implied directivity of the energy ejection (which has been well known for at least ten years) seems to demand an explanation as insistently as, for example, the spiral structure of galaxies, concerning which we received a degree of enlightenment earlier in this meeting.

References

BARDEEN, J.M.: Nature **226**, 64 (1970).
BLAKE, G.M.: Monthly Notices Roy. Astron. Soc. **156**, 67 (1972).
BLANDFORD, R.D.: Astron. Astrophys. **26**, 161 (1973).
BLANDFORD, R.D., REES, M.J.: Astrophys. Letters **10**, 77 (1972).
BRANSON, N.J.B.A., ELSMORE, B., POOLEY, G.G., RYLE, M.: Monthly Notices Roy. Astron. Soc. **156**, 377 (1972).
BURBIDGE, G.R.: Astrophys. J. **124**, 416 (1958).
CHRISTIANSON, W.A.: Monthly Notices Roy. Astron. Soc. **145**, 327 (1969).
CLAUDERC, P.: Ann. Astrophys. **2**, 271 (1939).
COLGATE, S.A.: Astrophys. J. **150**, 163 (1967).
DENT, W.A.: Science **148**, 1458 (1965).
DE YOUNG, D.: Astrophys. J. **167**, 541 (1971).
DE YOUNG, D.: Astrophys. J. Letters **173**, L. 7 (1972).
DE YOUNG, D., AXFORD, W.I.: Nature **216**, 129 (1967).
GUNN, J.E., REES, M.J.: Submitted to Astron. Astrophys. (1973).
KELLERMANN, K.I.: In D. O'CONNELL (ed.), Nuclei of Galaxies, Pont. Acad. Sci. Scripta Varia **35**, p. 217 (1971).
LOVELACE, R.V.E., BACKER, D.C.: Astrophys. Letters **11**, 135 (1972).
LE BLANC, J.M., WILSON, J.R.: Astrophys. J. **161**, 541 (1970).
LYNDEN-BELL, D.: Nature **223**, 690 (1969).
LYNDEN-BELL, D., REES, M.J.: Monthly Notices Roy. Astron. Soc. **152**, 461 (1971).
MACKAY, C.D.: Monthly Notices Roy. Astron. Soc. **154**, 209 (1971 a).
MACKAY, C.D.: Monthly Notices Roy. Astron. Soc. **151**, 421 (1971 b).
MILLS, D., STURROCK, P.A.: Astrophys. Letters **5**, 105 (1970).
MITTON, S.A., RYLE, M.: Monthly Notices Roy. Astron. Soc. **146**, 221 (1969).
MORRISON, P., CAVALIERE, A.: In D. O'CONNELL (ed.), Nuclei of Galaxies, Pont. Acad. Sci. Scripta Varia **35**, p. 485 (1971).
NORMAN, C.A., HAAR, D. TER: Submitted to Astron. Astrophys. (1972).
PIDDINGTON, J.H.: Monthly Notices Roy. Astron. Soc. **148**, 131 (1970).
REES, M.J.: Nature **211**, 468 (1966).
REES, M.J.: Nature **230**, 312 and 510 (1971).
REES, M.J.: In preparation (1972).
RYLE, M., LONGAIR, M.S.: Monthly Notices Roy. Astron. Soc. **136**, 123 (1967).
SALPETER, E.E.: Nature Phys. Sci. **233**, 5 (1971).
SHKLOVSKY, I.S.: Sov. Astron. **9**, 22 (1965).
SPITZER, L.: In D. O'CONNELL (ed.), Nuclei of Galaxies, Pont. Acad. Sci. Scripta Varia **35**, p. 443 (1971).
STURROCK, P.A., BARNES, L.: Astrophys. J. **176**, 31 (1972).

Discussion

OORT, J.H.:

The first desiteratum Professor REES had on his list was "evidence for intergalactic gas". Recently MILEY, VAN DER LAAN and VAN DER KRUIT at Leiden and Perola have presented data which to me appear to give rather strong evidence for the existence of rather dense intergalactic gas in some clusters of galaxies. In particular they have investigated structure and polarization of the "tails" shown by some radio galaxies in clusters. The most striking cases are in the Perseus cluster where RYLE and WYNDRAM first discovered this phenomenon in NGC 1265 and IC 310. In Westerbork another, fainter radio source with a tail was discovered: in this case the tail is directed roughly *towards* NGC 1275 instead of away from it as in the other two galaxies. The Leiden astronomers and Perola have interpreted these tails as due to motion of the radio galaxies through a fairly dense intergalactic medium. Similar "tails" have been observed in other clusters. It would appear that the interpretation given is a plausible one and gives in my opinion rather strong evidence for the presence of a considerable amount of intergalactic gas.

REES, M. J.:
I completely agree with Professor OORT that the work by MILEY and his collaborators is exceedingly important and probably provides the most convincing evidence so far that there actually is an intergalactic gas.

Perhaps I could also mention another piece of recent evidence which leads to the same conclusion. This comes from work by DE YOUNG who has compared the mean separation of the components of double sources inside and outside clusters. He finds that the separation is only half as great for the sources *within* clusters of galaxies. The most natural interpretation of this result is that the components are indeed being braked by intergalactic gas whose density is higher in the clusters than outside.

MCVITTIE, G. C.:
In the initial stage of your process it looked as if spherical "bubbles" were being produced in the gas. Can one envisage the collapse of such bubbles, with consequent violent effects, as in liquids?

REES, M. J.:
It would indeed be interesting if there were effects analogous to "depth charges" which gave rise to directed expulsion of energy from galactic nuclei. However, it seems to me that these effects are rather unlikely because the bubbles would collapse at a speed comparable with the external sound speed and this is very much slower than the rate at which the bubbles were inflated.

MALTBY, B.:
Could you tell me why you don't get radio radiation from outside the regions where the self-focusing waves are moving?

REES, M. J.:
Because the emission of radio waves requires the presence not only of relativistic particles, but also of low frequency radiation to accelerate them (unless there is also a non-oscillatory magnetic field).

Pulsars

(Invited Lecture)

By F. Pacini
Laboratorio di Astrofisica Spaziale, Frascati, Italy

Many outstanding problems of modern astrophysics involve the release of large amounts of non-thermal energy under the form of magnetic fields and relativistic particles. In our Galaxy, a typical example of this situation is the Crab Nebula, the debris of a star which exploded in 1054 A.D.

Polarization measurements have shown that the electromagnetic emission of the Crab Nebula (total output between the low radio frequencies and the hard X-rays $\sim 10^{38}$ erg s^{-1}) is due to the synchrotron process from electrons in the energy range 10^8 eV up to 10^{13}–10^{14} eV moving in a magnetic field 10^{-3}–10^{-4} gauss.

The total energy stored in the Crab amounts to about 10^{49} erg, probably in rough equipartition between particles and magnetic fields. The energy distribution for the nebular particles obeys a power law: the spectrum is rather flat at low energies $\left(N(E) \propto E^{-1.5}\right.$ at $\left.E \lesssim 10^{11}\,\mathrm{eV}\right)$ but tends to steepen at higher energies. Since the average electron energy is ~ 10 GeV, the Nebula contains at least 10^{51} electrons moving with relativistic velocity. If these electrons have been injected at an approximately constant rate over the last 10^3 years, some source injects electrons into the nebula at a rate $\sim 10^{40}\,s^{-1}$.

Several arguments have forced the astronomers to consider the possibility of a source providing continuously electrons to the Nebula. Prominent among these considerations is the short lifetime of the electrons emitting at high frequencies (optical and X-ray electrons). In order to maintain a steady-state, electrons should be injected at a rate which matches the nebular luminosity.

A similar problem exists for the origin of the nebular magnetic field which cannot be due to the expansion of the original stellar field or to a central hypothetical dipole (in both cases the required strength for the field source is unreasonable). As well known, the discovery of the pulsar NP 0532 inside the Crab has solved the basic problems of the energy source for the nebular activity. It is now generally agreed that both the relativistic particles and the system of currents generating the nebular field originate from the central rotating, strongly magnetized, neutron star. In the following we shall outline the basic processes around a rotating magnetized neutron star, with particular emphasis on those questions which still await an answer.

The basic physical principles which operate around a neutron star are very simple.

The first possibility is known in the Laboratory as the "unipolar inductor". If a magnetized sphere rotates and its pole is connected to the equator through a non-rotating circuit, an electromotive force arises and the circuit is traversed by a current density $\boldsymbol{J}$ (Faraday experiment). The torque $\boldsymbol{J}\times\boldsymbol{B}$ slows down gradually the sphere. In the case of a neutron star one cannot assume a priori the existence of an external conductor, since the very large gravitational field tends to squeeze the atmosphere into a few centimeters ($KT \ll$ escape energy ~ 50 MeV|nucleon). However one expects rapid rotation and very strong magnetic fields (if $B\sim 10^2$ gauss when $R = 1 R_\odot \sim 10^{11}$ cm, then $B\sim 10^{12}$ gauss when $R\sim 10^6$ cm!).

The spinning of this magnet causes a charge separation in the stellar interior and generates on the outside a large irrotational electric field. Indeed, inside the star (a good conductor) the electric force should vanish

$$\boldsymbol{E} - \boldsymbol{v}\times\boldsymbol{B} = 0$$

$$\boldsymbol{v} = \boldsymbol{\Omega}\times\boldsymbol{r}.$$

Therefore

$$4\pi\, e(n_- - n_+) = \operatorname{div}\left(\frac{\boldsymbol{v}}{c}\times\boldsymbol{B}\right)$$

or

$$n_- - n_+ = \frac{2\boldsymbol{\Omega}\cdot\boldsymbol{B}}{4\pi\, e\, c}.$$

This leads to a surface charge and to an external quadrupolar electric field. For a pulsar like NP 0532 with $P = 33$ m s, $E\sim 10^{12}$ volt cm^{-1} at the stellar surface if $R\sim 10^6$ cm and $B\sim 10^{12}$ gauss. This value is largely sufficient to overcome the gravitational field of the star and to extract from the surface electric charges, thus creating *on his own* a surrounding conducting medium. The fate of the extracted plasma is different according to its location on the star's surroundings. If it is connected with the field lines which close themselves before reaching the critical "speed of light distance" $= c/\Omega$, the plasma can corotate with the star. Charges will be pulled out from the star until they cancel the components of the exterior electric field parallel to $\boldsymbol{B}$. The local charge density $(n_- - n_+)$ is then determined by the Poisson equation. The corotating charge density is an important current source for the magnetic field at $r\sim c/\Omega$. A realistic self-consistent treatment of this problem is not yet available despite the efforts of several investigators.

On the other hand, it is clear that the plasma anchored to the field lines which go beyond the critical distance cannot corotate since corotation would imply velocities larger than c. Charges should therefore escape along these "open lines" after being accelerated electrostatically by the large existing potential differences. (Several authors have suggested ways in which charges escaping along these field lines could be responsible for the pulsed emission.) The distribution of electric potential is such that electric charges of different sign follow preferentially different sets of field lines and therefore give rise to an ordered system of poloidal currents which reach far away from the star. These poloidal currents in turn give rise to an extended toroidal magnetic field $\boldsymbol{B}_t$.

The slowing down rate of the Crab pulsar permits to estimate the braking torque and determines the strength of the stellar magnetic field. This turns out to be around 10^{12} gauss, in agreement with the expectations of flux conservation. Once the field and the period are both known, one can evaluate quantitatively the electromagnetic properties of the space surrounding the neutron star. For the Crab pulsar, one finds that it can accelerate particles to a top energy $\sim 10^{16}$ eV; also, the escaping charges would be able to generate a field $B \sim 10^{-4}$ gauss at ~ 1 parsec and this may very well be the origin of the magnetic field in the Crab Nebula. J. GUNN and M. REES have recently suggested a variance of this theory. In essence they note that the boundary of the Nebula represents naturally the conductive walls of a box which is continuously filled with currents and magnetic field by the unipolar mechanism. Inside the nebula one can therefore distinguish a central region $r < R_*$ where the outflowing energy dominates the ambient energy and an outer region $r > R_*$ where the opposite becomes true. The region $r > R_*$ is continuously fed with magnetic energy and the average spacing between lines of force becomes smaller than in region $r < R_*$. This is equivalent to a magnetic amplification where the nebular magnetic energy $W_B \sim B^2 R^3$ increases because of the "pulsar luminosity" L and decreases because of the adiabatic expansion at velocity v

$$\frac{d}{dt} W_B = L - \frac{v}{R} W_B .$$

The consequences of this model are presently under investigation.

There is another very simple process presumably also going on around a rotating neutron star, namely the emission of large quantities of magnetic dipole radiation at the basic rotation frequency Ω. The amount of energy lost by the star is simply given by the flux of the Poynting vector across the speed of light cylinder. Apart from a geometrical factor, this flux is given by

$$L \sim \frac{B_{\mathrm{cr}}^2}{8\pi} \times c \times 4\pi \left(\frac{c}{\Omega}\right)^2 .$$

In a dipole field $L \sim \frac{\mu^2}{c^3} \Omega^4$ (μ is the magnetic moment). Therefore one finds the usual expression for the magnetic dipole radiation losses. The same formula holds also for the losses due to the unipolar inductor mechanism but in one case (dipole radiation) μ indicates the component perpendicular to the rotation axis while in the other μ is the parallel component. There is a possible observational verification for this type of energy loss in a dipole field. Since the energy comes at the expenses of the stellar rotation, we have also $L = I \Omega \dot{\Omega} \propto \Omega^4$ (I is the stellar moment of inertia). The exponent n of the relation $\dot{\Omega} \propto \Omega^n$ can be determined observationally for the Crab Nebula pulsar and one finds $n \sim 2.6$ in approximate agreement with the expected value $n = 3$. It has been suggested that the nebular magnetic field is the low frequency wave field, rather than being a static field as it is generally assumed. In this case, however, the nebular radiation would not be ordinary synchrotron radiation but its equivalent in a wave field. M. REES and others have shown that one would then expect, for instance, a small percent of circular polarization in the nebular optical emission. Detailed recent measurements have failed to reveal this polarization and one can conclude that the waves do not constitute the dominant field over the nebular volume. Of course this leaves open two questions:

a) is it possible that the nebular field is the sum of a dominant static field (which causes most of the energy loss from the particles) and of a wave field (which keeps feeding energy to the particles)?

b) is it possible that the more internal electrodynamics at $r < R_*$ is dominated by the waves?

There is no answer as yet to these very important questions. Also, it is not clear at the present stage whether the plasma continuously escaping from the pulsar is sufficient to affect the propagation of the wave. This could happen in various ways:

1) by preventing the propagation at all of an electromagnetic phenomenon (the waves could then be of a magnetohydrodynamic type).

2) By changing the phase velocity of the waves.

3) By reducing very much the wave amplitude because of the energy transfer "waves into particles"–(in other words, the wave intensity would not only decrease because of the r^{-2} term but there could be additional attenuation). In all these cases, the energy losses from the star are likely to remain unchanged; however, the details of our understanding of the nebular electrodynamics could be radically affected.

In conclusion, it is clear that in the last few years one has started to understand the final energy source in at least one class of highly non-thermal objects. The basic mechanism appears to be deceptively simple: a rotating magnetized object gives rise to large scale, strong electromagnetic fields. In several ways, these fields are capable of accelerating particles and of creating a system of relativistic currents: these currents create large scale magnetic fields which interact with the high energy electrons and cause a conspicuous flux of non-thermal electromagnetic radiation. Several efforts are presently being made to understand along these lines not only the non-thermal activity of SN Remnant but also various aspects of the similar activity in violent extragalactic nuclei (see the paper of M. Rees). The understanding of all subtle complexities of the situation under investigation is likely, however, to require much more effort.

References

The present status of pulsar theory is also discussed in two review articles:

P. Goldreich, F. Pacini, M. Rees: Comments on Astrophysics and Space Science, Pulsar Theory 1: Dynamics and Electrodynamics, Vol. III, No. 5. Pulsar Theory 2: Radiation Processes, Vol. IV, No. 1.

An excellent, up-dated technical review of neutron stars and pulsar electrodynamics is:

M. Ruderman: Pulsars: structure and dynamics, Ann. Rev. Astron. Astrophys. 1972.

High Resolution 21-cm Observations of Tycho's Supernova Remnant

By R. G. STROM and R. M. DUIN
Sterrewacht, Leiden, The Netherlands

With 3 Figures

Abstract

High resolution observations of the remnant of Tycho's supernova show that the ring-shaped radio source is consistent with emission occurring in a spherical shell, provided that the magnetic field has a partially ordered radial component. Polarization observations suggest that a thin Faraday screen surrounds the source, and this may have been formed during compression of the interstellar medium swept up by the expanding remnant.

We have observed 3C10, the remnant of the supernova of 1572, with the Westerbork Synthesis Radio Telescope[1]. At its 1415 MHz observing frequency, this instrument (which was described earlier in Professor OORT's lecture) produces a synthesized beam whose half power dimensions (at the declination of 3C10) are 24.″3 arc in right ascension and 27.″1 arc in declination. The observations provide the information needed to produce the sky brightness distribution of the Stokes parameters I, Q and U as described by BROUW (1971).

Fig. 1 is a contour map of the Stokes parameter I, showing the nearly circular ring of emission whose steep outer boundary is unresolved in many places. Because the synthesis does not include a zero spacing measurement, there is a downward shift of all contours, the true zero having a negative value in the synthesized map. Taking this effect into account we find that although, in contrast with earlier measurements (e.g. BALDWIN, 1967), significant emission does come from the centre of the remnant, it is still not possible to reproduce the brightness distribution using a spherical shell where each volume element radiates isotropically. However, if the magnetic field has a partially ordered radial component, as suggested by short wavelength observations (KUNDU and VELUSAMY, 1971; WEILER and SEIELSTAD, 1971), the measurements are consistent with emission from a spherical shell whose thickness is a quarter of its outer radius. The required radial field would contain some 30% of the total magnetic energy density.

The linearly polarized intensity (in terms of Stokes parameters, $(Q^2+U^2)^{\frac{1}{2}}$) shown in Fig. 2 also has a ringlike appearance, but with more fine scale structure than the total intensity. A striking feature of this vector map is the presence of regions several beamwidths in size whose predominant polarization position angle is quite different from that of adjacent areas. There appears to be very little

[1] The Westerbork Radio Observatory is operated by the Netherlands Foundation for Radio Astronomy with the financial support of the Netherlands Organisation for the Advancement of Pure Research (Z.W.O.).

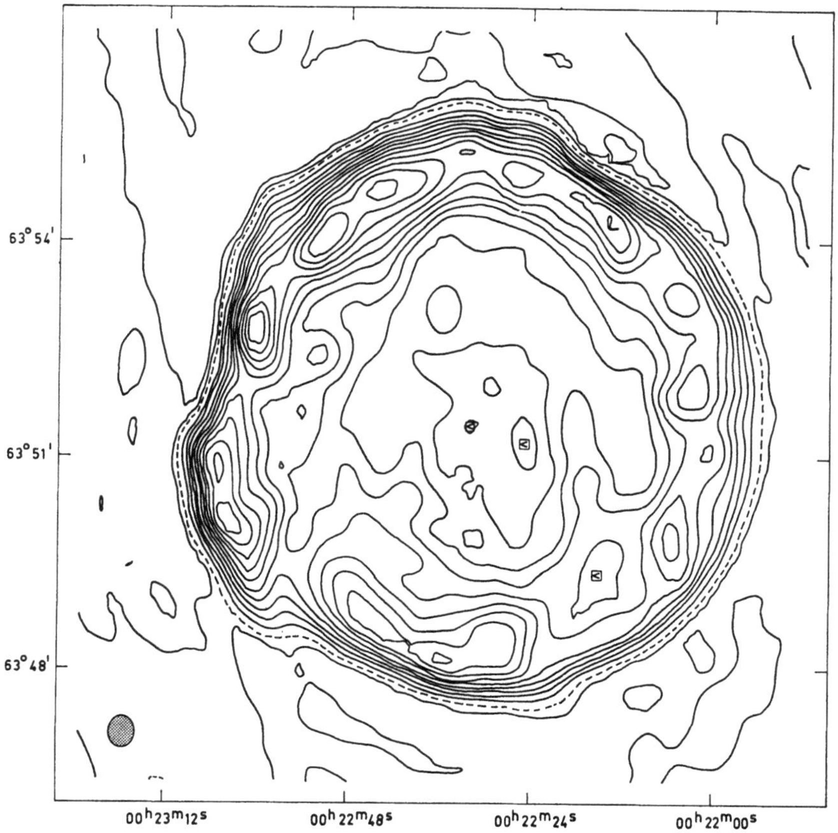

Fig. 1. A contour map of the total intensity, I. Above the apparent zero level (see text) which is dashed, the contour interval is 27.5 K; below the zero level, the contour interval is 13.75 K. The symbol "<" enclosed in a square indicates a local minimum. The synthesized half power beam is shown as a shaded ellipse

position angle variation within each region, though the boundaries between them are abrupt, generally unresolved. In view of the high degree of alignment which short wavelength observations indicate, this irregular structure is probably the result of Faraday rotation.

To quantify the effect, we have taken the absolute difference between the position angles of all neighbouring points in right ascension and declination. For each square of four adjacent points, we have computed the mean absolute difference and noted where it exceeds 10°. The resulting distribution of large position angle changes is shown superimposed on a contour map of the linearly polarized intensity in Fig. 3. In most cases, peaks in the polarized intensity are separated by features of large position angle change, suggesting that much of the observed fine scale structure is related to these features. We have generally not resolved transitions between adjacent regions and the depressions are probably the result of beamwidth depolarization, which is inevitable when large position angle changes occur over regions smaller than the beam.

Typically, the degree of linear polarization at individual peaks is 8%, though it exceeds 12% in several places. The degree of polarization at short wavelengths

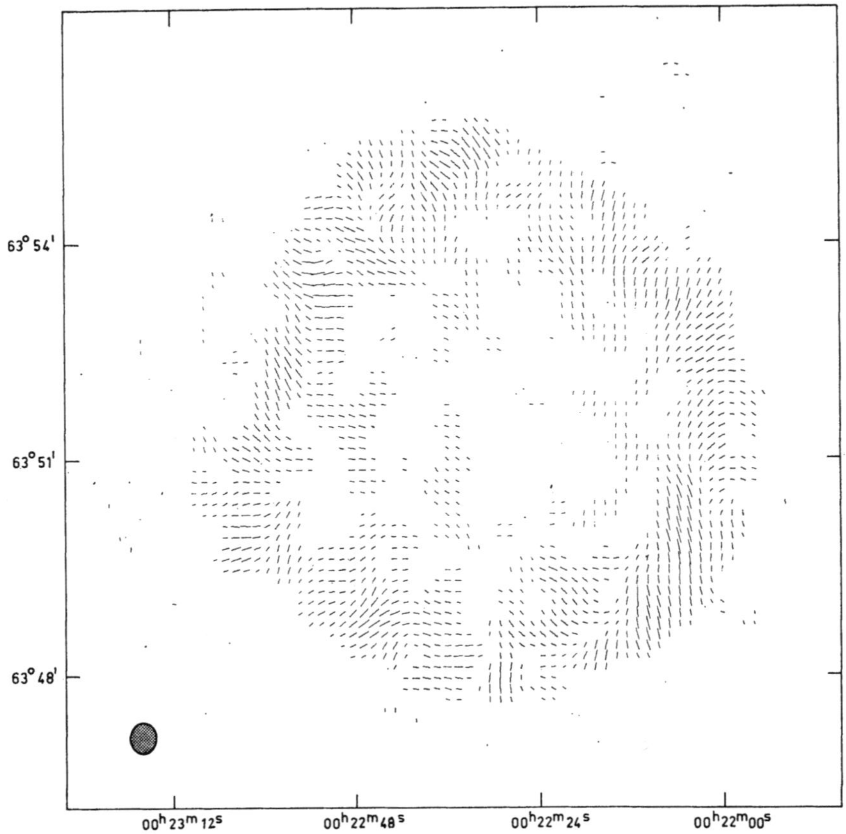

Fig. 2. A vector representation of the linearly polarized intensity, $(Q^2+U^2)^{\frac{1}{2}}$. All points where the polarized intensity exceeds 5.5 K have been included. The synthesized half power beam is shown as a shaded ellipse

(KUNDU and VELUSAMY, 1971; WEILER and SEIELSTAD, 1971) is similar, which makes it unlikely that the structure could be due to Faraday rotation *within* the radio emitting region, as that would result in considerable depolarization at our wavelength due to Faraday dispersion along the line of sight. Moreover, the fact that sharp boundaries separate regions of virtually constant position angle strongly suggests rotation in a thin Faraday screen.

The proper motion study of VAN DEN BERGH (1971) shows this remnant to have suffered considerable deceleration since the supernova outburst, and suggests the presence of a large quantity of swept up interstellar material. For a distance of 3 kpc and an interstellar hydrogen density near the remnant of 0.2 cm^{-3}, the total swept up mass would be some 2 $M_{\odot}$. This, combined with an interstellar magnetic field of about 7 μG, could account for the observed rotation, provided most of the material remains compressed in a shell about 0.5 pc thick surrounding the radio shell. The irregularities observed, whose scale size is about 1 pc, may reflect the inhomogeneity of the interstellar magnetoionic medium, as well as additional structure produced by the expansion dynamics.

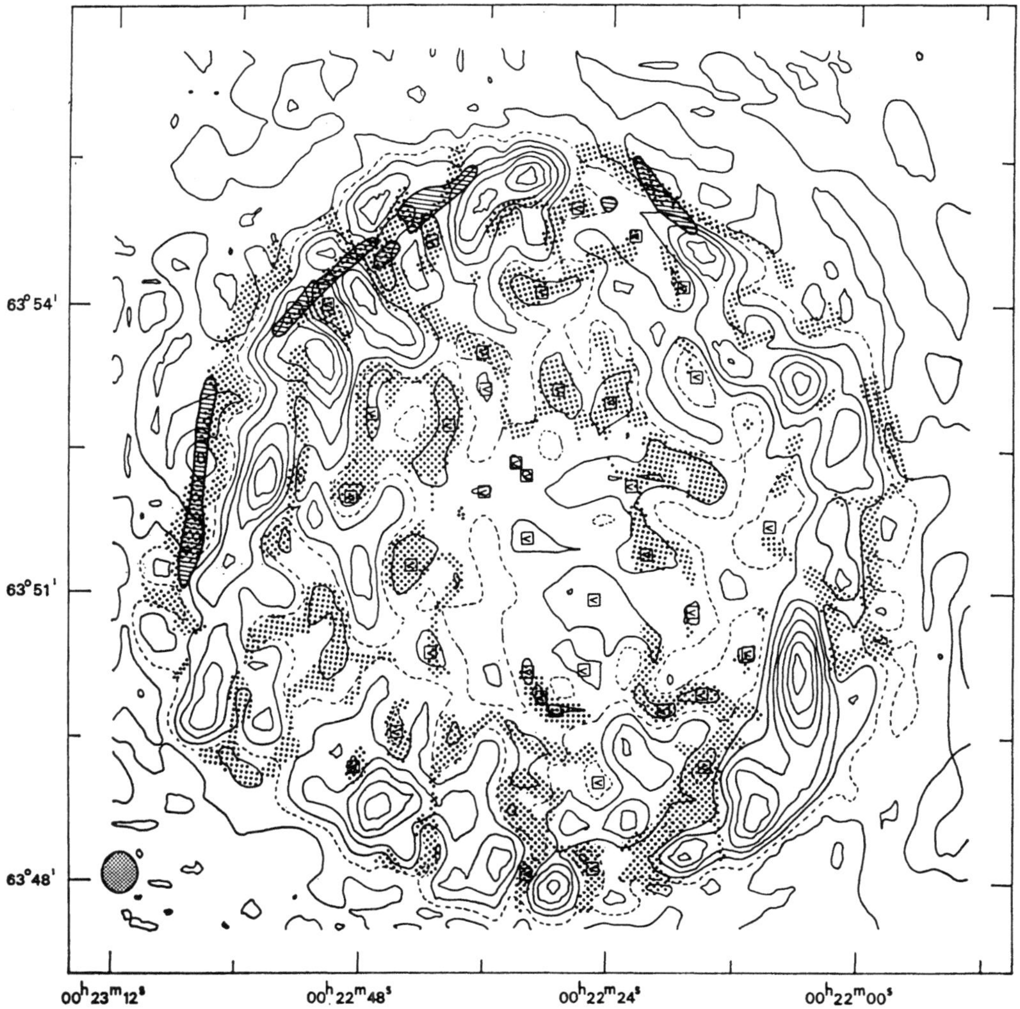

Fig. 3. A contour map of the linearly polarized intensity. The contour interval is 2.75 K with the contour whose value is 5.5 K shown dashed. Places where large changes in position angle occur (see text) and where the total intensity exceeds 55 K are shaded. Optical nebulosities associated with the remnant (VAN DEN BERGH, 1971) are shown by cross hatching. The symbol "<" enclosed in a square indicates a local minimum. The synthesized half power beam is shown as a shaded ellipse

Thus radio observations suggest that Tycho's supernova remnant consists of a shell of synchrotron particles moving in a partially radial magnetic field. Surrounding this is a screen of irregular magnetoionic material which has probably been swept up from the interstellar medium.

References

BALDWIN, J. E.: In H. VAN WOERDEN (ed.), Radio Astronomy and the Galactic System, Academic Press, London, p. 337 (1967).

BERGH, S. VAN DEN: Astrophys. J. **168**, 37 (1971).

BROUW, W. N.: Leiden University, (Ph. D. Thesis) (1971).

KUNDU, M. R., VELUSAMY, T.: Astrophys. J. **163**, 231 (1971).

WEILER, K. W., SEIELSTAD, G. A.: Astrophys. J. **163**, 455 (1971).

Radio Filaments in the Crab Nebula from High Resolution Maps

By A. S. WILSON
University of Cambridge, Cambridge, U. K.

With 6 Figures and 2 Plates

Abstract

High resolution maps of the Crab Nebula show radio emission from the bright optical filaments. The properties of this radiation are described and it is suggested that it is synchrotron in origin, the magnetic field being increased near the filaments. The increased field is caused by electric currents flowing along the filaments as suggested by WOLTJER.

I. Introduction

I shall be discussing high resolution radio maps of the total brightness and linear polarization distributions in the Crab Nebula which have been made with the Cambridge One-Mile Telescope. This instrument is presently operating simultaneously at 2.7 and 5 GHz and the half power beamwidth at the latter frequency is $6'' \times 16''$ arc (R. A. × Dec.). A description of the maps has already been published (WILSON, 1972, Paper I) and now I would like to concentrate on one particular aspect of the results. The maps show that bright optical filaments are emitting radio radiation ("radio filamentary radiation") and in this talk I shall discuss the properties of this radiation and then propose a theory for its origin.

II. Observations

It is well known that at optical wavelengths the Crab Nebula shows two distinct and very different components. Firstly, there is continuum radiation, believed to be produced by the synchrotron process, and secondly there is line emission from the gas ejected in the supernova explosion. The latter component is concentrated in long thin filaments. The use of filters enables the two components to be studied separately. Fig. 1 shows the isophotes of the continuum radiation obtained by WOLTJER (1957). A comparison of these isophotes with a photograph taken in a filamentary line shows that there is no apparent enhancement of *optical* continuum radiation by the filaments.

Now since both the radio and optical continuum radiation are believed to be produced by the same process, one might expect their brightness distributions to

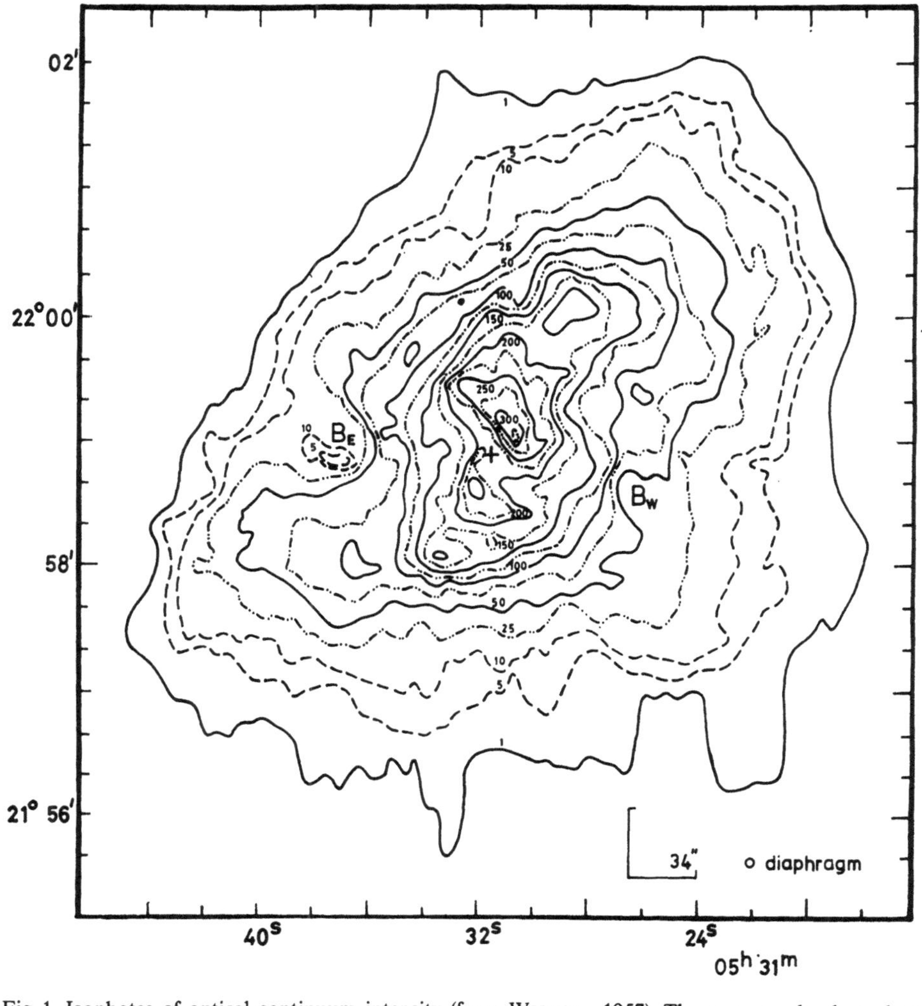

Fig. 1. Isophotes of optical continuum intensity (from WOLTJER, 1957). The cross marks the pulsar and the diaphragm size (5''.2) is shown in the bottom right hand corner. The coordinates are for epoch 1950.0 as throughout the paper. The figures and plates in this paper are reproduced by courtesy of Monthly Notices of the Royal Astronomical Society

be similar. This is true up to a point (Paper I) but, nevertheless, some of the radio emission is associated with the bright optical filaments. Plate I shows a superposition of our 5 GHz contour map of total intensity on a photograph taken in the filamentary lines $H\alpha$+[N II]. It is clear that many ridges of radio emission are associated with bright filaments.

Next we can consider the properties of the radio filamentary radiation. From Plate I we may note that:

a) The maxima of the radio ridges are very close to the $H\alpha$+[N II] maxima. There seems to be no tendency for the radio emission to be systematically on the inner or outer edges of the expanding filaments.

b) The widths of the radio ridges are larger than the widths of the associated optical filaments. It is, of course, likely that a long exposure photograph would show faint line emission around the bright filaments. However, the *half-intensity* sizes of the radio emission are clearly larger than the corresponding optical size.

c) The brightness of the radio filamentary radiation is typically 0.33×10^{-26} W m^{-2} Hz^{-1} (beam area)$^{-1}$ at 5 GHz which is equivalent to a brightness temperature of about 200 K.

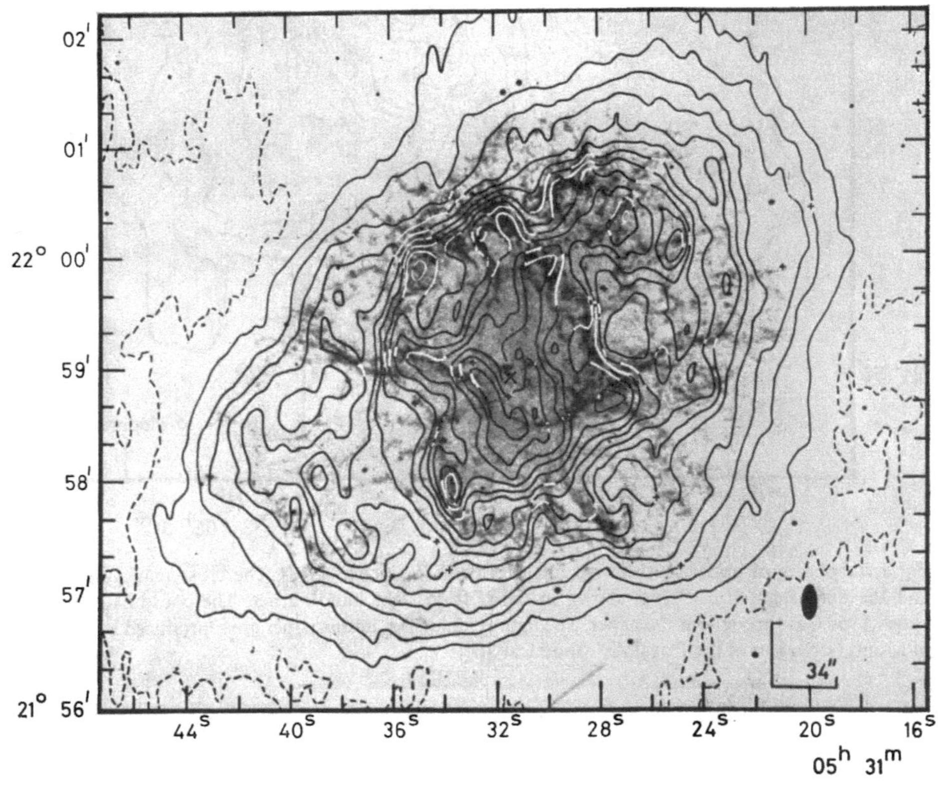

Plate I. A superposition of the 5 GHz map of total intensity and a photograph of the filamentary system taken in the lines $H\alpha$+[N II]. The cross (×) marks the pulsar and the other crosses (+) are stars and are drawn to aid the superposition. The photograph (TRIMBLE, private communication) is from the Hale Observatories and the radio beam is shown in the bottom right hand corner. The contour interval is 64 K (0.106×10^{-26} W m^{-2} Hz^{-1} (beam area)$^{-1}$)

We have noticed that there is no apparent enhancement of the optical continuum radiation by the filaments. It is then interesting to ask the question how many of the fine-scale differences between the radio and optical continuum distributions may be ascribed to radio emission from the filaments. To delineate these differences more clearly, the optical isophotes of Fig. 1 were convolved to the resolution of the 5 GHz map, normalised to the same maximum value and subtracted from it. The resulting distribution is shown superposed on the filamentary photograph in Plate II. The area H (at $\alpha = 05^h\,31^m\,30^s.6$, $\delta = 21°\,59'\,10''$) is a "hole" whereas the ring $JKLMN$ surrounding it is a "ridge". Further ridges clearly associated with filaments may be seen at $PQRS$ while regions which are low (e.g. $HTUVW$) tend to be free from filaments. The conclusion we can draw from this study is that most of the fine-scale differences between the radio and optical continuum intensity distributions are caused by radio emission from the filaments.

The radio filamentary radiation is correlated with the magnetic field structure. This may be seen in Fig. 2 where the contours of Plate II (i.e. the difference between the radio and optical continua) is superposed on the percentage polarization

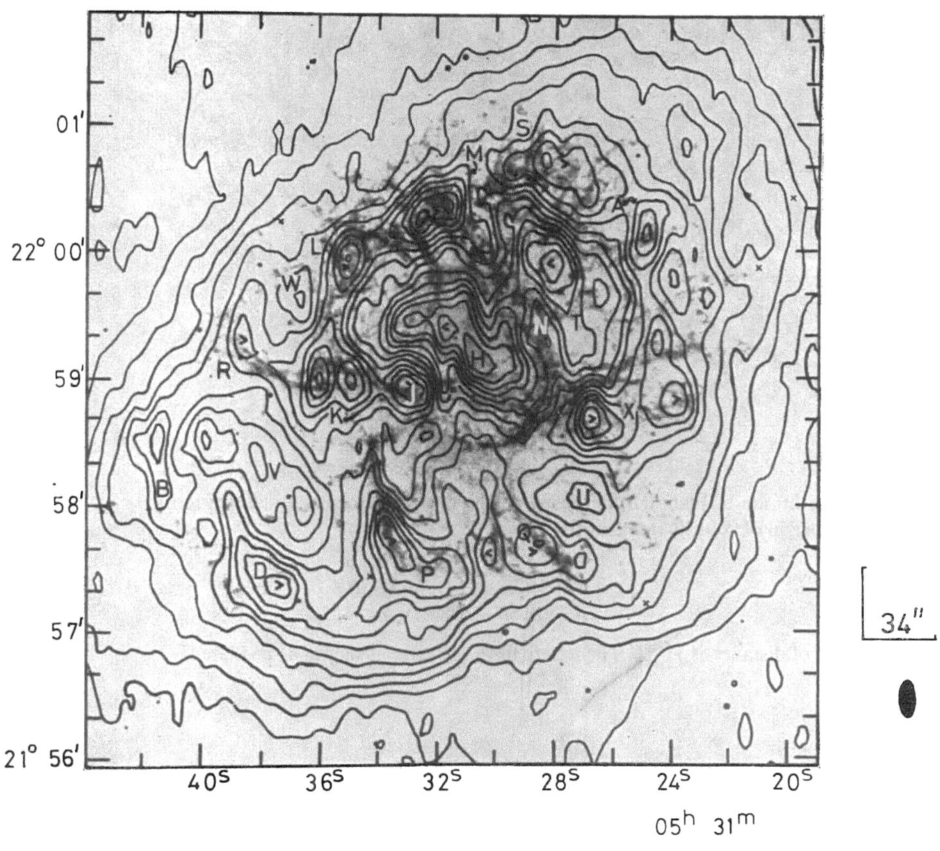

Plate II. The result of subtracting the smoothed optical continuum radiation from the 5.0 GHz map (see text) is shown superposed on the filamentary photograph of Plate I. The radio beam is shown in the bottom right-hand corner. The regions A, B, C etc. are discussed in the text

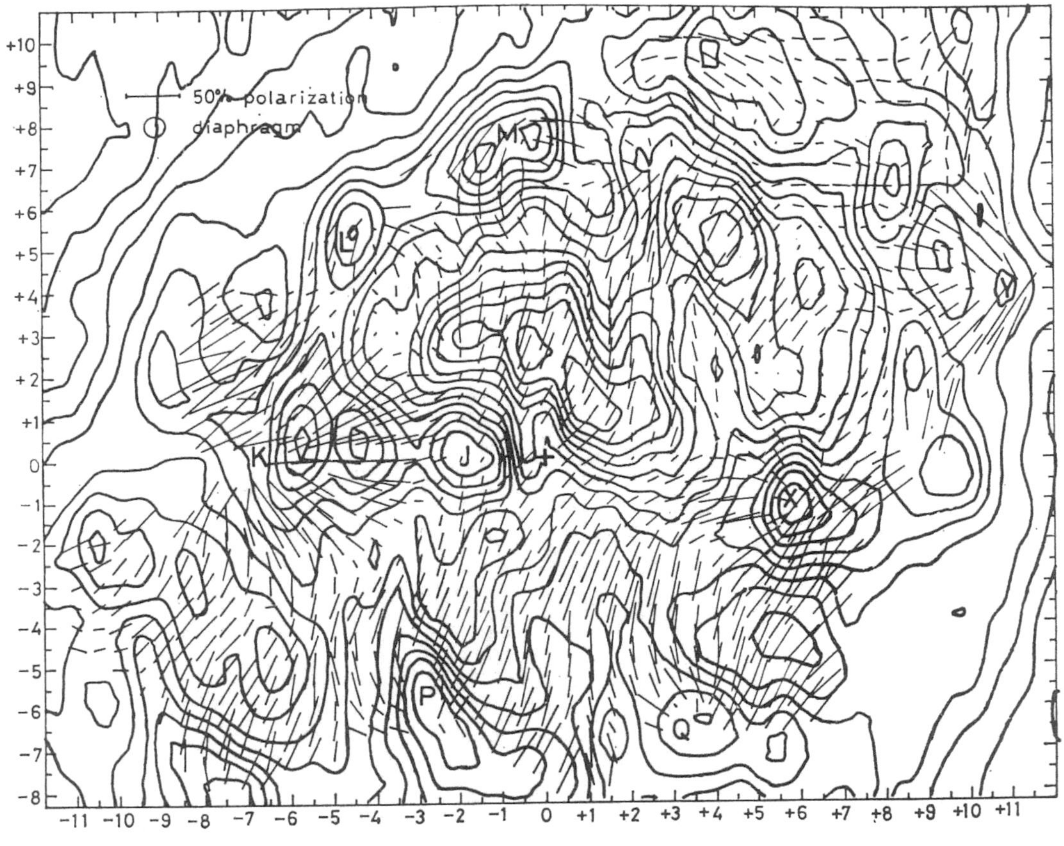

Fig. 2. The contours of Plate II superposed on the optical percentage polarization map of WOLTJER (1957). The origin of coordinates is at the pulsar

$$\left.\begin{aligned}\alpha &= 05^{h}\ 31^{m}\ 31^{s}.46\\ \delta &= 21°\ 58'\ 54''.8\end{aligned}\right\}\ 1950.0$$

and the unit of distance is 11''.16. The letters have the same meaning as in Plate II

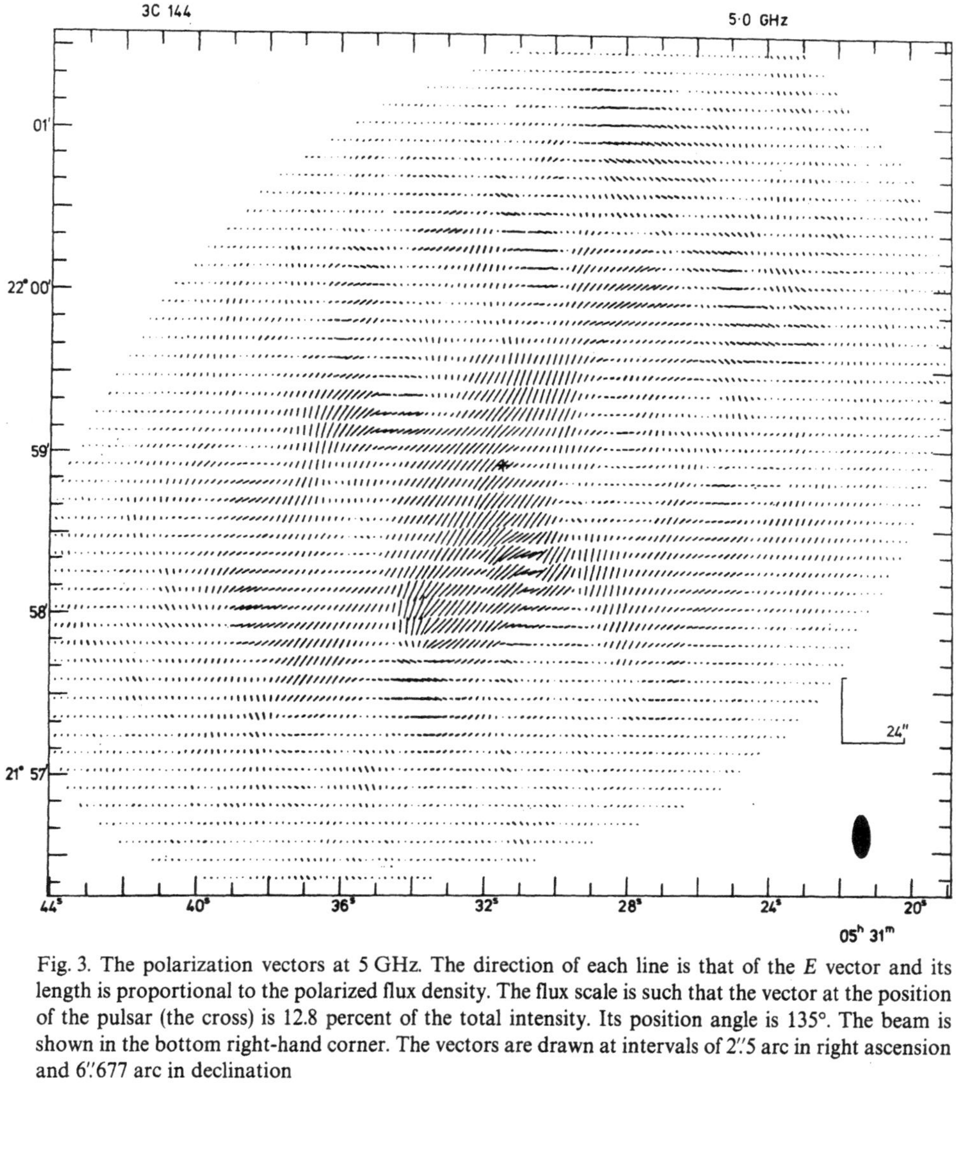

Fig. 3. The polarization vectors at 5 GHz. The direction of each line is that of the *E* vector and its length is proportional to the polarized flux density. The flux scale is such that the vector at the position of the pulsar (the cross) is 12.8 percent of the total intensity. Its position angle is 135°. The beam is shown in the bottom right-hand corner. The vectors are drawn at intervals of 2''.5 arc in right ascension and 6''.677 arc in declination

vectors of the optical continuum (WOLTJER, 1957). The vectors tend to diverge radially from the high regions marked by letters as in Plate II. The component of magnetic field perpendicular to the line of sight must therefore be circular in these regions.

Next we consider the available evidence about the polarization of the radiation. The distribution of polarized flux at 5 GHz is shown in Fig. 3. The radiation from the direction of the ridges is generally less polarized than are the central regions of the nebula. The tendency of the optical polarization vectors to diverge from the enhanged regions may also be seen in the 5 GHz polarization in positions corresponding to L and M of Plate II. This tendency suggests that the polarization observed in these regions is physically associated with the filament rather than being a background superposition. Some other regions, however, have very low polarization and here the enhanced radiation from the filament may itself be unpolarized.

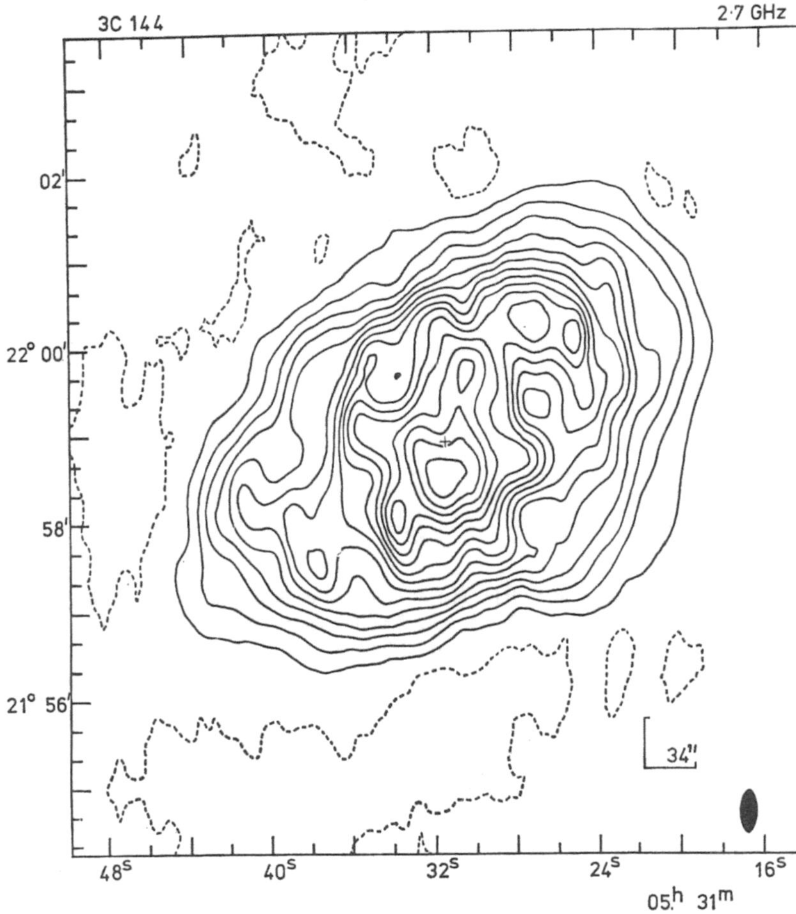

Fig. 4. Map of total intensity at 2.7 GHz. The half power beamwidth (11″ × 29″ arc) is shown in the bottom right-hand corner. The contour interval is 272 K $(0.431 \times 10^{-26}\ \mathrm{W\ m^{-2}\ Hz^{-1}\ (beam\ area)^{-1}})$ and the cross marks the pulsar

Finally we note that the ridges also appear on the total intensity map at 2.7 GHz (Fig. 4). Unfortunately there are no maps at other radio frequencies with sufficient resolution to distinguish them.

III. Discussion

WOLTJER (1958) has suggested that if the magnetic field is confined to a finite volume (that of the nebula itself), the boundary conditions require that it be surrounded by a surface current. He proposed that this current is, in fact, carried by the filaments because of the radially directed optical polarization vectors around some of them. Such filaments (e.g. *M* and *P* in Plate II and Fig. 2) are extended along the line of sight and the field is circular about them. Similarly TRIMBLE (1968) has shown that the regions *J* and *K* are a superposition of emitting material at the front and back of the object and it may be that filaments here thread through the nebula and give rise to the radial polarization vectors observed. Furthermore, MÜNCH (1958) noticed that some filaments run parallel to the optical polarization and presumably these are current carrying filaments extended in the plane of the sky. WOLTJER also suggested that the current will stabilise the filaments by the pinch effect.

If the filaments do, indeed, carry a current the magnetic field will be increased near them and the synchrotron volume emissivity enhanced. The radio filamentary radiation may thus be understood. The synchrotron radiation from filaments extended along the line of sight will be brighter than that from those in the plane of the sky because:

a) In the former case the magnetic field is always perpendicular to the line of sight (Fig. 5).

b) There will be a greater path length of emission from a filament extended along the line of sight.

This could explain the observed association of the most enhanced regions with the radial polarization vectors. The lack of enhanced optical continuum radiation from the filaments may be a consequence of depletion of the appropriate electrons by synchrotron losses. There must, however, be *some* "optical" electrons near the filaments in order to show the relationship between the optical polarization and the filaments. A more detailed and quantitative discussion of this interpretation may be found in WILSON (1973).

Finally, I speculate briefly about the origin of such currents. If there was to be relative motion between the magnetic field and a filament, the magnetic field could be distorted in the observed manner. The interaction with an "end on"

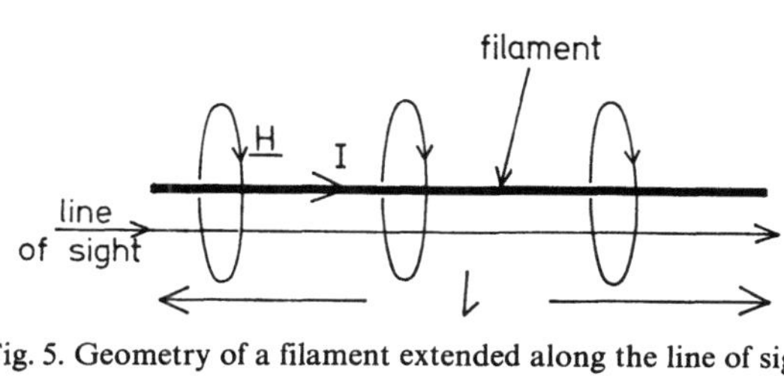

Fig. 5. Geometry of a filament extended along the line of sight

filament is shown schematically in Fig. 6. The field bends around the conducting gas (b) and may set up a current along the filament. Such a process could also explain the radial direction of the optical polarization vectors on the inner edge of the "bays", if the "bays" contain a large amount of conducting gas. KARDASHEV (1970) and GUNN and REES (1972) suggest that the field of the Crab Nebula has been "wound up" by the pulsar. GUNN and REES consider that when this field meets a filament it will wrap round it in the way indicated in Fig. 6.

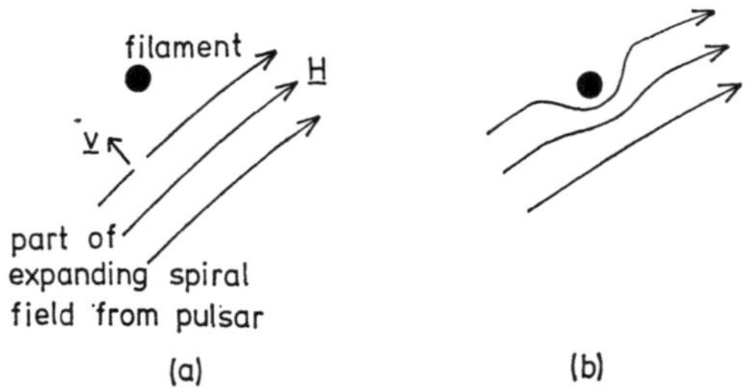

Fig. 6. Origin of relation of magnetic field to filaments

References

GUNN, J. E., REES, M. J.: (Private Communication) (1972).
KARDASHEV, N. S.: Soviet Astron. **14**, 375 (1970).
MÜNCH, G.: Rev. Mod. Phys. **30**, 1042 (1958).
TRIMBLE, V.: Astron. J. **73**, 535 (1968).
WILSON, A. S.: Monthly Notices Roy. Astron. Soc. **157**, 229 (1972).
WILSON, A. S.: Monthly Notices Roy. Astron. Soc. **160**, 373 (1972).
WOLTJER, L.: Bull. Astron. Inst. Neth. **13**, 301 (1957).
WOLTJER, L.: Bull. Astron. Inst. Neth. **14**, 39 (1958).

A Model of the Crab Nebula Derived from Dual-Frequency Radio Measurements

By K. W. Weiler
Radiosterrenwacht Westerbork, Post Hooghalen, The Netherlands
Netherlands Foundation for Radio Astronomy
and G. A. Seielstad
Owens Valley Radio Observatory, California Institute of Technology, Big Pine, California, USA

Abstract

The total intensity and linearly polarized emission from the Crab Nebula (Taurus A) were synthesized to a resolution of approximately 1 arc-minute at both 1420 and 2880 MHz. From these data were calculated the spectral index, rotation measure, intrinsic position angle, and depolarization ratio distributions. Then, combined with a source model, the physical conditions within the supernova remnant were established. The strength and orientation of both the homogeneous and random components of the magnetic field were determined and a measurement of the thermal electron plasma distribution obtained.

On the Possible Detection of an Energetic Photon Pulse from the Recent Supernova in NGC 5253

By H. ÖGELMAN and M. E. ÖZEL
Physics Department, Middle East Technical University, Ankara, Turkey

With 3 Figures

Abstract

A diffuse atmospheric fluorescence pulse having the proper signatures of an excitation caused by a fast burst of X-ray photons, incident from the general direction of the recent supernova in NGC 5253, was recorded on May 3, 1972 at $20^h\,48^m\,51^s$ U.T. Due to excessive background pulses of terrestrial origin, it is concluded that the evidence should be at best regarded suggestive. It is recommended that the data of satellite-borne X-ray detectors should be examined for anomalies at the indicated time.

I. Introduction

Motivated by the prediction of COLGATE (1968) that a short duration burst of X and gamma rays should accompany the very early stages of supernovae explosions, we have been operating an optical system designed to detect the fluorescence emission which would be produced in the atmosphere by such a pulse (ÖGELMAN and BERTSCH, 1970). Initial data revealed a wealth of light pulses in the night sky that were of terrestrial origin. In order to eliminate these terrestrial pulses we are presently using two detector systems separated by 500 km where simultaneous detection of similar pulses in both stations would be considered astrophysically interesting. Another such pair with a 3300 km baseline is being operated by Goddard Space Flight Center and the Smithsonian Astrophysical Observatory in the U.S.A. Up to now no supernova-like pulses have been observed in coincidence and the upper limits established are within the uncertainties of the predictions (BERTSCH *et al.*, 1972).

During May, 1972 a supernova in the galaxy NGC 5253 was discovered (KOWAL, 1972). Having an apparent photographic magnitude of 7.9 at its maximum around May 6 (FISHER, 1972), this supernova is the fourth brightest ever recorded and is certainly the brightest to occur within the operation time of the atmospheric fluorescence detector. Consequently we have examined our records carefully for the possible effects of this supernova. Unfortunately, during the interesting period prior to the maximum of the light curve only a single station in Ankara was operational and the data discussed in this report is limited to an interesting pulse recorded by this station on May 3, 1972.

II. The Detector System

The detector system consists of three 30 cm photomultiplier tubes with collimators that restrict their opening angle to 70° from the normal. Two of the tubes have wide band filters covering the useful atmospheric fluorescence region between 3200 Å and 4300 Å (denoted by V) and the third one responds to light between 4300 Å and 6500 Å (denoted by Y). A proper fluorescence pulse should have no amplitude in the Y tube. The two V tubes are tilted 30° from the zenith in opposite directions along the north-south axis, the Y unit is directed to the zenith. A 2 m^2 liquid-scintillator particle detector was also included in the system in order to investigate the existence of charged secondaries due to primary gamma-ray photons above 10^{13} eV accompanying the fluorescence pulses. A block diagram of the system is shown in Fig. 1. Simultaneous outputs of the V tubes integrated

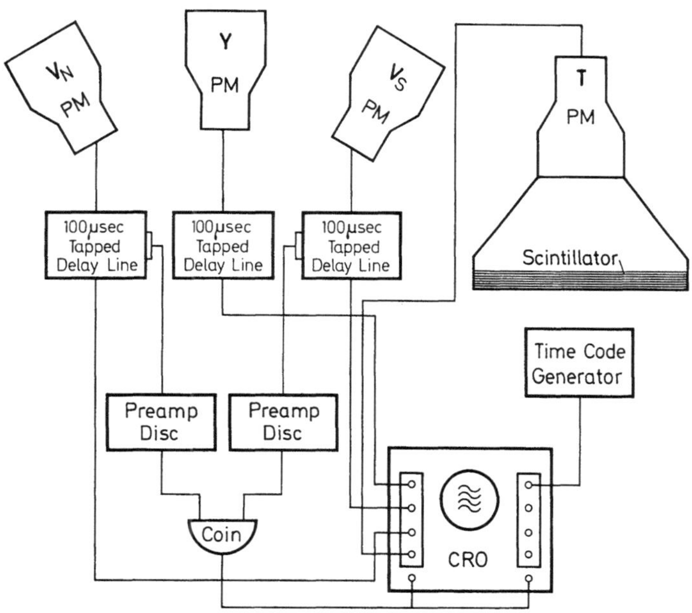

Fig. 1. Block diagram of the light detection system

over 50 microseconds trigger the system and the three tubes together with the scintillator tank pulses are recorded on film. The system is programmed to operate during moonless night sky conditions. The triggering sensitivity is around 10^2 photons cm^{-2} (50 microseconds)$^{-1}$. A surface energy density of 2×10^{-7} erg cm^{-2} in the form of 1 to 10 keV X-rays incident within a time scale shorter than 50 microseconds would suffice to trigger the system.

III. Data on Supernova 1972 in NGC 5253

The light curve of the supernova obtained from the data reported in the IAU Circulars is shown in Fig. 2 (Kowal, 1972; Wisniewski and Lee, 1972; Matchett, 1972; Detre, 1972; Ardeberg and De Groot, 1972; Fisher, 1972). From this

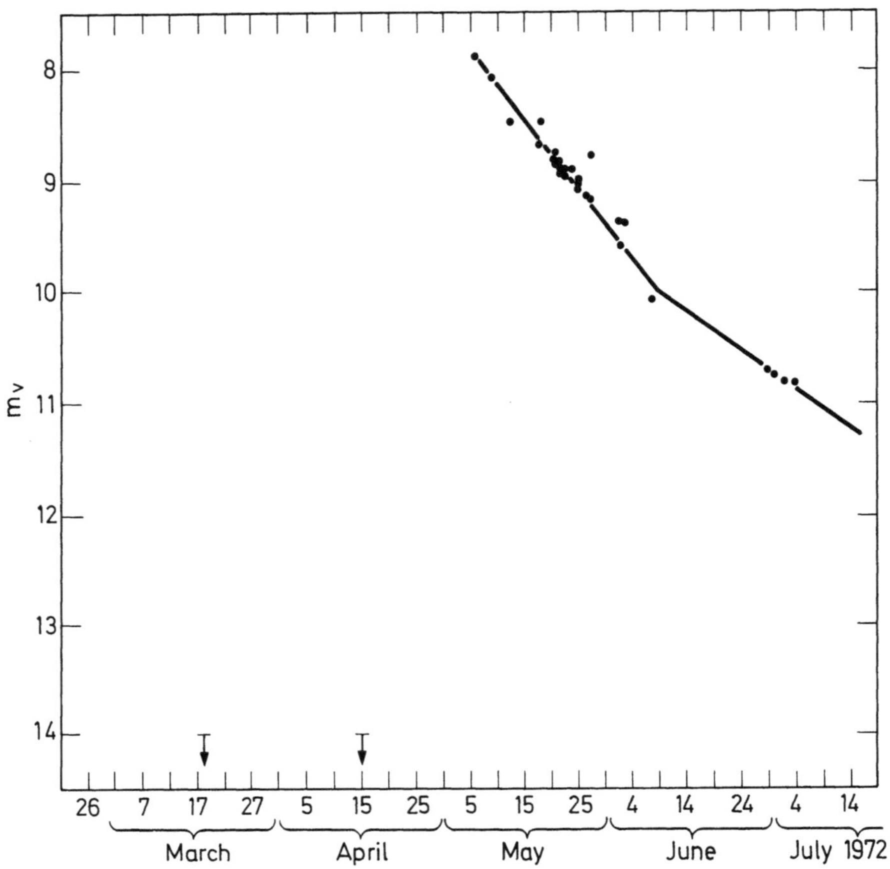

Fig. 2. The light curve of the supernova 1972 in NGC 5253 plotted according to the data reported in the IAU Circulars; 2405, 2407, 2409, 2411, 2421

information it appears that the maximum of the light curve occurred around May 6, 1972. Since this supernova's maximum apparent brightness is very similar to the supernova observed in NGC 5253 at 1895 we can adopt Minkowski's value of −19.8 as its absolute photographic magnitude corrected for absorption and classify SN 1972 in NGC 5253 as type I as well (MINKOWSKI, 1964). The distance estimate is about 3 megaparsecs (WELCH, 1970).

The energetic photon pulse is expected to occur during the very beginning stages of the supernova. The optical light emission is a more gradual process that has a rise time of between a few days and a few weeks. For this supernova the collapse of the central star should have taken place sometime between April 20 and May 5, 1972. During this interval the fluorescence detector was operational on the nights of May 3 and May 4 comprising a total "on" time of 8.6 hours.

The candidate pulse was recorded on May 3, 1972 at $20^h\,48^m\,51^s$ U.T.; the picture of this pulse is shown in Fig. 3. At this time NGC 5253 was 71° from the zenith towards the south, almost at the meridian of Ankara. The calculated response of the tubes to a hypothetical X-ray pulse with a very sharp wavefront (less than 10 microseconds wide) incident from the particular direction of NGC 5253

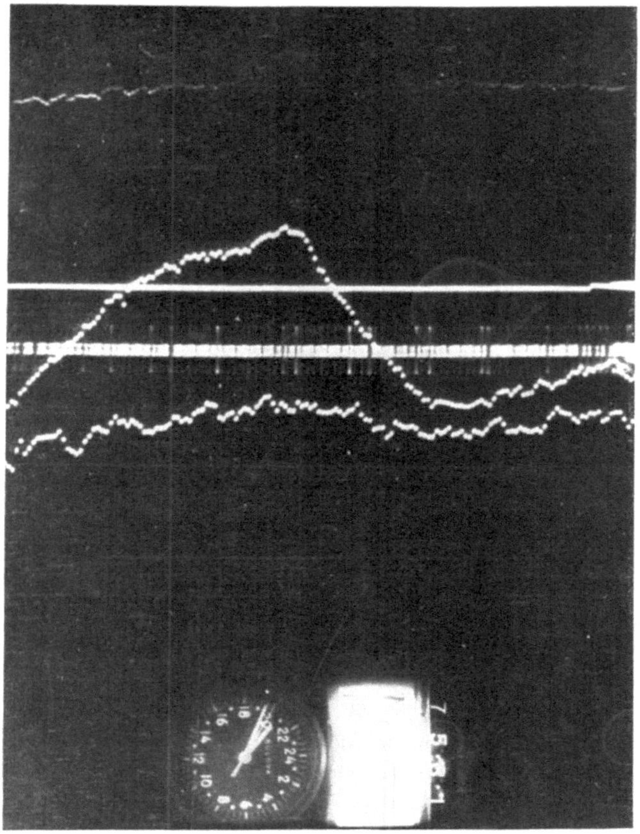

Fig. 3. The candidate pulse recorded on May 3, 1972 of $20^h 48^m 51^s$ U.T. Starting with the top first trace is the *Y* unit response (not too clearly visible); the second trace is the particle detector showing no response, the third trace is the NASA bit serial time code; the fourth trace is the south pointing *V* tube response which shows a large amplitude; the fifth trace is the north pointing *V* tube response which shows a broad small pulse. Full scale of time axis is 900 microseconds

at this time indicates that the south pointing *V* tube should have a bigger pulse with a rise time (to the maximum) of the order of 20 microseconds and a fall time (to half of the maximum) of about 150 microseconds while the north pointing tube should show a rise time of 250 microseconds and a fall time of 400 microseconds. The ratio of the maximum amplitude of the south pointing tube to that of the north pointing one should be around 8.

The candidate pulse conforms generally to the expected signature of having a larger and sharper response in the south pointing *V* tube, flattened and smaller response in the north-pointing *V* tube and no pulse on the *Y* tube; however the observed shapes are not in exact agreement with the idealized calculations. The spectrum and time profile of the incident pulse as well as the cloud conditions could alter the shapes sufficiently to make them compatible with the calculations.

IV. Discussion

If we assume that the very beginning of this supernova could have happened anytime between April 20 and May 5 and our system was 100% efficient in detecting this pulse if it occurred during the "on" time of the detector, then the probability of catching the energetic photon pulse is about 0.025. If we have truly recorded the very beginning of the supernova, on the basis of the low observations probability, we have to conclude that either we were lucky or that a multitude of such pulses may be produced by the supernova. Although a pulsar at the center may generate occasionally such pulses, it is difficult to reconcile how these photons could penetrate the envelope that is still optically thick.

To get an idea about the chance recording of a supernova-like pulse during this interesting interval, we have examined the 1600 hours of data taken between June 1970 and June 1972. Assuming that any pulse that shows no amplitude on the Y tube, irregardless of the shapes of the V tubes is a supernova-like event, the rate of such was measured to be 0.02 per hour. During 8.6 hours of the interesting observation time for the SN 1972 in NGC 5253, the probability of observing one or more supernova-like pulses is 0.16. If we put more stringent requirements on the shape of the V pulses such as having a larger amplitude on the south pointing tube, this probability reduces to 0.04.

If we assume that this pulse is caused by an X-ray burst in the 1 to 10 keV range from the supernova in NGC 5253 then the total energy released at the source would be 2×10^{47} erg which compares well with the earlier estimates of COLGATE (1968).

In view of the speculative nature of the experiment and the data presented above, it would be premature to claim positively the detection of an energetic photon burst associated with the supernova 1972 in NGC5253. However, it is worthwhile to expose these findings to the scientific community with the hope that at the time of this event other experiments such as satellite-borne X-ray detectors may find an interesting anomaly as well.

Acknowledgements

We thank M. AYDOĞDU and Ü. KIZILOĞLU for their help in maintaining the equipment and analyzing the data. This work is supported jointly by Goddard Space Flight Center and NATO Research Grants Programme.

References

ARDEBERG, A.L., GROOT, M.J.H. DE: IAU Circular, No. 2411 (1972).
BERTSCH, D., ÖGELMAN, H., FAZIO, G., WEEKS, T.: Bull. Am. Phys. Soc. **17**, Series II, 524 (1972).
COLGATE, S.A.: Can. J. Phys. **46**, 476 (1968).
DETRE, L.: IAU Circular, No. 2409 (1972).
FISHER, W.J.H.: IAU Circular, No. 2421 (1972).
KOWAL, C.T.: IAU Circular, No. 2405 (1972).
MATCHETT, V.L.: IAU Circular, No. 2407 (1972).
MINKOWSKI, R.: Ann. Rev. Astron. Astrophys. **2**, 247 (1964).
ÖGELMAN, H., BERTSCH, D.: Acta Phys. Acad. Sci. Hung. **29**, Suppl. 1, 35 (1970).
WELCH, G.A.: Astrophys. J. **161**, 821 (1970).
WISNIEWSKI, W., LEE, T.: IAU Circular, No. 2407 (1972).

Relevance of Electron Pair Production in the Interpretation of IR and X Extragalactic Sources

By S. Bonometto
Istituto di Fisica «G. Galilei», Università di Padova, Italy,
Istituto Nazionale di Fisica Nucleare – Sezione di Padova, Italy,
and F. Lucchin
Istituto di Astronomia, Università di Padova, Italy

With 2 Figures

Two possible explanations for X Radiation recently observed from a number of extra-galactic objects have been suggested: either the existence of a dense (10^{-28} g cm^{-3}) hot ($2 \cdot 10^7$ °K) gas is supposed, or Invese Compton (I.C.) Scattering of lower frequency radiation is supposed to take place.

In some cases, when the emitting region is very wide (e.g. Coma Cluster, see Perola and Reinhardt, 1972) the microwave background may be considered as a source for low-energy photons, but there are at least 3 cases (3C 273, NGC 1275, NGC 4151) where the X source coincides with a strong IR source (this is only partially true for NGC 1275, where a more extended X source seems also to be present). Then IR photons could be lower-energy photons providing a target for I.C.E.

This idea, firstly proposed by Takarada (1970) and later widely developped by Bergeron and Salpeter (1971), leads to the prediction of a γ-radiation from the same objects with an intensity given by

$$I_\gamma \simeq I_X^2 \, I_{IR}^{-1}.$$

As I_X and I_{IR} are still rather uncertain, one can obtain a non detectable (at present) γ-radiation.

Another interesting possibility was firstly outlined by Jelley (1967) who observed that the pair production reaction

$$\gamma + \text{lower energy photon} \rightarrow e^+ + e^- \qquad \text{(PP)}$$

could lead to the absorption of most γ photons.

As rather good arguments (Bergeron and Salpeter, 1971) can be put forward to set a peak for X radiation at $\omega_{0X} \simeq 10^{-3}$ m$_e$, while a peak for IR is likely to occur at about $10^{-7}-10^{-8}$ m$_e$ (i.e. 25–100 μ), a peak of γ-radiation ought to be expected at about 10 m$_e$.

As PP has a threshold, i.e. it occurs only provided that the condition (h$=c=1$ units are used)

$$\omega_\gamma \, \omega_{\text{target}} \geqq m_e^2$$

is fulfilled, target photons of $\simeq 50$ keV are needed to get it working. And, indeed, at frequencies ~ 50 keV the X spectra are generally obtained by extrapolating lower frequency data. Therefore a large uncertainty is involved, unless we assume that I.C. theory is correct and set the same spectral index both for IR (which is better known) and X spectra. Then assuming that α_X in

$$I_X(\omega) = A_X\,\omega^{-\alpha_X}$$

is 2, the mean-free-path λ of a γ of 10 m_e can be computed by means of the formula

$$\alpha(\omega) = \lambda^{-1}(\omega) \simeq \frac{4}{\alpha_X(\alpha_X+2)} A_X\,\sigma_T\,\omega^{\alpha_X} \tag{1}$$

(this formula is proven in Appendix 1) and compared with the size R of the source. The condition

$$R\,\alpha(\omega) > 1 \tag{1a}$$

must be fulfilled in order to get a relevant absorption effect. As α is increasing with ω, at high enough energy (1a) will be satisfied, but, for 10 m_e photons it will not, unless very small values of R are assumed; in effect it is

$$R\,\alpha(\omega) \simeq \frac{\alpha_X - 1}{\alpha_X(\alpha_X+2)} L_X\,\omega_{0X}^{\alpha_X - 1}\,\sigma_T\,\omega^{\alpha_X}\,\frac{1}{R} \tag{2}$$

(this formula is proven in Appendix 2), where L_X is the total X luminosity. In order to satisfy (a), for objects like 3C 273 or NGC 1275, $R \simeq 10^{14} - 10^{15}$ cm are needed.

Moreover none can say whether the extrapolation of $I_X(\omega)$, in the 100 keV range, is correct or not.

In any case, if the γ-radiation is absorbed, a lot of PP produced electrons will be input into the source, and they could lead to relevant modifications of IR and X spectra, by emitting their own energy (i.e. the γ-photons energy) because of the presence, inside the source itself, of different kinds of electromagnetic radiations.

With a more detailed analysis of this problem we have been able to show that, if we start supposing that γ-radiation is absorbed, we get out that X-radiation in the 50–100 keV range is widely emitted by PP produced electrons, leading to much shorter mean-free-paths for peak frequency γ photons. And, indeed, for sizes of $\sim 10^{16}$, 10^{17} cm, an almost complete self-consistence is reached.

The first point, in order to study this effect in detail, is to know which energy distribution the electrons created because of PP will have (this point is dealt with in Appendix 3). Then the emission due to these electrons can be calculated. But, indeed, as the absorption of γ's will be due to this emission, this problem requests a self-consistent solution, where the radiation produced by PP electrons is that one which has been used to absorb γ-radiation.

Then the connection between α_{XII} (spectral index of X-spectrum in the 50–100 keV region), γ (spectral index of the residual γ-radiation), β (spectral index of the equilibrium electron spectrum) and α_X (spectral index of X radiation) will have to be analysed.

We can show (Appendix 4) that three independent relations between α_X, β, α_{XII}, and γ hold that allow to calculate α_{XII} when $\alpha_X (= \alpha_{IR})$ is known.

The final result of this computation is the following one

$$\alpha_{\mathrm{XII}} = \frac{\alpha_{\mathrm{X}}+1}{2}.$$

When $\alpha_{\mathrm{XII}} (=\alpha_{\mathrm{IRII}})$ is known, if the total amount of radiation present in the XII spectrum is known, one can compute the constant A_{XII} in the formula

$$\mathrm{I}_{\mathrm{XII}}(\omega) = A_{\mathrm{XII}}\, \omega^{-\alpha_{\mathrm{XII}}}$$

which results to be

$$A_{\mathrm{XII}} \simeq \frac{1}{2r^2} A_{\mathrm{X}} \left(\frac{\omega_{0\mathrm{X}}}{\omega_{0\gamma}^2}\right)^{-\frac{\alpha_{\mathrm{X}}-1}{2}}$$

(this formula will be proven in Appendix 5) and use it to compute the mean-free-path of a γ photon, when owing to the XII spectrum. Then, using (1) and (2), one finds

$$\text{ratio} = \frac{\mathrm{R}\,\alpha(\omega)\,\{\text{with XII spectrum}\}}{\mathrm{R}\,\alpha(\omega)\begin{Bmatrix}\text{with extrapolation of}\\ \text{lower energy spectrum}\end{Bmatrix}}$$

$$= \frac{(\alpha_{\mathrm{X}}+1)(\alpha_{\mathrm{X}}+5)}{4\alpha_{\mathrm{X}}(\alpha_{\mathrm{X}}+2)} \frac{1}{2r^2} \left(\frac{\omega_{0\mathrm{X}}}{\omega_{0\gamma}^2}\right)^{-\frac{\alpha_{\mathrm{X}}-1}{2}} \omega^{-\frac{\alpha_{\mathrm{X}}-1}{2}}$$

(r is the ratio between L_{IR} and L_{X}).

For $\alpha_{\mathrm{X}}=2$, $\omega_{0\mathrm{X}}=10^{-3}\,\mathrm{m}_e$, $\omega_{0\gamma}=10\,\mathrm{m}_e$ one has, at $\omega=10\,\mathrm{m}_e$, ratio $\simeq 8$ (for $r=3$), ratio $\simeq 75$ (for $r=1$). Therefore, with radii of the order $10^{16}-10^{17}$ cm, for objects like 3C 273 and NGC 1275 an almost complete reabsorption of γ-radiation can take-place.

No such effect seems to be plausible for NGC 4151.

The structure of the X spectrum that is found in this way is shown in Fig. 1. It is worth noting that the same shape of spectrum should also appear in IR wavelenghts and it is somehow surprising that NGC 1275 (Kleinmann and Low, 1970) shows indeed this kind of feature in its IR spectrum (Fig. 2). Also other IR sources show similar (but less marked) features.

Our solution is a stationary one. If time dependence were introduced, one should probably expect bursts of γ-radiation to be emitted in correspondence with strong variations of IR and X fluxi, before the XII spectrum is "filled up".

Before concluding we wish to stress a point which was implicit in the calculation of A_{XII} (and is already outlined in Appendix 5).

It is well known that, in a source of a size $\sim 10^{16}-10^{17}$ cm, many electron sources must exist, as the mean-free-path of an electron – owing to its electromagnetic energy losses – will be $10^{12}-10^{13}$ cm. If, as is then quite reasonable, one supposes that a number of minor electron sources is scattered into the source itself, it might be quite reasonable to think that the magnetic field, giving rise to IR radiation, is practically confined in these minor sources. Therefore PP produced electrons, due to collisions between photons occurring everywhere inside the "big" source, will emit mainly because of the presence of IR and X radiations. We do not expect them to emit a substantial amount of their energy as Synchrotron radiation.

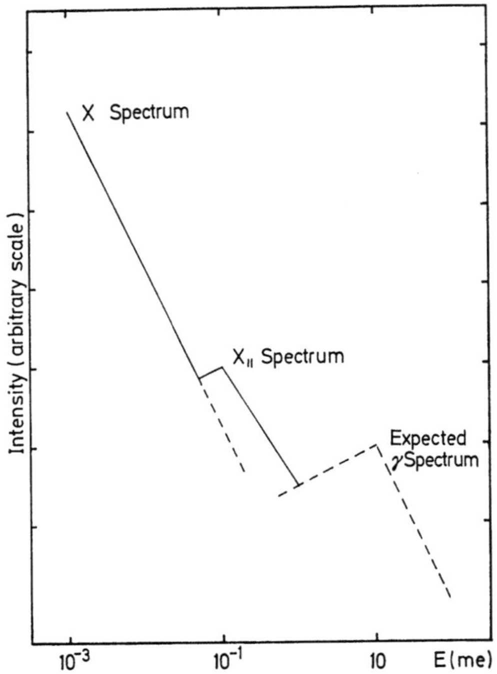

Fig. 1. Expected structure of X and γ spectral regions

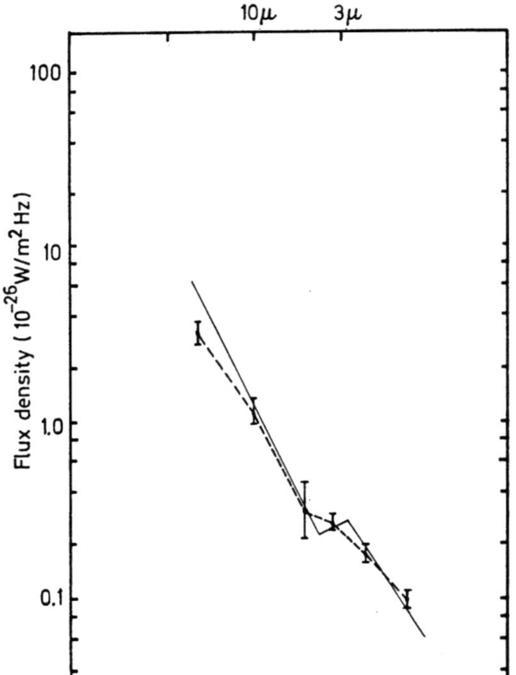

Fig. 2. The observed IR data on NGC 1275 are shown together with a fitting of our emission structure (for $r = 3$). The dashed line represents a tentative fitting of KLEINMANN and LOW in the paper where these data were first presented

The value of $A_{\rm XII}$ we have obtained is correct in the hypothesis that this picture is true. If a magnetic field, of almost constant intensity, is present everywhere into the "big" source, the values of $A_{\rm XII}$ (and, therefore, of "ratio") are to be divided by r. If, instead, $r \leqq 1$ no change will arise from this different picture.

Appendix 1

From GOULD and SCHRÖDER (1967)

$$\alpha(\omega) = \frac{4}{\omega^2} \int_1^\infty ds\, \sigma(s)\, s \int_{s\omega^{-1}}^\infty d\omega'\, \omega'^{-3}\, \mathrm{I}(\omega').$$

If one takes into account that most of the collision will occur near threshold (spectra are very steep), the approximation $\sigma(s) = \sigma_T$ can be made. Then the integration gets easy and we have

$$\alpha(\omega) = \frac{4}{\alpha_{\rm X}(\alpha_{\rm X}+2)} A_{\rm X}\, \sigma_T\, \omega^{\alpha_{\rm X}}.$$

Appendix 2

As

$$L_{\rm X} \simeq 4\mathrm{R}^2 \int_{\omega_{0\rm X}} \mathrm{I}_{\rm X}(\omega)\, d\omega = 4\mathrm{R}^2 \frac{1}{\alpha_{\rm X}-1} \omega^{-\alpha_{\rm X}+1} A_{\rm X}$$

one has

$$A_{\rm X} = L_{\rm X} \frac{\alpha_{\rm X}-1}{4} \omega_{0\rm X}^{\alpha_{\rm X}-1} \frac{1}{\mathrm{R}^2}$$

and then, owing to (1),

$$\mathrm{R}\,\alpha(\omega) = \frac{\alpha_{\rm X}-1}{\alpha_{\rm X}(\alpha_{\rm X}+2)} L_{\rm X}\, \omega_{0\rm X}^{\alpha_{\rm X}-1}\, \sigma_T\, \omega^{\alpha_{\rm X}} \frac{1}{\mathrm{R}}.$$

Appendix 3

The energy distribution of the electrons produced because of PP was studied in a previous paper by BONOMETTO and REES (1971) for different kinds of absorbing photon spectra. It is easy to verify that, when the mass of the electron becomes negligible with respect to its energy, the quantity $\alpha(\omega, E)$ defined to give the energy distribution of PP produced electrons, and obeying the normalization condition

$$\int \alpha(\omega, E)\, dE = \alpha(\omega)$$

has the following kind of dependence on ω and E

$$\alpha(\omega, E) = \alpha(\omega)\, \omega^{-1} \cdot K\left(\frac{E}{\omega}\right).$$

Moreover, if the absorption is due to a power-law photon spectrum with spectral index $-1{,}5$, a very good analytical approximation for $K\left(\frac{E}{\omega}\right)$ is

$$K\left(\frac{E}{\omega}\right) = 12\left[\frac{E}{\omega} - \left(\frac{E}{\omega}\right)^2\right].$$

As the solution we find then for $\alpha(\omega, E)$ is consistent with absorbed photons with spectral index $-1{,}5$, we can confidently take it as reasonable.

Appendix 4

The number-density of electrons of energy E produced per unit volume and time because of PP will be given by

$$\mathcal{N}_e(E) = \int_E n_\gamma(\omega)\,\alpha(\omega)\,\frac{1}{\omega}\,12\left[\frac{E}{\omega} - \left(\frac{E}{\omega}\right)^2\right] d\omega.$$

Now we shall take $n_\gamma(\omega)$ (number-density of photons of energy ω in a unit volume) to be

$$n_\gamma(\omega) = A_\gamma\,\omega^{-\gamma} \quad \text{for } \omega > \omega_{0\gamma}$$

$$n_\gamma(\omega) = \bar{A}_\gamma\,\omega^{-\frac{1}{2}} \quad \text{for } \omega < \omega_{0\gamma}.$$

Then it will be

$$\mathcal{N}_e(E) \propto E^{-\gamma + \alpha_{\mathrm{XII}} - 1} \quad \text{for } E > \omega_{0\gamma}$$

$$\mathcal{N}_e(E) \propto E \quad \text{for } E \quad \omega_{0\gamma}.$$

These formulae must be introduced into

$$\mathcal{N}_e(E) \propto \frac{d}{dE}\left(N(E)\,E^2\right).$$

Then, supposing $N(E) \propto E^{-\beta}$, one gets

$$\beta = \gamma - \alpha_{\mathrm{XII}} + 1. \tag{g1}$$

The spectral index of the photons emitted by these electrons will be

$$\alpha_{\mathrm{XII}} = \tfrac{1}{2}(\beta - 1). \tag{g2}$$

Moreover there must be equilibrium between production and absorption of γ's, i.e. it must be

$$n_\gamma(\omega)\,\alpha(\omega) = N_\gamma(\omega).$$

This relation leads to

$$\gamma = \alpha_{\mathrm{XII}} + \alpha_{\mathrm{X}} + 1. \tag{g3}$$

From (g1), (g2), (g3) one gets

$$\alpha_{\mathrm{XII}} = \frac{\alpha_{\mathrm{X}} + 1}{2}.$$

Appendix 5

It is

$$\mathrm{I}_{\mathrm{X}} = A_{\mathrm{X}} \int_{\omega_{0\mathrm{X}}} \omega^{-\alpha_{\mathrm{X}}}\,d\omega = \frac{A_{\mathrm{X}}}{\alpha_{\mathrm{X}} - 1}\,\omega_{0\mathrm{X}}^{-\alpha_{\mathrm{X}} + 1}.$$

Analogous formulae can be shown for IR, XII, IRII spectra. The contributions of lower energy tails are always neglected; they may be relevant, but are not expected to change the relative weights.

Then, if (without PP) it were

$$L_\gamma = \frac{1}{r}\,L_{\mathrm{X}}$$

this luminosity will be shared between IRII and XII spectra in a way proportional to L_{IR} and L_X respectively (we suppose the magnetic field, giving rise to IR, to be confined in small volumes, near the electron sources that ought to be scattered in many places through the source; this hypothesis is important only if $r > 1$).

Therefore it will be

$$L_{XII} \simeq \frac{1}{r^2} L_X .$$

Then

$$\frac{A_{XII}}{\alpha_{XII}-1} \omega_{0\,XII}^{-\alpha_{XII}+1} = \frac{1}{r^2} \frac{A_X}{\alpha_X - 1} \omega_{0\,X}^{-\alpha_X+1} .$$

From this relation, owing to

$$\alpha_{XII} = \frac{\alpha_X + 1}{2} \quad \text{and} \quad \omega_{0\,XII} \simeq \omega_{0\gamma}^2 \, \omega_{0\,X}$$

the relation

$$A_{XII} \simeq \frac{1}{2r^2} A_X \left(\frac{\omega_{0\,X}}{\omega_{0\gamma}} \right)^{-\frac{\alpha_X - 1}{2}}$$

can be easily obtained.

References

Bergeron, J., Salpeter, E.E.: Astrophys. J. **9**, L 121 (1971).
Bonometto, S., Rees, M.J.: Monthly Notices Roy. Astron. Soc. **152**, 21 (1971).
Gould, R.J., Schröder, J.P.: Phys. Rev. **155**, 1404 (1967).
Jelley, J.V.: Phys. Rev. Letters **16**, 479 (1967).
Kleinmann, D.E., Low, F.J.: Astrophys. J. **159**, L 165 (1970).
Perola, G.C., Reinhardt, M.: Astron. Astrophys. **17**, 432 (1972).
Takarada, K.: Prog. Theor. Phys. **43**, 303 (1970).

Author Index

Subject Index

MIX
Papier aus verantwortungsvollen Quellen
Paper from responsible sources
FSC® C105338

If you have any concerns about our products,
you can contact us on
ProductSafety@springernature.com

In case Publisher is established outside the EU,
the EU authorized representative is:
Springer Nature Customer Service Center GmbH
Europaplatz 3, 69115 Heidelberg, Germany

Printed by Libri Plureos GmbH
in Hamburg, Germany